COMMITTEE ON OPTICAL SCIENCE AND ENGINEERING

CHARLES V. SHANK, Lawrence Berkeley National Laboratory, *Chair*
ARAM MOORADIAN, Winchester, Massachusetts, *Vice Chair*
DAVID ATTWOOD, Lawrence Berkeley National Laboratory
GARY BJORKLUND, Optical Networks, Inc.
ROBERT BYER, Stanford University
MICHAEL CAMPBELL, Lawrence Livermore National Laboratory
STEVEN CHU, Stanford University
THOMAS DEUTSCH, Massachusetts General Hospital
ELSA GARMIRE, Dartmouth College
ALASTAIR GLASS, Lucent Technologies
JOHN GREIVENKAMP, University of Arizona
ARTHUR GUENTHER, Sandia National Laboratories
THOMAS S. HARTWICK, TRW (retired)
ROBIN HOCHSTRASSER, University of Pennsylvania
ERICH IPPEN, Massachusetts Institute of Technology
KRISTINA JOHNSON, University of Colorado at Boulder
DENNIS KILLINGER, University of South Florida
HERWIG KOGELNIK, Lucent Technologies
ROBERT SHANNON, University of Arizona
GLENN T. SINCERBOX, University of Arizona
BRIAN THOMPSON, University of Rochester
ELI YABLONOVITCH, University of California, Los Angeles

THOMAS BAER, Biometric Imaging Systems, *Special Consultant*

DONALD SHAPERO, Director, Board on Physics and Astronomy
ROBERT SCHAFRIK, Director, National Materials Advisory Board
SANDRA HYLAND, Senior Program Officer, National Materials Advisory Board
DANIEL MORGAN, Program Officer, Board on Physics and Astronomy

HARNESSING LIGHT

Optical Science and Engineering
for the 21st Century

NOTICE: The project that is the subject of this report was approved by the Governing Board of the National Research Council, whose members are drawn from the councils of the National Academy of Sciences, the National Academy of Engineering, and the Institute of Medicine. The members of the committee responsible for the report were chosen for their special competences and with regard for appropriate balance.

The National Academy of Sciences is a private, nonprofit, self-perpetuating society of distinguished scholars engaged in scientific and engineering research, dedicated to the furtherance of science and technology and to their use for the general welfare. Upon the authority of the charter granted to it by Congress in 1863, the Academy has a mandate that requires it to advise the federal government on scientific and technical matters. Dr. Bruce Alberts is president of the National Academy of Sciences.

The National Academy of Engineering was established in 1964, under the charter of the National Academy of Sciences, as a parallel organization of outstanding engineers. It is autonomous in its administration and in the selection of its members, sharing with the National Academy of Sciences the responsibility for advising the federal government. The National Academy of Engineering also sponsors engineering programs aimed at meeting national needs, encourages education and research, and recognizes the superior achievements of engineers. Dr. William A. Wulf is president of the National Academy of Engineering.

The Institute of Medicine was established in 1970 by the National Academy of Sciences to secure the services of eminent members of appropriate professions in the examination of policy matters pertaining to the health of the public. The Institute acts under the responsibility given to the National Academy of Sciences by its congressional charter to be an adviser to the federal government and, upon its own initiative, to identify issues of medical care, research, and education. Dr. Kenneth I. Shine is president of the Institute of Medicine.

The National Research Council was established by the National Academy of Sciences in 1916 to associate the broad community of science and technology with the Academy's purposes of furthering knowledge and of advising the federal government. Functioning in accordance with general policies determined by the Academy, the Council has become the principal operating agency of both the National Academy of Sciences and the National Academy of Engineering in providing services to the government, the public, and the scientific and engineering communities. The Council is administered jointly by both Academies and the Institute of Medicine. Dr. Bruce Alberts and Dr. William A. Wulf are chairman and vice chairman, respectively, of the National Research Council.

This project was supported by the Defense Advanced Research Projects Agency under Contract No. MDA972-94-1-0015, the National Science Foundation under Contract No. ECS-9414956, and the National Institute of Standards and Technology under Contract No. 50-SBNB-4-C-8197. Any opinions, findings, conclusions, or recommendations expressed in this material are those of the authors and do not necessarily express the views of the sponsors.

Cover: For photo credit and description, see p. 12.

International Standard Book Number 0-309-05991-7
Library of Congress Catalog Card Number 98-86525

First Printing, August 1998
Second Printing, January 1999

Preface

In July 1994, the National Research Council (NRC) issued a report titled *Atomic, Molecular, and Optical Science: An Investment in the Future* (National Academy Press, Washington, D.C.). The report found that optical science had become an integral part of a wide range of scientific disciplines and was a key contributor to economically important applications in many areas. Some aspects of optical science, however, and all of optical engineering, were beyond the scope of the 1994 report, which therefore recommended undertaking a more comprehensive assessment of the broad field of optical science and engineering.

A program initiation and planning meeting was organized by the Board on Physics and Astronomy in cooperation with the National Materials Advisory Board. This effort resulted in the formation of the Committee on Optical Science and Engineering in early 1995, under the auspices of the two boards and with funding from three federal agencies: the Defense Advanced Research Projects Agency, the National Science Foundation, and the National Institute of Standards and Technology.

The charge to the committee was as follows:

• Survey the field of optical science and engineering (OS&E). Define the technical scope and institutional structure of the OS&E community.

• Examine progress over the last decade and project the future impact of OS&E on societal needs in the short (3-5 years) and long terms (5-20 years). Focus on leading-edge developments. Develop a vision for the future and identify some "grand challenges" that could give the field direction and could focus efforts in areas that have potential for benefit to society.

• Identify technical opportunities and prioritize them in the context of national needs.

• Identify institutional and educational innovations that are needed to develop and organize the field in a more coherent fashion and to optimize the contributions of OS&E to addressing critical national needs.

• Determine how public policy influences the ability of OS&E to address national needs.

• Examine trends in private and public research activities and compare them with those in other countries.

The committee met for the first time in March 1995. It held six workshops over the course of the following year to gather technical input from the optical science and engineering community. There were also presentations and public forums at several professional society meetings, to inform the community about the study, to solicit further input, and to begin building a foundation of community support for the study process. Based on these inputs, additional inquiries by members of the committee, and extensive discussion and debate within the committee, this report was prepared to present the study's findings, conclusions, and recommendations.

The committee thanks the many members of the OS&E community who provided their assistance to the study by participating in the workshops and through other means (see Appendix B). Without such a broad range of input, no single group could have hoped to examine a field as broad and diverse as this one. Thanks are also due to Doug Vaughan of Lawrence Berkeley National Laboratory for his assistance in writing the Overview.

A final note on terminology: Many terms are used to describe this field and its various overlapping subfields. This report often simply uses the word *optics*, in its broadest sense, to include the whole spectrum of activity in the field, across all subfields, and from basic research to engineering.

Acknowledgment
of Reviewers

This report has been reviewed by individuals chosen for their diverse perspectives and technical expertise, in accordance with procedures approved by the National Research Council's (NRC's) Report Review Committee. The purpose of this independent review is to provide candid and critical comments that will assist the authors and the NRC in making their published report as sound as possible and to ensure that the report meets institutional standards for objectivity, evidence, and responsiveness to the study charge. The contents of the review comments and draft manuscript remain confidential to protect the integrity of the deliberative process. We wish to thank the following individuals for their participation in the review of this report:

Arthur Ashkin, AT&T Bell Laboratories (retired)
David H. Auston, Rice University
Arthur N. Chester, Hughes Research Laboratories
Anthony J. DeMaria, DeMaria ElectroOptics Systems
Paul A. Fleury, University of New Mexico
John L. Hall, JILA/University of Colorado
Wendell T. Hill, University of Maryland, College Park
William Howard, Scottsdale, Arizona
Daniel Kleppner, Massachusetts Institute of Technology
Paul W. Kruse, Infrared Solutions
Robert Laudise, Lucent Technologies
Jacques I. Pankove, University of Colorado at Boulder
Don W. Shaw, Texas Instruments (retired)
Watt W. Webb, Cornell University
and one anonymous reviewer

While the individuals listed above have provided many constructive comments and suggestions, responsibility for the final content of this report rests solely with the authoring committee and the NRC.

Contents

··

HARNESSING LIGHT

Executive Summary: Introduction to the Field and the Issues

The Field of Optics

We live in a world bathed in light. We see with light, plants draw energy from light, and light is at the core of technologies from computing to surgical techniques. The field of optics, the subject of this report, concerns harnessing light to perform useful tasks.

Light influences our lives today in new ways that we could never have imagined just a few decades ago. As we move into the next century, light will play an even more significant role, enabling a revolution in world fiber-optic communications, new modalities in the practice of medicine, a more effective national defense, exploration of the frontiers of science, and much more.

We are beginning to see the fruits of the scientific discoveries of the last three or four decades. The development of the laser in the 1960s produced light with a property never seen before on Earth: coherence. Coherent light can be directed, focused, and propagated in new ways that are impossible for incoherent light. This property of laser light has made possible fiber-optic communications, compact disks, laser surgery, and a host of other applications—in all, a multitrillion-dollar worldwide market. Applications of incoherent light abound as well, including optical lithography systems for patterning computer chips, high-resolution microscopes, adaptive optics for Earth-based astronomy, infrared sensors for everything from remote controls to night-vision equipment, and new high-efficiency lighting sources.

Although optics is pervasive in modern life, its role is that of a technological enabler: It is essential, but typically it plays a supporting role in a larger system. Central issues for this field include the following:

- How to support and strengthen a field such as optics whose value is primarily enabling; and
- How to ensure the future vitality of a field that lacks a recognized academic or disciplinary home.

The Report

This report is the product of an unprecedented effort to bring together all aspects of the field of optics in one assessment organized around national needs. The report reviews the field's status today, assesses the outlook for tomorrow, and considers what must be done to assure the field's future vitality.

Optics and National Needs

Because optics applications are everywhere, this report is selective rather than comprehensive. It highlights areas where breakthroughs are taking place, where rapid change is likely in the near future, and where national needs dictate special attention. The field is largely defined by what it enables; as a result, applications drove the structure of the study and the report, which is organized around seven major areas of national need: (1) information technology and telecommunications; (2) health care and the life sciences; (3) optical sensing, lighting, and energy; (4) optics in manufacturing; (5) national defense; (6) manufacturing of optical components and systems; and (7) optics research and education. Some issues connected with these areas are outlined here. Some specific recommendations may be found in the Overview. More detailed conclusions on each area will be found in the chapters dedicated to those areas.

Information Technology and Telecommunications

In *information technology*, progress during the past decade has been extraordinary. For example, optical fiber for communications is being installed worldwide at a rate of 1,000 meters every second, comparable to the speed of a Mach 2 aircraft. Just 10 years ago, only 10 percent of all transcontinental calls in the United States were carried over fiber-optic cables; today 90 percent are. Meeting the computing and communications needs of the next 10 to 20 years will require advances across a broad front: transmission, switching, data storage, and displays. Many capabilities will have to advance a hundredfold. Although institutions have access to this rapidly growing, high-speed global telecommunications network, the infrastructure is not yet in place to provide individual consumer access that fully exploits the power of the system.

Health Care and the Life Sciences

In *medicine*, optics is enabling a wide variety of new therapies, from laser heart surgery to the minimally invasive knee repairs made possible by arthroscopes containing optical imaging systems. Optical techniques are under investigation for noninvasive diagnostic and monitoring applications such as early detection of breast cancer and "needleless" glucose monitoring for people with diabetes. Optics is providing new *biological research* tools for visualization, measurement, analysis, and manipulation. In *biotechnology*, lasers have become essential in DNA sequencing systems. Optics is playing such an important role in the life sciences and medicine that organizations concerned with these disciplines need to recognize and adjust to these developments.

Optical Sensing, Lighting, and Energy

Advances in *lighting* sources and light distribution systems are poised to dramatically reduce the one-fifth of U.S. electricity consumption now devoted to lighting. Innovative *optical sensors* are augmenting human vision, showing details and revealing information never before seen: infrared cameras that provide satellite pictures of clouds and weather patterns; night-vision scopes for use by law enforcement agencies; infrared motion detectors for home security, real-time measurements of industrial emissions, on-line industrial process control, and global environmental monitoring. High-resolution digital cameras are about to revolutionize and computerize photography and printing, and improvements in photovoltaic cells may permit *solar energy* to provide up to half of world energy needs by the middle of the next century. These developments will affect energy and environmental concerns on a national scale.

Optics in Manufacturing

Optics has had a dramatic economic influence in *manufacturing*, particularly since the advent of reliable low-cost lasers and laser-imaging systems. Optical techniques have become crucial in such diverse industries as semiconductor manufacturing, construction, and chemical production. Every semiconductor chip mass produced in the world today is manufactured using optical lithography. Just making the equipment for this business is a $1 billion industry, and it ultimately enables a $200 billion electronics business. Other applications include laser welding and sintering, laser model generation, laser repair of semiconductor displays, curing of epoxy resins, diagnostic probes for real-time monitoring and control of chemical processes, optical techniques for alignment and inspection, machine vision, metrology, and even laser guidance systems for building tunnels. Optics can play an important role in ensuring a healthy U.S. manufacturing enterprise.

National Defense

In national *defense*, optical technology has become ubiquitous, from low-cost components to complex and expensive systems, and has dramatically changed the way wars are fought. Sophisticated satellite surveillance systems are a keystone of intelligence gathering. Night-vision imagers and missile guidance units allow the U.S. armed forces to "own the night." Lasers are used for everything from targeting and range finding to navigation, and may lead to high-power directed-energy weapons. The Department of Defense has a significant stake in optics.

Manufacturing of Optical Systems and Components

As the impact of optics has increased, changes have become necessary in how optical components and systems are designed and made. The manufacture of mass-market optics is now dominated by companies in Asia, but some recent developments are enabling U.S. industry to recapture selected market segments. One example is the emergence of new classes of numerically controlled optical grinding and polishing machines. Another is a better understanding of the characteristics of optical materials, from glasses to polymers to metals, thus permitting broader use of these automated technologies. Advanced optical components cannot be considered commodity items, and even though they represent only a small fraction of the value of the optical systems they enable, their availability is essential for the success of new high-level applications that rely on those systems. The U.S. optics industry is currently strongest in the design and manufacture of high-performance specialty products. A key U.S. strength is in optical design, which is being revolutionized by the development of fast and affordable ray-tracing software. The United States can preserve a presence in world markets for optical components and systems by focusing on areas where domestic capabilities are strong and by addressing the process by which international standards are set.

Education and Research

Underpinning the explosive growth of optics are investments in *education and research*. Research continues to lead to extraordinary discoveries. Although the field is growing rapidly and its impact is both pervasive and far-reaching, it remains a "multidiscipline" with components in many university departments and government programs. The presence of optics in these diverse programs reflects its pervasiveness but also reveals an Achilles' heel. Trends and developments in optics can easily be missed in such a disaggregated enterprise. Educational and research organizations will need to pay close attention to ensure that the field develops in a healthy way that ensures continuing benefits to society.

Overview

··

The role of light in our lives is both pervasive and primordial. Ultraviolet light probably had a role in the very origins of life, and light-driven photosynthesis underlies all but the most primitive of living things today. For humans, sight is the most crucial of the senses for perceiving the world around us. Indeed, the highly evolved vertebrate eye is one of the most exquisite light detectors ever created. Yet light is influencing the way we live today in ways we could never have imagined just a few decades ago. As we move into the next century, light will play an even more critical role—often the central role—in the ways we communicate, in the practice of medicine, in providing for the nation's defense, and in the tools we use to explore the frontiers of science. Optical science and engineering—or, more conveniently, just optics—is the diverse body of technologies, together with their scientific underpinnings, that seek to harness light for these and other tasks. This report addresses a broad range of issues pertinent to this field: its status today, the outlook for tomorrow, and what must be done to ensure its future vitality.

In a broad sense, optics has a long history. Mirrors were already in use thousands of years ago. By the early seventeenth century, lenses were being ground for microscopes and telescopes, and in 1704, Isaac Newton published his classic text *Opticks*, which set down the fundamental principles of reflection and refraction. Yet optics is also a thoroughly modern field, a young and vital one whose character has been revolutionized by technological advances over the last few decades.

The development of the first laser in 1960 produced a kind of light never before seen on Earth—light with the property of *coherence*. This coherence permits light to be directed, focused, and propagated in new ways impossible for incoherent light. Laser light has thus made possible

OPTICS DEFINED

Optics is the field of science and engineering encompassing the physical phenomena and technologies associated with the generation, transmission, manipulation, detection, and utilization of light.

fiber-optic communications, compact disks, laser surgery of the retina, and laser welding. Other developments in optics have perhaps been less obvious, but their impact can be equally well seen in the commodities and conveniences of our world. Examples include optical lithography systems for making computer chips, high-resolution microscopes, adaptive optics for ground-based astronomy, infrared sensors for a multitude of applications, and highly efficient lighting sources. The sidebar on page 7 suggests some of the ways in which these and other optical technologies affect our everyday lives.

Optics: A Pervasive Enabler

Not surprisingly, then, optics is rapidly becoming an important focus for new businesses in the global economy. In the United States, both large and small businesses are significant players in emerging optics business activity. Optics-related companies number more than 5,000, and their net financial impact amounts to more than $50 billion annually. More significant than this, however, is the role of optics as an *enabler*. Just as a lens in a pair of glasses enables clear vision, so an investment of a few hundred million dollars in optical-fiber technology has enabled a trillion-dollar worldwide communications revolution. A mere six laser transmitters are used in a transatlantic undersea telephone transmission system that can carry 40 million simultaneous conversations. The cost of the lasers is a tiny fraction of the cost of the system or the revenue it generates, but without them the system would be useless. Indeed, in his report to Congress on July 22, 1997, Federal Reserve chairman Alan Greenspan alluded to this enabling role: "We may be observing . . . a number of key technologies, some even mature, finally interacting to create significant new opportunities for value creation. For example, the applications for the laser were modest until the later development of fiber optics engendered a revolution in telecommunications."

As another instance, a compact disk player incorporates hundreds of intricate electronic and mechanical parts, all working together and all absolutely dependent on a single laser costing less than a dollar to illuminate the spinning disk. The following pages contain dozens of additional examples. Often, perhaps even usually, those who developed the enabling optical technologies never imagined their ultimate applications. In this report, the committee has thus sought to address the pivotal question, *How does one support and strengthen a field such as optics whose value is primarily enabling?*

The remarkable breadth of optics' enabling role is both an indicator of the field's importance and a source of challenges. Virtually every

ILLUMINATING OUR DAILY LIVES

Optics has a pervasive impact on our daily lives, but that impact is rarely noticed because the products of optical technology are, ironically, often invisible and because we accommodate so swiftly to modern technology. Today we pay as little attention to infrared remote controls, light-emitting diodes, and laser printers as to the mirrors that have been with us since antiquity. Here is a brief story to remind us of some of these pervasive optical technologies.

John reached over and shut off the *alarm clock*. He turned on the *lights* and got up. Downstairs, he began to make his morning coffee and *turned on* the *television* to check the *weather forecast*. Checking the time on the kitchen *clock*, he poured his coffee and went to the *solarium* to sit and read the *newspaper*.

Upstairs, the kids were getting ready for school. Julie was listening to a favorite *song* while getting *dressed*. Stevie felt sick, so his mother, Sarah, checked his *temperature*. Julie would go to school, but Stevie would stay home.

John drove to work in his new *car*, a high-tech showcase. He drove across a *bridge*, noticing the *emergency telephones* along the side of the freeway. He encountered *traffic signals*, *highway signs*, and a police officer scanning for *speeders*.

Awaiting John in his office were several *telephone* messages and a *fax*. He turned on his *computer*, checked some reference data on a *CD-ROM*, and *printed* it to look at later. After *copying* some last-minute handouts, he went to the conference room to make a *presentation*.

Meanwhile, Julie was walking to school. As she passed the neighbors' house, a *security light* came on. On the next block she passed a *construction* site for a new apartment building, then a block of *medical* offices. A few blocks away was the *factory* where her uncle worked.

At school, Julie's first class was biology. The students looked for *microbes* in water samples they had collected on a nature walk the previous day. On the walk they had also done some *birdwatching* and taken still and video *pictures* of the plants and wildlife. The teacher put on her *glasses* to read Julie's lab report.

At lunchtime, John left his office to do some grocery shopping. At the *checkout counter* he paid with a *credit card*. Among his purchases were a bag of *apples*, a *bottle* of wine, and a *carton* of milk. Each was labeled with a *bar code*.

At home, Stevie was watching a *movie* on the *large-screen television*. With her sick son occupied, Sarah connected her *laptop computer* to the *office network*. Modern technology let her do her work, despite having to stay home with the child—and at least John was stuck doing the shopping.

light-emitting diode (LED) displays
energy-saving compact fluorescent lamps
infrared remote controls
optical fibers for distributing cable television
satellite-based optical weather imaging
liquid crystal displays (LCDs)
temperature-moderating window coatings
phototypesetting
compact disks
laser fabric cutting
infrared noncontact "ear" thermometers
infrared automobile security systems;
  optical monitors for antilock brakes; LED, LCD, and
  optical fiber dashboard displays; LED taillights
optical-fiber sensors to monitor bridge integrity
solar power for emergency services
LED traffic lights
high-reflectivity surfaces for highway signs
laser traffic radar
optical fiber telephone cables
optical scanners and fax machines
photolithography for making computer chips
optical data storage
laser printers
photocopiers
overhead projectors, slide projectors, laser pointers
infrared motion sensors for home security
laser range-finders and surveying equipment
laser surgery, optical tools for medical diagnosis
laser welding and cutting, optical stereolithography
  for rapid three-dimensional prototyping
microscopes, magnifying lenses
binoculars
cameras, videocameras
eyeglasses
supermarket bar-code scanners
credit card holograms to prevent counterfeiting
image recognition for produce quality control
optical inspection to ensure clean bottles
optical inspection for labeling and packaging
bar-code readers for inventory control
videodisks and videodisk players
television displays
active-matrix displays for computers
optical fiber local area networks

scientific discipline uses optics in some way, so (perhaps unsurprisingly) optics courses in most universities are taught in several science departments, as well as engineering departments and schools of medicine. As a result of such organizational structures as these—in spite of, or perhaps because of, its pervasive importance—optics tends to be an orphan, owned by no one. A second, related question thus emerges: *How does one nurture a field that lacks a recognized academic or disciplinary home?*

Optics is thus an invaluable means, rarely an end in itself—a field often seamlessly integrated with electronics and materials science (see inset below) and an enabling presence in the university research lab, in our daily lives, and in countless businesses. In the diversity and pervasiveness of optics lies great strength, but these same qualities similarly pose a daunting hurdle to concise assessments and simple prescriptions. The central challenge of this study was to overcome that hurdle in providing a coherent picture of optical science and engineering today and in pointing the way to the field's continuing vitality.

Optics encompasses a broad set of technologies and techniques for exploiting the properties of light, and as suggested earlier, its applications are to be found everywhere. As a consequence, this report cannot hope to be comprehensive. The intent of the committee has been to highlight those areas in which breakthroughs are taking place, in which

....

MATERIALS: AN ENABLER FOR OPTICS

Just as optics has enabled scientific advances in diverse disciplines and spawned entire industries, so advances in materials science and engineering have enabled many of the advances in optics described in this report. Many of the devices and systems of modern optics are based largely on classical optical principles that could be fully exploited only after the discovery of new materials and the invention of efficient processes for bringing these new developments to the factory and the marketplace. In the 1870s, Alexander Graham Bell patented the "photo phone," which allowed a conversation to be transmitted through the atmosphere by a beam of light. Practical optical communications, however, awaited the discovery of a materials system capable of guiding light with negligible losses over distances of many kilometers—fiber optics. This practical development was the result of an intimate collaboration between materials and optical scientists. In other instances, new materials enabled the emergence of entirely new branches of optics. Nonlinear optical materials, such as lithium niobate and potassium niobate tantalate, and laser hosts such as yttrium aluminum garnet, were "molecularly engineered" through a synergism between optical physicists and engineers on the one hand and, on the other, solid-state chemists and materials scientists with a deep understanding of how chemical bonding and crystallographic structure determine optical properties. Understanding of the basic principles of crystal growth and phase equilibria, coupled with novel preparation techniques and factory engineering, paced the development and introduction of these new materials. In recent years, both materials science and modern optics have flourished in an atmosphere of interdisciplinary research and engineering, and together they have changed our modern world.

rapid change is likely in the near future, and to which national needs dictate special attention. The report is organized along the lines of seven major areas of national need: (1) information technology and telecommunications; (2) health care and the life sciences; (3) optical sensing, lighting, and energy; (4) optics in manufacturing; (5) national defense; (6) manufacturing of optical components and systems; and 7) optics research and education.

Trends and Developments

To motivate the report's key recommendations and underscore the dynamic state of optics today, it is useful to look at a few critical applications in which the developments have been most dramatic and the trends point to revolutionary change in the coming years.

Information Technology and Telecommunications

The explosive growth in many areas of optical technology constitutes an optics revolution. In information technology and communications, for example, progress during the past decade has been extraordinary. Around the world, optical fiber is currently being installed at a rate of 1,000 m every second, comparable to the speed of a Mach 2 aircraft. By the year 2005, about 600,000 km of fiber-optic cable will cross the oceans, enough to encircle Earth 15 times. Just 10 years ago, only 10 percent of all transcontinental calls in the United States were carried over fiber-optic cables; today the number is 90 percent.

To suggest the ramifications of this burgeoning capability for high-data-rate communications, a vision was recently proposed for the future of information technology: the "tera era." This vision articulates the need for cost-effective networks of enormously broadened bandwidth. The rapid growth of several key information technologies promises to meet the requirements of this vision: Fiber transport capacity, computer processing power, and magnetic storage density are all advancing by factors of 100 every 10 years. This implies that today's "giga" (10^9) performance will improve to "tera" (10^{12}) performance within 15 years. Some of the elements of the tera era vision are summarized on page 10.

Central advances toward the tera era will involve optical technologies. All elements of information transport are likely to require optical fibers and lasers, including 100-gigabit-per-second access networks, 10-gigabit-per-second local area networks, and even 1 gigabit per second to the desktop. Information processing is likely to require advances in both electronics and optics to achieve tera-era rates, and information storage is likely to rely on both volume optical storage and advanced

THE "TERA-ERA" VISION FOR INFORMATION TECHNOLOGY IN 10 TO 15 YEARS

TRANSPORT	Terabit-per-second backbone, long-haul networks • Access networks operating at hundreds of gigabits per second • Local area networks operating at tens of gigabits per second • 1 gigabit per second to the desktop
PROCESSING	Teraoperations-per-second computers • Terabit-per-second throughput switches • Multigigahertz clocks • Interconnections operating at hundreds of gigabytes per second
STORAGE	Terabyte data banks • Multiterabyte disk drives • Tens-of-gigabit memory chips

The tera era is a 10- to 15-year vision for the needs of the information age, as articulated by Joel Birnbaum of Hewlett-Packard in October 1996. Projections for switching and details for storage have been added. This vision demands hundredfold improvements in many central capabilities. For example, clock speeds for most information processing tasks were measured in hundreds of megahertz in 1997; to achieve the goal of the tera era, they must increase to several gigahertz by 2010.

• •

magnetic storage. Many capabilities will have to advance a hundred-fold to achieve this vision.

With the ubiquity of broadband networks and information services, displays are also increasing in importance. Text, graphics, images, and video are the outputs of the Information Age. Historically, the workhorse of display devices has been the cathode-ray tube (CRT). CRTs are cheap, they are easy to manufacture, and they have benefited from the mass market for televisions. In 1968, however, a new kind of display was invented—the liquid crystal display (LCD). These displays are lightweight, draw little current, and are thus ideal for portable devices such as watches, small televisions, and laptop computer displays. LCDs now command a large and growing share of the display market (see figures on page 11).

Health Care and the Life Sciences

Optics has had a dramatic impact on American health care, changing the practice of medicine and surgery and offering new approaches to both therapy and diagnostics. Some fields, such as ophthalmology,

have completely integrated lasers into clinical practice. Lasers have also enabled a number of new therapies, ranging from the treatment of kidney stones to the removal of skin lesions. Optics has also enabled the introduction of fiber-optic endoscopes, which allow convenient viewing of the body's interior and have led, in many cases, to the replacement of open surgery by such minimally invasive therapies as arthroscopic knee repairs.

Arguably, the broadest impact of optics in health care is in the fabrication of eyeglass frames, lenses, and contact lenses. In 1994 the market was estimated to be $13.2 billion, extending to 55 percent of the entire U.S. population. More important to the future of health care will be the application of a variety of more complex optical systems and techniques to a broad array of health needs. Lasers are approved by the Food and Drug Administration for refractive surgery to eliminate the need for glasses and are being used with light-activated drugs to treat cancers. Flow cytometry, an optically based diagnostic technique, is a critical tool for monitoring viral loads in AIDS patients and for guiding their therapy. Optical techniques are also under active investigation for noninvasive applications ranging from "needleless" glucose monitoring for the control of diabetes to the early detection of breast cancer.

Optical techniques are providing new tools for biology in the areas of visualization, measurement, analysis, and manipulation. Confocal laser-scanning microscopy has presented us with highly detailed three-dimensional pictures of biological structures. Two-photon techniques have not only enhanced the capabilities of fluorescence microscopy but

One future market expansion for displays is in flat, light, low-volume visual interfaces for personal digital assistants, Web browsers, augmented and virtual reality engines, and portable and wearable computers. The display shown here is a full-color VGA-sized device that carries nearly the information content of a personal computer monitor. (Courtesy of Planar America.)

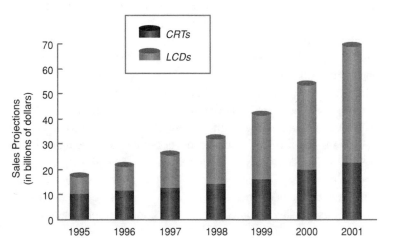

Worldwide market for electronic displays. Over the next several years, LCD sales are expected to grow rapidly.

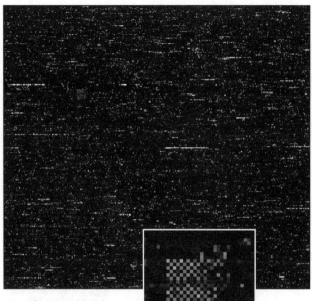

also opened up new possibilities for performing highly localized photo-chemistry within cells. Near-field techniques have enhanced resolution to beyond the diffraction limit.

In biotechnology, lasers have become essential parts of all the systems used for DNA sequencing, ranging from commercially available systems to more experimental capillary electrophoresis systems. Other applications of optics to biotechnology range from sophisticated systems using "DNA chips" (at left) to simpler systems using transmission probes.

Finally, laser tweezers are an example of the novel turns optics can take. Normal optical forces are minuscule on the scale of larger organisms, but they can be significant on the scale of macromolecules, organelles, and even whole cells. Laser tweezers thus afford an unprecedented means for manipulations at microscopic scales. They can tow a bacterium through water faster than it can swim, halt a swimming sperm cell in its track, or arrest the transport of an intracellular vesicle.

Part of a "DNA chip," showing fluorescently labelled DNA bound to an 8,000-site GeneChip® probe array. (Courtesy of Affymetrix, Inc., Santa Clara, Calif. Copyright © Affymetrix, Inc. All rights reserved. Affymetrix and GeneChip are registered trademarks used by Affymetrix, Inc.)

Optical Sensing, Lighting, and Energy

Advances in lighting sources and light distribution systems are poised to bring about a profound change in the way we use energy for lighting. Currently, lighting accounts for almost one-fifth of the electrical power used in the United States each year. However, recent developments in light sources now offer a dramatic new set of options. Among them are high-efficiency metal halide, sulfur-dimer, and light-emitting diode (LED) sources that can be up to 10 times as efficient as standard incandescent light bulbs.

Innovations in optical sensors have been equally dramatic. With the development of new optical and infrared sensors, it is now possible to greatly augment the normal human visual process and, in some cases, to show details and reveal information never before seen (see figure on page 13). A broad range of newly developed sensors have already taken their places in our daily lives: infrared cameras that provide satellite pictures of clouds and weather patterns for the evening news, infrared night-vision scopes for use by law enforcement agencies, infrared sensors in home security motion detectors, and supermarket bar-code scanners.

Less apparent in our everyday lives, open-path optical sensors are beginning to have a major impact on real-time measurements of several environmental gases, as well as emissions at industrial plants, and are just beginning to find applications to on-line industrial process control. Sensitive optical-imaging cameras and spectroscopic instruments are being used to monitor green biomass (plant density), as well as the ozone hole, and similar instruments have flown aboard spacecraft to measure the composition and atmospheric makeup of Mars, Jupiter, and several comets. Recent advances in real-time atmospheric turbulence compensation techniques ("rubber mirrors") now allow ground-based telescopes to achieve optical resolutions as good as the orbiting Hubble telescope. For the future, high-resolution, optical-imaging digital cameras are poised to revolutionize and computerize the photographic and printing industries, and improvements in photovoltaic solar cells may permit renewable energy sources to provide as much as 50 percent of the world's energy needs by the middle of the next century.

Infrared thermal image of a motor and drive assembly, showing the increased heat generated in the bearing as a result of misalignment between the drive-shaft and the bearing. This real-time imaging technique is used for testing motors and machinery in industrial plants while in operation and under load. (Courtesy of FLIR Systems, Inc.)

Optics in Manufacturing

Apart from the more recent acceptance and growth of optics and photonics in information technology, nowhere has optics had a more dramatic economic influence than in manufacturing, particularly since the advent of reliable low-cost lasers and laser-imaging systems. Optical techniques, applied both directly to manufacturing and as process control and diagnostic tools, have become crucial in such diverse industries as semiconductor manufacturing, civil construction, and chemical production. To cite the most obvious example, every semiconductor chip mass produced in the world today is manufactured using optical lithography. Manufacturing the necessary optical processing equipment alone is a $1 billion industry, enabling in turn a $200 billion electronics business. In addition, the semiconductor industry's vitality is a powerful stimulus for research and development: Shrinking feature sizes to 0.1 μm will demand new resist materials, new types of optics, new ultraviolet sources, and improved metrology for mask alignment.

Laser materials processing is another rapidly growing technology and one that promises to significantly reduce manufacturing costs in the automobile and aerospace industries, among others—especially where hostile environments may be present. Laser welding and sintering and

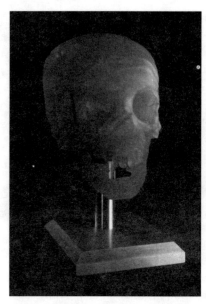

Lasers can be used to make three-dimensional prototypes of mechanical parts under the direct control of computer-aided design software. This skull is a demonstration of the complex shapes that can be made. (Courtesy of 3-D Systems, Inc.)

three-dimensional laser modeling hold great promise as new parts of the agile and flexible digital factories of the future. Laser model generation (at left), for example, lends itself to rapid prototyping, which is emerging as a new paradigm for reducing time and cost by eliminating many of the steps between engineering design and manufacturing. Laser repair and customization of high-cost items (e.g., semiconductor displays) are two other important cost-saving manufacturing techniques.

Optics can also be found in a host of chemical-related situations, for example, in the curing of epoxy resins in flat-panel displays or in diagnostic probes for the monitoring and control of on-line chemical processes. Optical sensors for real-time control of chemical processes have been shown to double the productivity of some existing manufacturing facilities. In the future, we may even see the routine industrial use of lasers to promote bond-selective chemical reactions.

As part of real-time process control, optical techniques for alignment and inspection have been shown to reduce variation and shorten the launch time for new product designs. Machine vision, for example, has enormous potential for enhancing automated manufacturing processes, but its use is often limited by the requirement for sophisticated, customized software algorithms. Customization is also frequently required in metrology applications, but the economies inherent in non-contact, real-time process feedback control often warrant the expense—for example, in guaranteeing the accurate assembly of complex structures in aircraft manufacturing. Optical inspection of finished or intermediate parts for quality control and statistical process control is especially critical in the electronics industry.

Much less sophisticated applications—but nonetheless highly useful ones—are found in laser guidance systems for the construction industry, where such systems have greatly reduced costs and allowed more rapid and more precise alignments in building, tunneling, and surface grading.

National Defense

Television images of precision laser-guided bombs zeroing in on and obliterating a military headquarters building in Baghdad during

Operation Desert Storm highlighted the central role of optics in modern warfare. Optical technologies play a key enabling role in modern battle plans and have dramatically changed the way wars are fought.

The roles of optics in defense are pervasive and ubiquitous. These range from low-cost components to complex and expensive systems. Among the latter, sophisticated satellite surveillance systems are a keystone of our intelligence gathering (see photograph above). Lower-cost night-vision imagers and missile guidance units have recently allowed the United States to "own the night" and to dominate the battlefield. Not surprisingly, lasers also have a place. In applications analogous to the use of radio waves in radar, laser radiation propagating in free space can be used for targeting and range finding. Recent developments have also led to such novel applications as laser gyros for navigation. In the future, laser weapons based on gas or chemical lasers that generate more than 1 MW of power will open the door to directed-energy weapons that may find both terrestrial and space-based applications. Developments in optics also promise a host of less obvious payoffs for defense—many with likely civilian spin-offs as well. For example, Department of Defense investments in new high-leverage optical technologies such as photonics and chemical agent detection will provide a unique military advantage on the battlefield of the future, as well as nondefense payoffs.

Manufacturing of Optical Components and Systems

As the impact of optics has increased in recent years, major changes have been necessary in the way optical components and systems are designed and fabricated. As recently as a decade ago, most optical components were made in small specialty optics shops, with a premium on

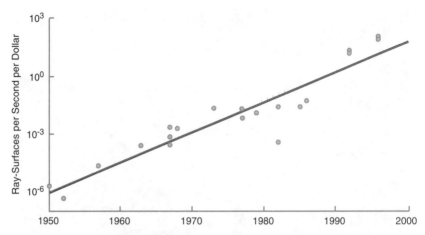

There has been a dramatic revolution in both the speed of ray tracing and the capital cost of the computers required (note the logarithmic scale).

craft work. This work was driven by the needs of the military, government agencies, and scientific research. Today, as military budgets and lot sizes decrease and the use of optics grows in applications as diverse as optical fibers for communication, large space optics, and high-performance short-wavelength aspheric optics for integrated-circuit fabrication, the demand is growing for cheaper, faster, more flexible optics manufacturing with increased capabilities.

In the face of these needs, the manufacture of mass-market optics is now dominated by Asian countries—potentially a cause for concern. On the other hand, developments are enabling U.S. industry to recapture some segments of the optical component fabrication business. One example is the emergence of new classes of numerically controlled optical grinding and polishing machines. Another is a better understanding of the characteristics of optical materials from glasses to polymers to metals, thus permitting broader use of these automated technologies. Advanced optical components cannot be considered commodity items, and even though they represent only a small fraction of the value of the optical systems they enable, their availability is essential for the success of new high-level applications that rely on such systems. To take full advantage of new fabrication methods, flexible manufacturing, assembly, and testing techniques have to be integrated. The U.S. optics industry is currently strongest in the design and manufacture of high-performance specialty products. A key U.S. strength is in optical design, which is being revolutionized by the development of fast and affordable ray-tracing software (see figure above).

In looking to the future, affordable infrared systems for military uses and high-quality components for deep-ultraviolet wavelengths in the semiconductor industry are two of the major challenges within reach of today's optical industry. General aspherics, diffractive and gradient-index optical elements, and conformal optics also offer new optical design and packaging possibilities. In particular, industry support for deep-ultraviolet aspherics is required for chip making. Fiber devices such as fiber lasers and amplifiers, fiber gratings, and dispersion compensators are also growing rapidly in importance, driven by information technology applications.

Photonic materials are another growth area. Continued advances in photonic networks will depend on continued reductions in the cost of photonic components, increased functionality, and increased levels of integration. Computer-assisted design tools, which are becoming essential for shortening design cycles and increasing functionality, are not available for designing photonic components. Process improvements are also needed to ensure the future practical use of photonic materials such as cadmium telluride, silicon carbide, gallium nitride, and aluminum nitride.

Research and Education

Research continues to yield extraordinary discoveries. Light is now being used to control atoms, with applications in laser cooling and the engineering of quantum states (see figure below). New fluorescent dyes allow the detection of single molecules, with important implications for chemistry, biology, and medicine. New techniques using ultrashort, high-peak-power laser pulses are being made possible by the development of femtosecond optics. Advances in semiconductor lasers are dramatically reducing costs and increasing utility. Research on nonlinear optical materials promises to open up new approaches for future optical devices. High-frequency sources and optical components are enabling microscopy and lithography to move into the extreme ultraviolet

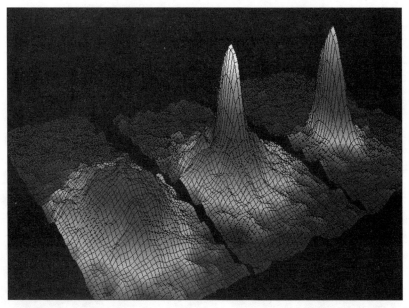

The momentum distribution of atoms during Bose condensation of a laser-cooled dilute gas. (Courtesy of M. Matthews, JILA.)

and even the x-ray wavelength regime. These are just a few highlights of the exciting research now under way in optics.

There are only a few formal postgraduate educational programs in optics. In most universities, optics is taught in a wide range of departments—biology, physics, materials science, electrical engineering, computer science, and others—as befits the span of the field's applications. Academic research in optics is no less diverse, and in their diversity, universities have made indispensable contributions to furthering optical technologies. In information technology, for example, academic programs are particularly strong in the basic sciences and the development of device technologies, but weak in the corresponding systems, packaging, and applications areas. Accordingly, universities are relatively ineffective in transferring technologies to U.S. industry and in keeping pace with the needs of newly developing systems.

Key Conclusions and Recommendations

Along with explaining the state of the art in optics and presenting a vision of its future, an important goal of the study that produced this report was to identify actions that could facilitate the field's future technical development and enhance its contributions to society. Each chapter of the report draws conclusions that bear on this goal and offers recommendations aimed at reaching it.

As this report shows, optics is an extraordinarily strong and dynamic field. Its diverse applications are making important contributions to society in areas that range from telecommunications to medicine to energy to national defense. In most cases, the field's vitality is sufficient to keep up the rate of advance without major changes in public policy or other major intervention. In a few important cases, however, action is needed to overcome barriers that might slow the present pace of rapid progress—to ensure consumer access to the dramatically increasing capacity of optical-fiber communications, for example, or to take full advantage of the potential of noninvasive optical methods for medical monitoring and diagnosis. Key recommendations aimed at addressing these concerns are summarized below, along with brief synopses of the arguments that underlie them.

Information Technology and Telecommunications

Optics is now the preferred technology for the transmission of voice, data, and video information over long distances. Already the dominant communications technology throughout the world, fiber-optic systems continue to be deployed at an increasing rate, linking cities and continents

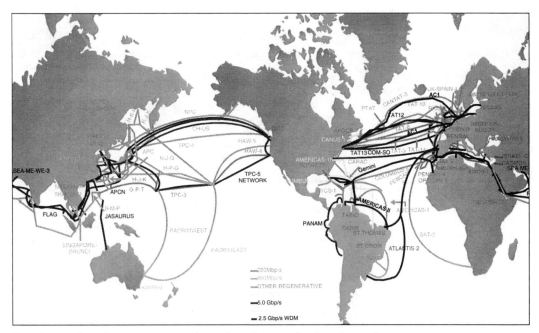

(see figure above). As one example, only a thousand voice channels spanned the Atlantic Ocean in 1975; in 1998 10 million are in place or under construction. This technological shift is due primarily to the huge long-distance information capacity made possible by the use of optical fibers and laser transmitters. Data rates exceeding 1 terabit per second per fiber have been demonstrated in the laboratory, and systems capable of 20 gigabits per second are being deployed commercially.

Worldwide undersea network of fiber-optic cable for telecommunications. (Courtesy of P.K. Runge, Tyco Submarine Systems Laboratories.)

As technological advances reduce the cost of optical components, it is becoming cost-effective to use optical fiber communication systems over shorter and shorter distances. Today's long-distance fiber networks extend only as far as the local telephone office, where signals are transferred to metal wires for transmission to and from individual homes and offices. It will eventually be feasible to extend these fiber networks all the way to the end user, dramatically increasing the bandwidth available to individual users for voice, video, and data services.

Demand for high-bandwidth services has grown tremendously over the past few years. The extraordinary growth in use of the Internet, especially the explosive growth of the graphics-intensive World Wide Web, has been the main driving force. Continued rapid expansion of the Web, along with increased demand for audio and video transmissions, will accelerate this growing demand. There are ways to greatly increase transmission capacity while still using the twisted-pair wires that now link homes and offices to the telephone network, but optical connections are much faster than wires will ever be. Just a few years ago, even today's demand for Internet access and other high-bandwidth services would have seemed extraordinary. Optical fiber to the home will provide for the expected rapid growth in these services. It will also "future-proof" the network against the bandwidth needs of future services that do not even exist yet but that recent history indicates are sure to emerge.

Fiber also has a number of technical advantages over metal wiring. For example, it is corrosion resistant and immune to electromagnetic interference, and it can eliminate the need for electrical power between the central office and the home, with considerable potential savings in the operation and maintenance of the network. A number of technical challenges remain to be overcome, however, before the ultimate goal of fiber to the home can be achieved. Chief among these is the development of new technologies that reduce the costs of optical components, packages, and systems. Also to be considered is the combined technical-regulatory issue of lifeline power (i.e., enabling communication during emergencies if the customer's premises lose power).

The rapid emergence of broadband telecommunication networks and broadband information services will have an enormously stimulating impact on commerce, industry, and defense worldwide. Congress should challenge industry and the federal regulatory agencies to ensure the rapid development and deployment of a broadband fiber-to-the-home information infrastructure. Only by beginning this task now can the United States position itself to be the world leader in both broadband technology itself and its use in the service of society.

Health Care and the Life Sciences

Developments in optics and especially in lasers have made dramatic contributions to health care. Rigid and flexible viewing scopes allow minimally invasive diagnosis and treatment of numerous sites inside the body, such as the colon, the knee, and the uterus. Accordingly, lasers have become accepted and commonly used tools for a variety of surgical applications in fields ranging from ophthalmology to gynecology (see figure on page 21). In contrast to the progress in therapeutics, however, noninvasive diagnostic methods are less well developed, despite their great potential in the medical laboratory (more accurate blood tests, for example), in the clinic (adjuncts to x-ray mammography), and for home care (noninvasive glucose monitoring). Particularly in the area of monitoring basic body chemistries, the fundamental science is often incomplete; for example, the optical signatures of some human biological processes and substances have yet to be determined. Although the National Institutes of Health (NIH) does have a modest effort in biomedical optical technology, the disease-oriented structure of NIH does not encourage the growth of biomedical optical technology programs.

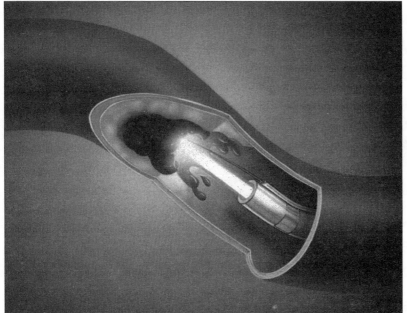

Schematic diagram illustrating laser thrombolysis, the use of a laser to destroy a blood clot. (Courtesy of K. Gregory, Oregon Medical Laser Center.)

To hasten progress in this area, mechanisms should be sought to encourage increased public and private investment in the development of noninvasive optical monitoring of basic body chemistries. A clearer separation between the roles of the public sector (support for basic science and proof-of-principle demonstrations) and those of the private sector (device development) is needed.

Optimal utilization of the fluorescence-based techniques so essential to biology depends on the continual development of new, specific, and inexpensive molecular probes. Scientists, engineers, and technicians with cross-disciplinary training will enhance the transfer of optical science to biology and medicine.

The National Institutes of Health should recognize the importance of optical science in biomedical research aimed at understanding human disease by establishing a study section dedicated to this area. NIH should raise the priority for funding innovative optical technologies for medicine and medical research. An initiative should be launched to identify the optical signatures of human biological processes and substances for application to noninvasive monitoring.

Optical Sensing, Lighting, and Energy

The economic impact of lighting the United States has already been noted. Lighting accounts for about 19 percent ($40 billion) of the total annual electricity use in this country, but higher-efficiency light sources and distribution systems are now available that might dramatically change these numbers. Among the new sources are high-efficiency

LIGHT SOURCE EFFICIENCY

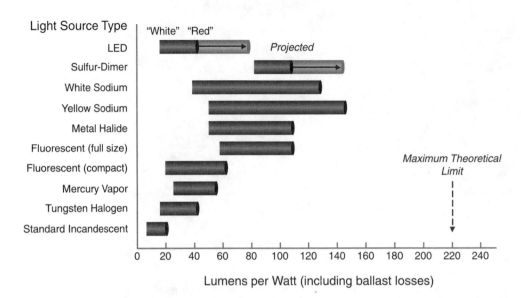

Efficiency of selected types of light sources. The output of several new lighting sources exceeds that of conventional incandescent bulbs three- to tenfold.

metal halide lamps, sunlight-spectrum sulfur-dimer lamps, and white LED sources that can be three to 10 times more efficient than conventional lightbulbs (see figure above). Accordingly, these new light sources have the potential to reduce consumer electricity bills in the United States by tens of billions of dollars each year. For instance, the ongoing replacement of red traffic lights with new red LED lights is expected to save about $175 million annually in the United States within the next few years.

The Department of Energy, the Environmental Protection Agency, the Electric Power Research Institute, and the National Electrical Manufacturers Association should coordinate their efforts to create a single program to enhance the efficiency and efficacy of new lighting sources and delivery systems, with the goal of reducing U.S. consumption of electricity for lighting by a factor of two over the next decade, thus saving about $10 billion to $20 billion per year in energy costs.

National Defense

Optical technologies have played and continue to play an indispensable role in national defense. Post–Cold War Department of Defense (DOD) policy has not substantially reduced the need to develop and acquire optical systems, but it has affected the R&D process. Acquisition reform that actively leverages the use of commercial off-the-shelf technology to reduce procurement and R&D costs is difficult to apply widely to optics because commercial optical products require special adaptation or improvement for DOD use. For example, computer displays cannot withstand the conditions of vibration, shock, abrasion, and so on that are common on the battlefield. Nonetheless, current DOD policy, which stresses the development of new systems with low risk at low cost, has led to greater reliance on existing or adapted commercial technology and, concomitantly, to less aggressive funding of new optical technology than in the past. The Department of Defense is also seeking to limit its support of manufacturing and other core optical competencies, a position inconsistent with DOD's special operational requirements for systems to be produced at low volume.

The war-proven effectiveness of optics ensures that DOD will continue to require highly capable optical systems for surveillance and for surgical strikes with precision guided munitions, to mention only two conspicuous examples. Downsizing must be done very carefully to ensure that an adequate manufacturing infrastructure is maintained and supported. Similarly, in light of a past climate in which DOD aggressively drove optics development, future efforts must be carefully planned to extract maximum benefit from lower discretionary R&D budgets, expended mainly by the Defense Advanced Research Projects Agency (DARPA). Certain R&D and manufacturing areas deserve special attention and emphasis by the Department of Defense, especially those that offer near-term DOD payoffs, as well as potential commercial economic returns. These areas include the low-cost manufacturing of precision aspheric, diffractive, and conformal optics, discussed below under "Manufacturing of Optical Components and Systems."

Despite the reality of Department of Defense downsizing and acquisition reform, DOD should stress investment in R&D on key optical technologies such as photonics, sensors, and high-power tunable lasers to gain maximum defense competitive advantage. Special attention should also be given to investment in low-cost manufacturing of precision aspheric, diffractive, and conformal optics.

Optics in Manufacturing

Optical techniques are used in a wide variety of manufacturing environments, from photolithography for making semiconductor chips to optical sensors for measuring the temperature of molten steel. These applications fall into two broad classes: (1) the use of light to perform manufacturing, including photolithography, laser materials processing such as welding and machining, and optical methods for rapid prototyping and manufacturing; and (2) the use of optics to control manufacturing, including metrology, machine vision systems, and a wide variety of sensors. Optical lithography is an essential part of the semiconductor, flat-panel, and CRT display industries. In light of this leading role, there is a compelling need for—and great advantage to be gained from—

further advances in ultraprecise optical lithography. Other related fabrication approaches, such as microelectromechanical systems, are rapidly developing as future technologies for control systems, sensors, displays, and other emerging applications. In this broad area, several entities, including DARPA and SEMATECH, not only play key, high-level oversight and coordination roles, but also demonstrate an effective cooperation among government, industry, and academia. A notable example of such cooperation is the Precision Laser Machining Consortium. This government-led consortium has brought together a wide variety of companies, from aerospace to automobiles, to solve their common problems. No single company would have independently devoted the necessary R&D resources to solving these problems, but by coming together, the consortium is making significant progress that is of value to all participants.

Participation in the DARPA-sponsored Precision Laser Machining Consortium should be extended to other optically assisted manufacturing areas by establishing a test facility as a service center.

Manufacturing of Optical Components and Systems

The optical fabrication industry is fractionated, with each company generally being quite specialized. Much of the U.S. high-precision optics supplier base comprises small innovative companies, which can often be flexible and dynamic in their approach to optical system design and fabrication.

In today's fertile and competitive environment, the several professional and trade organizations that represent different portions of the optical industry have, until recently, contended rather than cooperated. (This situation is changing, partly through the formation of the Coalition for Photonics and Optics.) In the face of such specialization—even disunity—industry needs a central voice and source of developments if it is to grow and prosper. Several collaborative programs with significant government participation offer sufficient incentives for modernization of optical design and fabrication.

Active government and industrial participation and support for strong, applicable international standards for optical components are important in cementing the links among the diverse components of the optical industry. Virtually all recent standards have been produced by overseas industry. The failure of U.S. industry and the federal government to support these activities has led to U.S. industry being a follower in adapting to new international opportunities.

The National Institute of Standards and Technology should become a leader in the development of international optics standards by coordinating the efforts of U.S. industry and the domestic and foreign standards-setting communities.

Traditionally, lens systems have been based on the use of spherical refracting surfaces; however, significant design simplifications and improvements in image quality can be obtained by using general aspheric surfaces. Currently, the United States has a significant lead in the low-volume production of high-precision aspheric components. The future of new, efficient imaging systems depends on developing a cost-effective method of producing such surfaces in volume. Other opportunities in manufacturing can be expected from the seamless integration of lens systems and active optical components. These findings lend additional support to the recommendation in Chapter 4 regarding the low-cost manufacturing of aspherics.

Research and Education

As an enabler, optics affects many fields and supports a vast array of applications. However, the funding of such cross-disciplinary areas of research and development is often hindered by the structure and organization of federal agencies. Accordingly, the National Science and Technology Council has been charged to identify promising areas of science and technology that cut across disciplines and would benefit from coordinated initiatives involving multiple agencies.

Optics is an enabling multidisciplinary technology that has a significant and growing impact in many areas of our lives. Multiple agencies should form a working group to support optics as a crosscutting initiative similar to the recent initiative in high-performance computing and communications systems.

Materials advances have been an integral part of the progress in optical devices and systems, from the demonstration of the first laser to the invention and installation of low-loss optical fiber. Progress in materials is a recurring theme in this report, from information technologies to the manufacturing of optics and the demand for sensors to withstand extreme environments. Research on new materials and materials processing methods thus remains critical to advances in optics. However, the development of materials for future applications is especially challenging in the United States, where research funding is rarely consistent with the decade-long time frames required to bring a new material to the level of commercial readiness.

Progress in materials science and engineering is critical to progress in optics. DARPA should therefore coordinate and invest in research on new optical materials and materials processing methods with the goal of maintaining a stream of materials breakthroughs.

In 1995 the National Science Foundation (NSF) announced a new multidisciplinary research initiative in optical science and engineering.

The call for preliminary proposals elicited more than 600 pre-proposals, which were reduced by selection to 70 submitted full proposals, of which only 18 could be funded. These numbers indicate both the extremely competitive nature of funded research and the nationwide interest in optics research. The NSF initiative also incorporated educational aspects that meet many of the needs of optics for interdisciplinary education, with an emphasis on teamwork and systems understanding. The initiative included undergraduate students in the research programs, a key stimulus to student interest in graduate research. Despite its apparent success, however, this NSF initiative was a short-lived venture.

NSF should develop an ongoing, agency-wide, separately funded initiative to support multidisciplinary research and education in optics. Examples of research and education opportunities include fundamental research on atomic, molecular, and quantum optics; femtosecond optics, sources, and applications; solid-state laser sources and applications; and extreme ultraviolet and soft x-ray optics.

Universities should encourage multidisciplinarity in optics education, cutting across departmental boundaries, and should provide research opportunities at all levels, from the bachelor's degree to the doctorate and from basic science to applied technology.

Despite its essential enabling role—or perhaps because of it—optics remains an ill-defined educational program at most institutions. Progress in optics has not obviously suffered from this situation, but a greater commitment to professional education in optics would serve well the continuing expansion in optics research and development. Wider visibility for existing educational programs and more systematic information about them would also greatly benefit the next generation of optical scientists and engineers.

X-ray microscope images of a red blood cell show features of a malaria infection site observed at a resolution not obtainable with an optical microscope. The left image shows a new infection. The right image was taken after 36 hours. (Courtesy of C. Magowan and W. Meyer-Ilse, Lawrence Berkeley National Laboratory.)

Professional societies should continue to expand their commitment to professional education in optics. Accordingly, they should work to

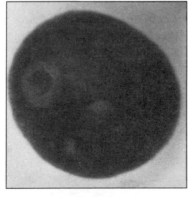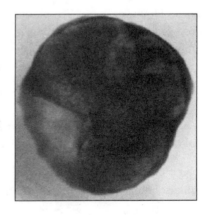

strengthen optics as a recognized crosscutting area of science and technology through the recently established Coalition for Photonics and Optics. Professional societies should also evaluate optics programs and jointly produce an annual guide to educational programs in optics.

A Vision of the Future

As we peer into the next century, we foresee developments in optics that will change our lives in ways that today we can hardly even contemplate. In almost every major area described in this report, we expect optics to change our world. The future will undoubtedly surprise us, but here is one possible vision:

We imagine the entire world linked together with high-speed fiber-optic communications, as ubiquitous as today's telephone system, made possible by advances in optical materials that enable the mass production of inexpensive, very high quality optical components and systems. This will result in the growth of very high speed Internet data and video transmission and other new broadband communications services.

In health care, the development of optical ways to monitor human processes could have an enormous impact on diagnosis and treatment. We dream of a day when people have personal health monitors that can monitor their health cost-effectively and noninvasively by evaluating the optical properties of their blood and tissue. We also foresee a growing impact of optics in many other areas of diagnostic and therapeutic medicine, biomedical research (see photos on page 26), and our quality of life.

Facing a world enveloped in greenhouse gases, we will have to consume energy more wisely. Highly efficient lighting technologies will significantly reduce the energy it takes to illuminate the world. Solar cells will reduce our dependence on fossil fuels by making electricity from the light of the sun.

In industry, optical sensors and infrared imagers will make significant inroads into process control for manufacturing and materials processing. Factories will employ optical sensors extensively in the manufacture of everything from textiles to automobiles, and digital cameras will substitute for film in printing and photography. In the electronics industry, which relies on photolithography to create circuit patterns on chips, producing features smaller than 0.1 μm will require optical steppers that use soft x-ray or extreme ultraviolet light; optical components for these machines will have unprecedented optical figure and atom-level surface smoothness.

The battlefield of the future, of which the Gulf War gave us but a glimpse, will see optics used in virtually every aspect of battle, from weapons targeting to the detection of chemical and biological warfare agents. This omnipresence will depend critically on the availability of low-cost optical systems, many of them developed for commercial use; unique military needs for performance and reliability, unmet by the commercial marketplace, will continue to require targeted investments in optics research.

The role of optics in research, which already cuts across nearly all fields of science and technology, will be limited only by our imagination. High-power laser systems will make possible the construction of particle accelerators that extend the energy frontier for experiments in particle physics. Lasers will manipulate individual atoms in light traps. Laser interferometer experiments may unravel the mysteries of gravity. Femtosecond visible and x-ray sources will provide new tools for understanding the dynamics of materials.

As the importance of optics grows, colleges and universities will be challenged to meet the educational needs of a growing work force. In time, we expect the field of optics to become a discipline, as computer science has over the past few decades, and to become recognized as such in educational institutions around the world.

1

....

Optics in Information Technology and Telecommunications

The information industry, including the services it provides, is growing rapidly worldwide. Its annual revenue is estimated to exceed $1 trillion, which, at an average revenue of $200,000 per employee, translates into 5 million jobs.

The demand for new information services, including data, Internet, and broadband services, has combined with the supply of innovative information technology to move us rapidly into the information age (Figure 1.1). The information age has recently been called the "tera era" because its technology demands are terabit-per-second information transport, teraoperations-per-second computer processing power, and terabyte information storage (see Box 1.1). Because of the growing importance of image and video information, there is also demand for

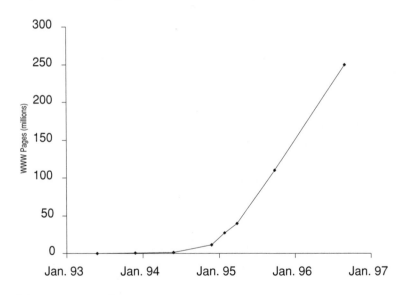

FIGURE 1.1 *The growth of the Internet and the World Wide Web is among the factors driving the growth of information technology. Although not all sources agree on the exact statistics of this growth, all agree that it is a spectacular phenomenon. Optics is an important enabler for information technology. (Courtesy of P. Shumate, Bellcore.)*

BOX 1.1 THE "TERA ERA" VISION FOR INFORMATION TECHNOLOGY IN 10 TO 15 YEARS

TRANSPORT | Terabit-per-second backbone, long-haul networks
- Access networks operating at hundreds of gigabits per second
- Local area networks operating at tens of gigabits per second
- 1 gigabit per second to the desktop

PROCESSING | Teraoperations-per-second computers
- Terabit-per-second throughput switches
- Multigigahertz clocks
- Interconections operating at hundreds of gigabytes per second

STORAGE | Terabyte data banks
- Multiterabyte disk drives
- Tens-of-gigabit memory chips

The tera era is a 10- to 15-year vision for the needs of the information age, as articulated by Joel Birnbaum of Hewlett-Packard in October 1996. Projections for switching and details for storage have been added. This vision includes the need for cost-effective networks of virtually unlimited bandwidth. Note that the roadmaps of several key information technologies promise to meet the requirements of this vision: fiber transport capacity, computer processing power, and magnetic storage density are all advancing by a factor of 100 every 10 years. This implies that giga (10^9) performance will improve to tera (10^{12}) performance within about 15 years.

••

strong advances in display technology. Optics and electronics are partnering and complementing each other in meeting this demand for information technology and thus enabling the information age.

There are five major technology segments in which optics plays a major role or has the chance to do so in the future. One of them is information *transport* over long distances, through large networks under the ocean, across continents, and in the local networks of the telephone and cable television systems. For this, optical fiber transmission is already the technology of choice, with a clear edge in cost and performance over competing technologies such as coaxial cable or satellite communications. Optical *processing*, including switching and networking, is another segment in which worldwide R&D hopes to open up new markets. A third segment is the *storage* of information, where technology has to meet rapidly increasing demands for more and more storage capacity. Optical techniques are a strongly growing complement to traditional magnetic storage. A fourth segment is the *display* of information, for which optics is the intrinsic, unavoidable link between the human eye and the electronics of a television or computer. The fifth important segment is the interface between electronic machinery and information recorded on paper, which includes *printers, scanners, and copiers.* This segment is not covered in this report since most of its

R&D is conducted in industry, which—for competitive reasons—keeps these subjects quite proprietary; very little work is done in this area in university or government laboratories.

The field of information technology provides many examples of the enabling role of optical technology. Often these enablers are small in size or cost but have an impact on a grand scale in large systems and applications. A tiny semiconductor laser, for example, enables the building of an optical transmitter, which enables a transmission system, which enables the construction of a telecommunications network, which enables the delivery of information age services such as multimedia or the Internet. Another example is the optical fiber, which enables the construction of an optical cable, which enables the construction of a network, and so on. A third example is the liquid crystal, which enables the flat-panel display, without which the laptop computer could not exist. These chains of enablers make it difficult to place firm dollar values on individual component technologies; components such as lasers or fibers are relatively inexpensive, but the service revenues of telecommunications networks are in the hundreds of billions of dollars. Box 1.2 gives a more detailed illustration of the enabling devices for long-haul information transport systems.

As for optics in general, optical materials play an important enabling role. Chapter 6 gives further details of this topic (see Box 6.2, "Photonic Materials").

A few rather obvious but critical points about *high-tech mass markets* and *low-cost manufacturing* should be more widely understood and appreciated. They clearly apply to the mass markets for optical information technologies:

BOX 1.2 ENABLING PHOTONIC DEVICES FOR PRESENT AND FUTURE LIGHTWAVE LONG-HAUL SYSTEMS

Semiconductor lasers
- High-power pumps for fiber optical amplifiers
- Integrated laser-modulator transmitters
- Multiwavelength transmitters
- Tunable transmitters
- Soliton sources

Semiconductor optical amplifiers
- Power amplifiers in integrated transmitters
- 1.3-μm optical amplifiers
- Optical switches and switch arrays
- Wavelength converters and other nonlinear functions

Photodetectors and OEIC receivers
- High-speed (>10 gigabit-per-second [Gb/s]) and high-sensitivity OEIC receivers
- High-speed (>10 Gb/s) avalanche photodiodes

Planar integrated waveguide components
- Couplers
- Filters (fixed and tunable)
- Wavelength multiplexers and routers
- Modulators and switches
- Doped-waveguide devices
- Dispersion compensators

Fiber-type components
- Doped-fiber devices (amplifiers and lasers)
- Couplers
- Grating filters
- Wavelength multiplexers
- Dispersion compensators
- Switches (mechanical)
- High-power lasers and brightness converters
- Isolators and circulators
- Nonlinear devices

(Source: T. Li, AT&T Laboratories.)

- When a technology creates a mass market, its products become commodity items whose production requires a capability for low-cost manufacturing.
- Suppliers that have prepared and invested to develop competence and capability in low-cost manufacturing by the time the mass market emerges are likely to gain the largest market share.
- The revenues generated by a large market share provide a significant source of R&D funds. This fact results in a rapid learning curve for the technology and its manufacturing, which drives costs down further and creates barriers to market entry for other suppliers.

Three established mass markets in optical information technology serve as excellent illustrations of the above points: (1) compact disk (CD)-based optical storage, (2) liquid crystal displays, and (3) cathode-ray tube displays. Each market has global revenues in excess of $20 billion per year, and in each market, offshore manufacturers planned early and were ready with timely low-cost manufacturing. As a result, more than 90% of the manufacturing of these high-tech, mass market products now occurs offshore.

At least four emerging optical information technologies—fiber to the home, optical data links, small and miniature displays, and projection displays—have an excellent chance of creating global mass markets in the not too distant future that are similar in magnitude to the three listed above. We should learn from history and make sure that U.S. industry captures a good share of the emerging mass markets.

The remainder of this chapter summarizes key trends in the four major technology segments described above, including the size of their markets and the technologies they compete with. It also outlines key challenges in this area for science, technology, education, and international competitiveness. The bulk of the chapter is based on the expert presentations and position papers listed in Appendix B.

Information Transport

Optics is now the preferred technology for the transmission of information over long distances. This technology, based on semiconductor junction lasers and optical fibers, is the technology of the information superhighways used to transmit voice, data, and video information. The major market segments include undersea transmission between continents, terrestrial long-distance telephone trunking between states and cities, local telephone exchange links between the central office and the home, and the networks of the cable television companies. Satellite-to-satellite links also offer exciting possibilities. Table 1.1 illustrates typical distance scales for these market segments.

TABLE 1.1 Typical Distance Scales for Selected Information Transport Applications

Application	Distance (km)
Satellite links	50,000
Undersea transmission	1,000-10,000
Terrestrial long haul	20-1,000
Cable television links	10-20
Fiber in the local exchange	10-20

The optical fiber medium offers several advantages over earlier transmission media such as coaxial cables and copper wire pairs. The most significant among these are large transmission capacity (high bandwidth), large repeater spacings, small cable size, low cable weight, and immunity from electromagnetic interference. These advantages have led to large-scale installations of optical fiber all over the world. It is estimated that more than 100 million kilometers of fiber are now installed. Figure 1.2 provides more details on U.S. installations.

Among the grand challenges that face the R&D community in this area are (1) the development of a cost-effective wavelength division multiplexing technology for transmitting multiple signals on a single fiber, and

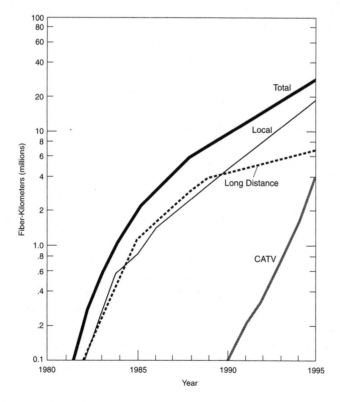

FIGURE 1.2 Nearly 40 million kilometers of fiber have now been installed in the United States, and the number is still growing strongly. As the figure shows, long-haul installations (by both long-distance and local telephone companies) started the trend around 1982. Since 1990, fiber installations by cable television (CATV) companies have grown strongly. It is anticipated that fiber-to-the-curb or fiber-to-the-home installations by local telephone companies will follow. Note that a single fiber cable typically contains 20 to 40 optical fibers. (Courtesy of T. Li, AT&T Laboratories.)

(2) the development of a low-cost technology for large-scale deployment of fiber-to-the-home systems that can deliver broadband services.

An issue common to all optical information transport technologies is the lack of systems education at U.S. universities. They offer excellent programs on the required materials, device, and component aspects of the field, but they seem not to have found a way to integrate into their programs the worldwide trend toward a greater emphasis on software and systems.

Four major application areas of optical information transport are discussed in this section: (1) long-distance transmission, (2) fiber to the home, (3) analog transmission, and (4) optical communications in space.

Long-Distance Transmission

Undersea systems and terrestrial long-haul systems have been early large-scale users of optical technology. Rapid advances in both photonics and electronics are producing a wealth of new technologies that continue to significantly increase the performance of these systems. Recent progress includes the development of optical fiber amplifiers, wavelength division multiplexing technology, photonic integrated circuits, and video compression. The vast increase in single-fiber transmission capacity illustrates the rapid pace of progress and serves as a key technology roadmap for the field (see Figure 1.3). It shows exponential growth, with capacity increasing by a factor of about 100 every 10 years. Transmission at 1 terabit per second (Tb/s) in the research laboratory was reported in 1996. Commercial systems appear to parallel the trend of laboratory demonstrations, with a lag of 3 to 7 years.

To appreciate this staggering information capacity, recall that an optical fiber is just a thin strand of glass, about as thick as a hair. Contemplate one of your hairs and note that a terabit is a million megabits. This means that at the recently demonstrated capacity of 1 Tb/s, a hairlike fiber can transmit as many as 40 million data connections (at 28 kilobaud), 20 million conventional digital voice telephony channels, or half a million compressed digital television channels.

Undersea Transmission

Undersea fiber systems have become the major information highways between continents (see Figure 1.4). The first large system of this kind, linking the United States and Europe, was placed in service in 1988, as the successor to the transatlantic telegraph cable, which was installed in 1858, and the transatlantic telephone cable, which began operation in 1956. The old system used coaxial cables and carried information in analog form. It had a capacity of about 40 voice circuits. The new fiber systems use digital technology, which (as in computers and CDs) provides a considerable improvement in quality; at

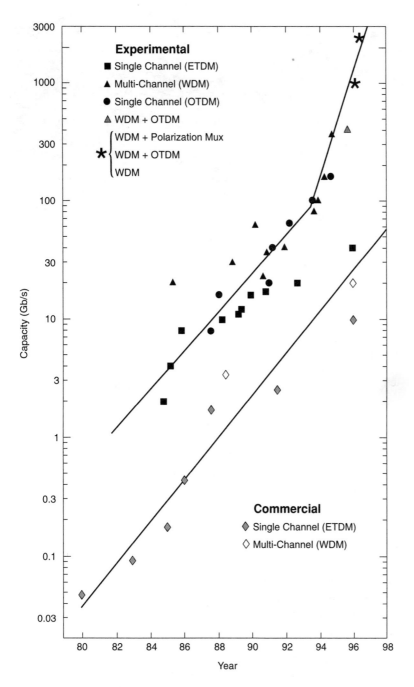

FIGURE 1.3 A technology roadmap for lightwave systems, showing the transmission per fiber achieved in leading long-distance optical fiber systems over the past 15 years. The graph also indicates the progression of technologies developed for this accomplishment.

Note the logarithmic scale for capacity on the vertical axis, which means that the straight trend line implies an exponential growth in fiber capacity. The growth rate is approximately a factor of 100 every 10 years, about the same as the rate of increase in the processing power of computers.

Trend lines are shown both for demonstrations of experimental systems in research laboratories and for the first introduction of commercial systems. The time lag between the two is 3 to 7 years.

The major multiplexing systems used are marked ETDM, OTDM, and WDM: ETDM systems use only electronic time-division multiplexing to achieve high capacity by interleaving pulses of a multiplicity of signals; OTDM systems use high-speed optical techniques to accomplish the same interleaving; WDM refers to wavelength-division multiplexing, in which different TDM signals are transmitted by light of different wavelengths. The asterisks mark a set of recent experiments that demonstrate terabit capacity using combinations of the above techniques. (Courtesy of C.A. Murray, Lucent Technologies.)

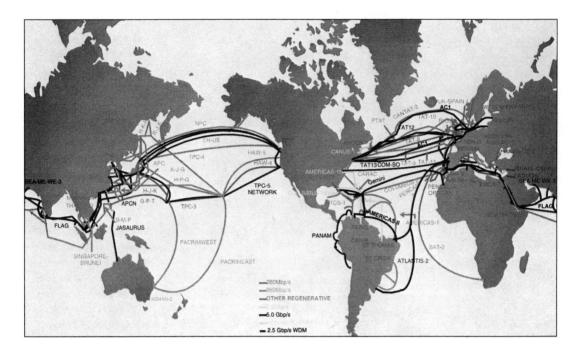

FIGURE 1.4 A map of the current global undersea optical fiber network. By the end of 1996 more than 300,000 km of optical fiber cable were installed in this network. Note that this map was totally empty before 1988, when the first large optical undersea system was deployed.

Installations have been given names, such as "Columbus," or acronyms, such as TAT (for TransAtlantic Transmission), TPC (TransPacific Cable), and FLAG (Fiber Link Around the Globe). (Courtesy of P. Runge, Tyco Submarine Systems Laboratories.)

the same time, their information capacity is at least 1,000 times larger. The rapid introduction of optical fiber into global networks is illustrated by the fact that in 1991, just three years after the first optical system was introduced, fiber already carried more international digital information traffic than satellites did.

More than 300,000 km of undersea lightwave cable had been installed by the end of 1996. Typical system lengths are about 6,000 km in the Atlantic and 9,000 km in the Pacific. The rapid pace of progress is reflected in the fact that we can already distinguish three different generations of the technology, as described in Box 1.3.

The dramatic increase of undersea cable capacity, enabled by optical fiber technology, has led to an equally dramatic decrease in the cost of a voice circuit across the oceans. The key reason for this is that the cost of constructing an undersea system, particularly of laying an intercontinental undersea cable, has remained almost the same, even though a single cable now carries at least 1,000 times more voice circuits than it did 40 years ago. An example of this dramatic cost decrease is given in Figure 1.5, which shows the cost of AT&T transatlantic systems and indicates a cost reduction by more than 1,000 times over the 40-year period.

Terrestrial Systems

Optical fiber systems for long-distance transmission on land (also called "trunking") provide the links between metropolitan telephone offices, between cities, and across the continent.

BOX 1.3 THREE GENERATIONS OF UNDERSEA LIGHTWAVE TECHNOLOGY

The 1988 first-generation undersea technology uses conventional single-mode fiber operating at a wavelength of 1.3 μm. The system uses optoelectronic regenerators with semiconductor lasers and p-i-n photodetectors both made of InGaAsP. The bit rate of the optical signal stream in each fiber is 280 megabits per second (Mb/s), and the typical regenerator spacing is 70 km.

A second generation of technology was used in the 1991 installations. It too employs conventional fibers, but with the operating wavelength moved to 1.55 μm to benefit from lower fiber losses. Single-frequency distributed-feedback lasers are used in optoelectronic regenerators, together with avalanche photodetectors. The bit rate per fiber is doubled to 560 Mb/s, and the typical repeater spacing is 150 km.

The third-generation technology was prepared for installation in 1995. It uses optical repeaters with optical amplifiers. The systems operate at 1.55 μm and use dispersion-shifted fiber to provide near-zero dispersion in the operating range. The design bit rate per fiber is 5 gigabits per second (Gb/s).

Year	Technology Generation	Voice Channels per Cable
1955	COAX 1	48
1963	COAX 2	140
1970	COAX 3	840
1976	COAX 4	4,200
1988	FIBER 1	8,000
1991	FIBER 2	16,000
1995	FIBER 3	122,880

The above table shows the information-carrying capacity of these three lightwave systems (FIBER 1-3) along with that of the preceding four generations of coaxial systems (COAX 1-4). Note that the data shown are for cable capacity without voice processing. Current undersea systems use digital voice processing to take advantage of silent periods in speech. This increases effective capacity by another factor of 5, enabling the third-generation lightwave system to carry more than 500,000 voice channels.

• •

Early metropolitan applications, such as those in the San Francisco Bay area, highlighted optical transmission's advantages. The Bay Area system, introduced in 1980, provided digital transmission at 45 megabits per second (Mb/s) with a repeater spacing of 7 km. This spacing was already sufficient to allow links between telephone offices to be direct, with no need for electronics outside the buildings. The optical system also helped to save space in metropolitan cable ducts since a single fiber replaced the earlier T1 carrier's 28 pairs of copper wires, with a capacity of 1.5 Mb/s per pair.

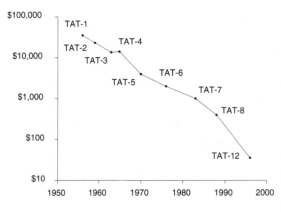

FIGURE 1.5 *The falling cost per circuit of transatlantic telephone (TAT) systems. (Note the logarithmic vertical scale.)*

The first major lightwave installation in the United States was the Northeast Corridor system, which linked Washington with New York in 1983 and New York with Boston in 1984, a total length of 747 miles. Its technology carried 90 Mb/s per fiber, the equivalent of 1,344 digital voice channels. This system confirmed the inherently higher quality of digital technology, but it also brought out another point: the excellent technical and economic synergy between the digital lightwave system and the digital telephone switches along the route. No analog-to-digital conversion was needed to interconnect the fiber system to the 23 existing digital electronic switching systems along the route, which resulted in considerable cost savings.

The Northeast Corridor was soon followed by more large-scale fiber installations, using rapidly advancing technology. More than 100 million kilometers of fiber had been installed around the world by the end of 1996, about one-third of the total being in the United States (see Figure 3.2).

Terrestrial fiber systems, like undersea systems, have evolved through several technology generations. (See Figure 1.3 for more detail on this rapid progress.) Advanced systems now employ optical amplifiers and wavelength-division multiplexing (WDM). WDM is a multiwavelength technique that increases the number of digital information channels that can be sent over a fiber simultaneously by transmitting each channel on its own distinct optical wavelength, as described in the section "Optical Networking and Switching" below. Terrestrial long-haul networks will benefit significantly from amplified multiwavelength transmission systems designed to access the large inherent bandwidth in the installed fiber. Capacity will increase by a factor of 10 to 50, not only providing for ample and graceful growth, but also allowing flexibility in network architecture design and in the management of network restoration in case of a failure (e.g., a cable cut). The new networking flexibility will be exploited to enable novel routing of traffic. WDM system experiments involving more than 100 channels already have demonstrated ultrahigh-capacity transmission over long distances, and large-scale experimental projects are under way to explore the potential of WDM technology for high-capacity networking. Amplified 16-channel WDM systems at 2.5 gigabits per second (Gb/s) per channel

TABLE 1.2 Estimated Global Markets for Long-Distance and Interoffice Lightwave Systems (U.S. dollars)

	1995	1996
Undersea	$2 billion	$2.5 billion
• Cable and deployment: ~70%		
• Lightwave equipment: ~30%		
Terrestrial	$2 billion	$4 billion
• SONET (OC-48) and SDH (STM-16)		
• Transmission equipment only		
• Lightwave: ~20%		
• Electronics: ~80%		

NOTE: SONET = Synchronous Optical Network, SDH = Synchronous Digital Hierarchy. (Source: T. Li, AT&T Laboratories.)

have been developed and are being manufactured for massive deployment in the embedded terrestrial network. These revolutionary system solutions will meet the demand of envisioned broadband services for many years to come and thus make lightwave communications the principal component of the global information services infrastructure.

Table 1.2 gives estimates of the global markets for undersea and terrestrial lightwave systems. There are strong systems vendors in Europe, Japan, and North America. Because of the early large-scale deployment of WDM systems in the United States, North American vendors have gained an early market lead in WDM systems.

Fiber to the Home

Although most experts agree that in the future, fiber will be installed all the way from the telephone company central office to the home, opinions vary widely as to when this will happen. Developments in the United States and elsewhere are beginning to suggest that it may happen sooner than commonly thought.

The technical benefits of fiber to the home (FTTH) include its well-known capacity for transmitting incredibly high bandwidths at relatively negligible losses. This "future-proofs" the resulting network against demands for rising bandwidths, which history shows will indeed be necessary. Since fiber lasts 30 years or more, labor-intensive installation of cable and passive components has to be done only once because no electronics are installed in the outside plant between office and home. Upgrades take place on the premises of the service provider and the customer. Nearer-term benefits of fiber include its small size and weight compared with metallic cable, especially coaxial cable; its total immunity to both inward and outward leakage of electromagnetic interference; and its resistance to corrosion. Systems with active components at only the ends of the network ensure high reliability. As important and practical as these advantages are, more practical benefits may help drive fiber to the home now. These include solving the high costs of drop-cable maintenance and electrical powering.

The advantages of completely replacing metallic telephone and cable television cables with fiber were recognized in the 1970s. Numerous trials in Japan, Europe, Canada, and the United States, in both the telephone and the cable television industries, established the technical feasibility, but the costs for either replacement or new construction were prohibitive. To move ahead with fiber, the telephone and cable industries refocused on less expensive fiber-to-the-curb and hybrid fiber-coax solutions, both of which realize many of fiber's advantages while sharing costs among many customers. However, these systems still complete the final connection to the customer with metallic cables (see Figure 1.6), which requires the supply of power in the outside plant on the way from the telephone office to the home.

Recent advances in key FTTH technologies have driven costs down and removed technical barriers. These advances include low-cost lasers, solutions for delivering video, and network topologies that share costs. Equally important, there are a growing number of bandwidth-thirsty services that make the high-bandwidth capability of FTTH more compelling. These include high-speed Internet access, telecommuting, home offices, and high customer satisfaction with digital video; an all-fiber network can satisfy these needs simultaneously. Perhaps most important, there is now a much clearer understanding of the cost savings that result from bringing fiber to the home. The time appears right for seriously reconsidering FTTH, at least for operators whose initial focus is on providing telecommunications services.

The most recent and perhaps most significant steps toward reducing FTTH costs are in the areas of lasers and laser packaging. New strained-layer, multi-quantum-well lasers have been developed recently that operate reliably at high power levels throughout the required wide range of temperatures (–40°C to +85°C). Some of the most recent structures also integrate a beam expander to simplify coupling to the output fiber and laser packaging.

Electrical powering has always been a major issue and has often been referred to as the Achilles' heel of FTTH. In the United States, telephone service is nearly always available during power outages, and the public expects uninterrupted "lifeline" service. Since fiber does not conduct electrical current as copper wires do, lifeline service has been an issue for all fiber-based networks. The challenge is to provide inexpensive, high-reliability power from the customer's premises. (Wireless technologies have the same requirement.)

Most of the recent progress in FTTH has been achieved in Japan and Europe. Manufacturers have developed equipment for NTT in Japan and for trials in England, Belgium, Germany, and Denmark. A promising development is a recent international initiative among ten network operators, including two from the United States, to define complete specifications for "full-service" access networks, including FTTH, all

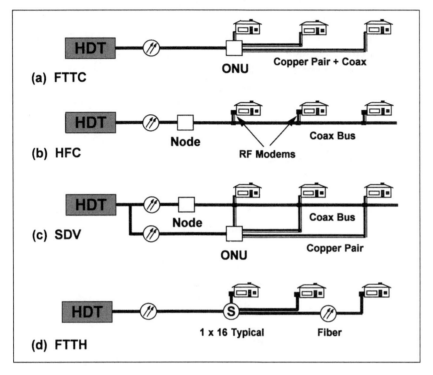

FIGURE 1.6 Four leading architectures for broadband information transport to the home from the telephone central office or a remote host digital terminal (HDT). A typical central office serves about 10,000 homes. Typical distances are 10 km from the HDT to the node or the optical network unit (ONU) and less than 1 km from these to the home. Fibers are marked by the symbol ⊘.

(a) In the fiber-to-the-curb (FTTC) architecture, the fiber is terminated at an ONU on a street curb in a neighborhood or at a large building, and transmission continues via copper wire or coaxial cable. This technology has been used for the large-scale OPAL installation in Germany.

(b) The hybrid fiber-coax (HFC) architecture uses coaxial cable from the node to the home. It carries broadcast analog television channels, as well as digital telephony and data channels, and is very similar to the technology used by the cable television industry.

(c) The switched digital video (SDV) architecture is a combination of (a) and (b) that carries digital and analog services separately. It links to very few (4 to 24) homes per ONU, which allows the provision of switched digital video services.

(d) The fiber-to-the-home (FTTH) architecture uses a passive optical splitter (marked Ⓢ) to share the long stretch of fiber from the telephone office to the neighborhood among 16 to 32 homes, which reduces the cost per home. Note that FTTH is the only one of these four technologies that requires no power supply in the outside plant between the telephone office and the home. (Courtesy of P. Shumate, Bellcore.)

Chapter 1

of which can easily evolve to FTTH. Their efforts should result in new FTTH products becoming available in 1998. Cost is still cited as the major barrier to wider early deployment, but recent analyses show that in rural areas and for some high-end suburban cable television situations, FTTH can meet first-cost objectives now. Imaginatively reducing costs even further, however, is a major challenge for the optical R&D community, in both industry and academia.

The rapid emergence of broadband telecommunications networks and broadband information services will have an enormously stimulating impact on commerce, industry, and defense worldwide. **Congress should challenge industry and its regulatory agencies to ensure the rapid development and deployment of a cost-effective broadband fiber-to-the-home information infrastructure.** Only by beginning this task now can the United States position itself to be the world leader in both the broadband technology itself and its use in the service of society.

Analog Lightwave Transmission

The lightwave systems discussed above are digital systems, in which information is transmitted as pulses of light. In the analog systems that are the subject of this section, a waveform such as a radar signal or an analog television signal is modulated directly onto a laser beam. This avoids costly digital-to-analog conversion at the receiver but generally limits the transmission distance to about 20 km.

Cable Television

By far the most widespread application of analog lightwave systems has been in the cable television industry. Analog lightwave systems in cable television architectures link the cable headend office to a fiber node, where a receiver extracts the radio frequency (RF) signal and transmits it to between 500 and 2,000 households using a tree-and-branch coaxial cable plant [see Figure 1.6(b)]. Fiber's success in the cable television industry arises from the fact that a single analog lightwave system with only two active components—a transmitter and a receiver—can replace a trunk coaxial cable with 12 to 30 active trunk amplifiers while providing better end-of-line performance. Virtually absent in the early 1980s, fiber now exists in more than 25% of cable systems, and fiber deployment by cable television operators accounted for 34% of all fiber deployed in North America in 1994.

A cable television signal consists of 40 to 80 analog video channels, each of which requires a 45-decibel (dB) signal-to-noise ratio at the subscriber's television set. Because of the subsequent coaxial cable plant, the signal-to-noise ratio of the lightwave system must be even higher than this 45 dB, so the linearity and noise requirements on the enabling lightwave components are very demanding. The first successful lightwave transmitter of multiple video channels was a directly modulated 1300-nm semiconductor distributed feedback (DFB) laser. This technology, introduced in the late 1980s, remains a workhorse of the cable television industry, and advances in DFB lasers have increased both the average output power and the allowable channel load (from 40 to 100 channels). Additional enabling technologies now under development, aimed at further performance improvements, include 1550-nm DFB lasers, external modulators, and optical amplifiers.

The market for analog lightwave systems for cable television-like signals is expected to remain strong. Cable television operators, feeling the threat of competition from wireless cable and direct broadcast satellite (DBS) systems, are looking to upgrade their plants to improve both the quantity and the quality of their services. This typically means more fiber-rich architectures, in which each fiber node serves fewer homes. As cable television operators look to enter the telephony and data network market, they are installing fiber ring architectures that improve the reliability of their networks. The arrival of digital television channels, such as those used by the DBS providers, breathes added life into the analog lightwave industry. These digital channels are coded onto subcarrier frequencies, which casts them into a format readily carried by an analog lightwave system.

Remote Antennas

The second most common application of analog lightwave systems is in links to remote antennas. The signals carried are for mobile radio, such as cellular telephones or personal communication services (PCS), or for microwave or millimeter-wave applications.

Mobile radio is the largest potential market for antenna remoting. The explosive growth of the cellular telephone market has led to a need for more frequency reuse of the radio spectrum. As cells shrink to microcells, the economic case weakens for putting a complete base station at each antenna, requiring analog lightwave backhaul from multiple mobile radio antennas. Mobile radio signals are higher frequency than cable television signals (900 MHz for cellular and 2 GHz for the new U.S. PCS licenses) and occupy less bandwidth, but they require very high dynamic range because of the "near-far" problem. In other words, the signal-to-noise ratio must be maintained for all mobile users, whether near the antenna or far from it, even though the RF power received by the antenna depends significantly on the distance to the user. Required breakthrough technology includes much cheaper laser transmitters and more compact (perhaps integrated) laser-modulator packages.

In the future, analog lightwave links may be used for antenna remoting of stationary microwave or millimeter-wave radio systems.

At microwave frequencies, direct modulation of semiconductor lasers or modulators is possible. At millimeter-wave frequencies (30 to 100 GHz), more sophisticated techniques are needed. In some respects, optical fiber is the only medium that can transport millimeter waves over any appreciable distance, because the attenuation of free space for millimeter waves is very large compared with the attenuation of fiber.

Competing technologies for antenna remoting include signal processing at the remote antenna so that a digital link can be used, line-of-sight microwave links, and coaxial cable. The choice will be based on both performance and economics.

Optical Space Communications

In the past decade, satellite communications have developed into a booming commercial market, global in nature but dominated by the United States. Currently valued at about $15 billion per year, this market is expected to grow to at least $30 billion per year in the next decade, or even more if mobile satellite communication services are deployed successfully. Fixed satellite services are relatively mature; their projected growth is a modest 10% per year. Direct broadcast satellites will see more rapid growth of at least 50% per year. Governments have long used satellites for a variety of communications missions (Figure 1.7), and the Department of Defense (DOD), National Aeronautics and Space Administration (NASA), and other government agencies will continue to launch and maintain dedicated satellite links, but they are expected to use commercial links more and more to save costs.

Most current satellite communications systems use simple, single-satellite relays between ground terminals. If it is necessary to span a longer distance, two relays may operate in series through a relay ground station. An alternative—direct cross-links between satellites—eliminates a relay ground station. Though not now commonly employed, this approach could be important in future satellite networks, especially for mobile communications. In addition, the utilization efficiencies of sensor satellites in low Earth orbit can be improved significantly by the use of a readout cross-link to a relay satellite in high Earth orbit. The capacity of the French SPOT-4 satellite, for example, will increase six-fold with its planned optical readout link to the geosynchronous relay satellite ARTEMIS. Furthermore, reading out low-Earth-orbit Earth resources satellites such as LANDSAT via cross-links to high-Earth-orbit relay satellites can increase sensor coverage, improve access times, and eliminate expensive global tracking networks.

Cross-links have already been operated, such as NASA's data relay satellite TDRSS (Tracking and Data Relay Satellite System), but their use is not routine. The networking choice and the particular cross-link hardware have to meet stringent cost-risk requirements for a new system design.

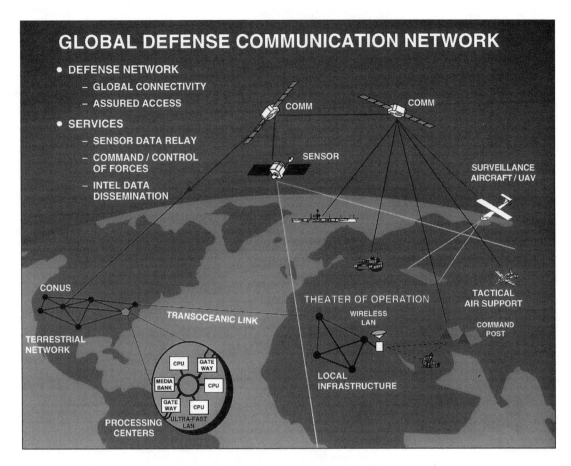

GLOBAL DEFENSE COMMUNICATION NETWORK

- **DEFENSE NETWORK**
  - GLOBAL CONNECTIVITY
  - ASSURED ACCESS
- **SERVICES**
  - SENSOR DATA RELAY
  - COMMAND / CONTROL OF FORCES
  - INTEL DATA DISSEMINATION

COMM

COMM

SENSOR

SURVEILLANCE AIRCRAFT / UAV

CONUS

THEATER OF OPERATION

TACTICAL AIR SUPPORT

TRANSOCEANIC LINK

WIRELESS LAN

COMMAND POST

TERRESTRIAL NETWORK

CPU

GATE WAY

MEDIA BANK

CPU

GATE WAY

CPU

ULTRA-FAST LAN

LOCAL INFRASTRUCTURE

PROCESSING CENTERS

Intersatellite cross-links can be either optical or radio frequency. The shorter wavelength of optical systems allows modest telescope sizes and transmission at a high data rate. Significant weight, power, and size advantages are realized over RF systems of similar performance, especially at very high data rates. However, optical space communications is still an emerging technology, with a checkered history. Many tough technical issues are yet to be resolved, and their solutions must be demonstrated before the technology is mature enough for deployment. Among these critical technology and system issues are transmitter and receiver technology; spatial acquisition and tracking of very narrow beams; optical-mechanical-thermal engineering of high-precision optical systems for space use; and a good understanding of system architectures and techniques for design, fabrication, integration, quality assurance, and risk mitigation. These are mainly engineering issues.

The United States has a history of major disappointments in this field. More than $1 billion has been spent, without yet producing a working optical link in space. The failed attempts can be traced to unsuccessful technology development, poor understanding of system

FIGURE 1.7 A schematic sketch of the global defense communication network. An essential element is the network in the sky consisting of interlinked communication satellites and sensor satellites. Extensive satellite networks are also planned for commercial mobile telephone and data service applications. The text discusses the capacity advantage of optical cross-links between the satellites of the network. (Courtesy of V. Chan, MIT Lincoln Laboratory.)

engineering, lack of satellite payload integration experience, and lack of creativity of the technical teams assigned to the programs. The know-how needed to successfully deploy optical communications in space does exist in U.S. universities and federally funded laboratories, but an effective mechanism must be devised to transfer the technology to the U.S. aerospace industry. Specifically, U.S. satellite communications companies, although they collectively possess many of the critical technical ingredients, seem unwilling to make the major investment necessary to aggressively pursue the development of an optical cross-link payload comparable in quality and lifetime with RF links. Most U.S. companies are studying optical links, but their strategies stress RF.

As technology advances, optical cross-links look more attractive, but detailed analyses of costs, benefits, and risks require the development of actual space-qualified optical communications payloads that are competitive with RF cross-links. As the Federal Communications Commission encourages the construction of global satellite networks, the requirement for several cross-link terminals per satellite, which is difficult for large RF antennae, is expected to become a strong motivation for optical cross-links. So far, however, neither industry nor government has made the financial commitment necessary to fund such an effort in the United States. As a result, if the market develops, the United States is likely to enter it late.

European and Japanese companies are not as reluctant as U.S. companies to proceed with a commercial payload. In Europe and Japan, government and industry have joined forces in long-term R&D in this area for the past 6 to 7 years. Notably, the European Space Agency's Semiconductor Intersatellite Laser Experiment (SILEX) represents a payload investment of approximately $250 million, to be launched soon. Japan is planning to launch the Optical Intersatellite Communication Engineering Test Satellite (OICETS) to link with Europe's ARTEMIS in 1998 and later with Japan's own geosynchronous relay COMET. The commercial low-Earth-orbit satellite constellations for mobile phone and small-terminal data services will be a lucrative outlet for such technology.

Information Processing

Of the four major application areas discussed in this chapter—optical transmission, storage, display, and processing of information—the first three have succeeded in large commercial markets. In these applications, light performs functions that cannot be done by electronics. By contrast, for most applications in information processing, electronic technology is excellent and sets a high standard of performance. Thus, the use of optics in information processing remains a research topic, still seeking a clear competitive advantage.

In this report, the definition of optical information processing is taken to include the use of optics in data links, telecommunications switching, both analog and digital computing, and image processing. In essentially all of these applications, silicon-based electronics is presently the technology of choice. However, as cheaper and more practical optical and optoelectronic devices become available that can be used in suitable systems, optics will play an increasingly important role. This section describes optical information processing technologies from the most advanced to the most speculative.

Optical Data Links

Optical fibers are an excellent transmission medium and, as seen in the first section of this chapter, they reign undisputed in long-distance transmission links. At short distances, in local area networks (LANs) that link computer workstations around a campus or from desk to desk within a building, the opportunities for optics are growing rapidly, as costs come down and bandwidth needs continue to grow. Optical data links will be important, however, only as they become sufficiently inexpensive to compete with electronics, particularly in low-end applications such as connecting desktop computers. Optical data links also require extraordinary reliability, since they transmit computer data and images.

In many datacom applications, the cost per channel is reduced by using a high degree of parallelism, that is, by using arrays of lasers and detectors, connected by fiber ribbons (see Figure 1.8 and Table 1.3). Parallel optical links not only lower costs, but also reduce cable congestion, board area, and bandwidth demands on sources, detectors,

FIGURE 1.8 Motorola's Optobus system. Motorola's optical data link package contains a 10-fiber ribbon up to 100 m long, driven by arrays of lasers and detectors. The lasers are short-wavelength vertical cavity surface emitting, connected to an optical interface unit, which is a molded waveguide with direct chip attached to the optoelectronic array and to the fiber ribbon. Other parallel optical data links have been developed by Hewlett-Packard and Hitachi, among others. By 2005, typical prices are projected to drop to $10-$15 per duplex channel at 622 Mb/s. (Courtesy of R. Nelson, Motorola.)

TABLE 1.3 Features of Some Selected Optical Communication Application Segments

Target	Distance	Data Rate	Modem Cost Target
Telecom			
Long distance	5-100 km	0.6-2.5 Gb/s	$10,000
Short distance	1-10 km	50-622 Mb/s	1,000
Datacom			
High performance			
Campus	300-2,000 m	200-1,000 Mb/s	100
Interbuilding			
Telecom switches			
Low cost			
Desktop	< 100 m	< 200 Mb/s	10
Backbone	100-500 m	< 200 Mb/s	

(Source: R. Nelson, Motorola.)

electronics, and fiber. They also allow many alignments in one assembly operation and eliminate the high cost of electronic multiplexing. The result is a lower packaging cost per channel and the elimination of delays caused by electronic signal processing.

In computer applications, data links are present in the local area network that connects workstations within the computing cluster, as well as to and from file servers, data communication adapters (telecom and satellite), display and printer adapters, and data storage systems. Optical data links have shown modest market penetration, with copper wiring usually being cheaper at present. Success in penetrating the market will require leverage of the enabling optical technologies. The technology driver is primarily cost; obstacles to progress include the costs of connectors and cables, packaging, and installation and alignment. Improved molding or plastic packaging as well as modeling and simulation tools for high-frequency, low-cost packages are needed.

Figure 1.9 indicates the large future markets for optical data links, as found by several market studies. The United States is in a strong position to capitalize on these markets. Companies will invest in the necessary R&D, but industry needs good standards and improved manufacturing. University research can help by working on advanced devices and low-cost fiber technologies.

The United States is competitive in high-performance datacom, but cost reductions are needed. In low-performance datacom, the United States is strong. The challenge is to enable research in the new R&D environment, in which low-cost, high-volume global markets mean low marginal profits.

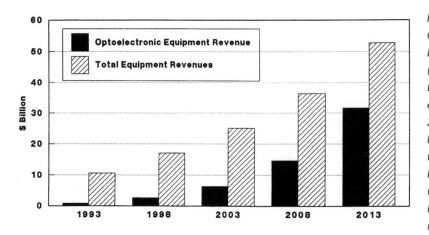

FIGURE 1.9 The Optoelectronics Industry Development Association (OIDA) predicts $25 billion in total data communications equipment revenues by 2003, of which $7 billion will be in optoelectronics. The total market should double by 2013, of which $30 billion will be optoelectronics. OIDA has produced a detailed roadmap that lays out those technologies that will be inserted in the marketplace for specific applications. Restricting the consideration to the worldwide market in parallel optical links, ElectroniCast predicted in July 1995 that the market will be $1 billion by 2000 and $3.3 billion by 2005, roughly half in computers and half in telecommunications, with 10% military. (Courtesy of OIDA.)

Optical Networking and Switching

High-bandwidth optical telecommunication systems are digital and transmit many simultaneous telephone calls or data channels with their digits interleaved in a process called multiplexing. Today's systems perform this multiplexing electronically, beginning at the local exchanges and increasing the bandwidth and number of simultaneous channels as the signals approach the long-distance nodes (of which AT&T, for example, has 250 around the country). Commercial telecommunication systems currently do all their routing electronically. At switching nodes, receivers convert optical signals from the fibers into electrical signals, which are switched and routed to the appropriate output fiber so that each phone call gets where it is intended.

In an all-optical network, light signals would remain light signals throughout, and the switching would be optical. The architecture of the system determines the approach and specifies both the hardware and the software used for multiplexing and switching. The key technological task is to identify the capabilities of optics that enable it to play crucial roles in the assembly, management, and distribution of large numbers of signals that do not all start and end at the same two distinct points.

Switching

With the reduction in the cost of information transmission, switching is becoming an ever larger fraction of communications costs (see Figure 1.10). Switching is therefore both a large challenge and a large market; today, just one fiber can saturate the switching capability of the largest switch ever made. Since transmission cost is falling exponentially, the primary system cost will be in switching and networking, unless network architectures change to use longer spans of fiber between switches.

The current practice is electronic switching, in which a transceiver detects optical signals and electronic logic sends them where they belong, conditions them, retimes them, and then modulates a laser that transmits them down the appropriate fiber. Electronic switches use logic functions to route the signal spatially from one channel to another. This is in contrast to switches that physically change the routing, such as the patch cords used by the first telephone operators.

The WDM systems (see Box 1.4) currently under development are all optical. They take advantage of the fact that the transmitted information is in the form of light, physically routing it via optical cross-connects. This keeps signals optical throughout switching and routing, and it avoids expensive optical-electronic conversions. In essence, it provides "transparent pipes" that enable transmission of any bit rate, any packet length, any transport format, and any modulation format including SONET (Synchronous Optical Network) and ATM (asynchronous transfer mode).

Another approach is to replace only selected electronic components with optical ones, where the optical components have a distinct advantage. For example, taking advantage of the interconnection ability of optics can enable very high-capacity switching machines, such as might be needed for terabit-per-second ATM switches. Such devices have commonality with the advanced optical technologies developed for image processing (see below). Their use of optics solves most of the physical problems inherent in high-density electrical interconnections: Optics has no frequency-dependent loss or cross talk and is intrinsically a very high-bandwidth medium; there is no distance-dependent loss or degradation; and photons allow for electrical isolation and immunity to electromagnetic interference. Meanwhile, the logic in such systems remains electronic, where it is most efficient.

FIGURE 1.10 Transmission costs per bit have seen dramatic reduction, driven in large part by the ever-improving optical fiber transmission technology. Switching technology has not kept pace; the information sent on one fiber is more than enough to saturate the largest telecommunications switch in existence. Switching cost is likely to be an increasing fraction of the cost of telecommunications. This shows up in telephone calls costing the same whether the destination is in the same state or across the country. (Courtesy of P.E. White, Bellcore.)

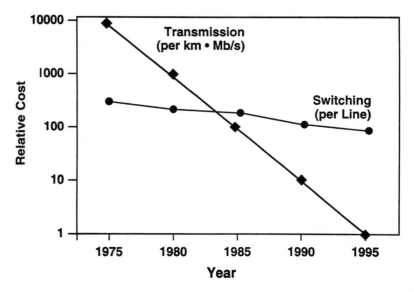

BOX 1.4 MULTIPLEXING: SPACE, TIME, WAVELENGTH

Multiplexing, or placing many simultaneous calls on a single line, may occur by dividing up the signals in time, space, or wavelength. When signals are multiplexed, they can travel either through fixed circuits or in separately addressed packets. The system architecture determines when and how much the signals in different channels are multiplexed.

In *time-division* multiplexing, pulses corresponding to bits from different signals are placed in different time slots. The format that has become standard in long-distance telephone systems is called SONET. SONET consists of digital frames that are sent at a bit rate of 155 MHz and each contain an address that sends them in the proper direction.

In *space-division* multiplexing, different signals travel on different fibers. Each fiber can be routed in a different direction, connected to the various nodes of the system. This approach reduces the need for switching.

Wavelength-division multiplexing (WDM) is the newest approach. In WDM, each channel is transmitted in a slightly different color of light. Color-sensitive switches or receivers determine the routing and which call is received at which station.

For mixed-format multimedia signals, in the time-division-multiplexed mode a new standard is being developed, called the asynchronous transfer mode. In ATM, short packets of data (typically 53 bytes) are used for all kinds of traffic, including voice, video, data, and multimedia. Each packet has a header that carries the address, and each packet is transmitted independently of every other one. ATM requires high-speed switches with fine granularity that can switch rather small packets automatically.

••

WDM Networks

In a WDM network, the use of many wavelengths permits huge network capacity, and the technology is scalable, modular, evolvable, and reconfigurable. WDM is economically viable because successful optical fiber amplifiers have been developed that can simultaneously amplify a large number of channels of different wavelengths without cross talk or interference with one another.

To build up a reconfigurable WDM network, wavelength-dependent cross-connects are needed. The ideal cross-connect would allow any wavelength on any input fiber to be connected to any wavelength on any output fiber. This requires wavelength converters, a technology that has not yet proven feasible for commercial exploitation, but cross-connects without wavelength conversion could be sufficient for a significant segment of the market.

Box 1.5 provides some background information on the considerable number of R&D projects on WDM networking now under way worldwide. These projects are providing a growing consensus: Point-to-point WDM is happening now at 8, 16, and perhaps 32 wavelengths, separated by 200, 100, or 50 GHz each, respectively. Applications are in long-distance transmission and in exhausted regional routes. WDM sources, wavelength-selective switches, and WDM amplifiers have all proven feasible. Layered WDM network architectures provide graceful evolution with increasing demand. If commercial feasibility is to be evaluated realistically, network management and control systems have to be included in these study projects.

The enabling technologies for all-optical networks are optical amplifiers, waveguide switches, pump lasers, and fiber. The emerging technologies are WDM passive components, wavelength add-drop components, multiple-wavelength lasers, photonic integrated circuits, and cross-connects. Technologies that are not here yet and are speculative include low-cost, single-mode waveguide device packaging and wavelength converters. Finally, mode-locked ultrafast laser sources may push WDM capabilities even farther than in present systems studies.

European researchers have made tremendous progress with RACE (Research Advanced Communications, Europe) and ACTS (Advanced Communications Technologies and Services) and are well coordinated,

BOX 1.5 WORLDWIDE PROGRAMS ON WDM NETWORKING

There are several major optical networking programs under way worldwide. An early WDM networking demonstration was by British Telecom in 1991, which ringed London with five nodes connecting 89 km route lengths at 622 Mb/s, using erbium-doped fiber amplifiers and WDM routing. RACE (Research Advanced Communications, Europe) had a joint research project among eight companies and two universities that developed a Multi-wavelength Transport Network. The demonstration in Stockholm in 1994-1995 investigated optical network layering, transparency, evolutionary paths, standards, and economic modeling.

The program following this in Europe is ACTS (Advanced Communications Technologies and Services). Its first project is to develop METON (Metropolitan Optical Network), designed with an ATM optical transport layer to be cost-effective in a metro area net and a local access network. It uses optical cross-connects and optical add-drop, using fiber grating, silica-on-silicon, InP, and optoelectronic integrated circuits. The ACTS program has several projects under way similar to METON, utilizing collaboration among industry, government, and universities from a number of countries in Europe. This project includes all major European industrial interests, all major telecom network operators, leading broadcasters and cable television operators, and key European equipment manufacturers. It is the largest set of linked trials and demonstrations of new telecommunications services in the world. It will provide a trans-European Information Infrastructure, with fiber, cables, radio, and satellite links. There will be up to 10,000 businesses and 1 million individual participants in service trials.

The Photonic Transport Network Project under way at NTT in Japan is also aimed at a WDM network demonstrator.

In the United States, the Defense Advanced Research Projects Agency has funded several projects on WDM networking:

with a hard push toward commercialization and local access application. The Japanese program is in its early phases but has a highly developed WDM technology base and rapid progress is expected. In China, university laboratories are in unusually close contact with industry, which increases their access to technical and business ideas and helps to guide university research. The U.S. programs are roughly comparable in scope and quality to European and Japanese programs. WDM standards are being pushed hard in Europe and are needed for commercial progress.

High-Density Optical Switching

Physical cross-connects can route entire optical channels transparently, independent of the bit rate or the transmission format. They need not switch at the bit rate (i.e., reroute the signal at each signal pulse), but since the control (electrical) and signal (optical) are not in the same form, such switches cannot control the network or retime,

• •

• AON (All-Optical Network) is a consortium among four industrial and university laboratories to achieve a scalable, universal all-optical network. It includes three service types: circuit-switched, scheduled time-division multiplexing, and unscheduled datagram. The technologies used for AON are integrated passive optical WDM routers and broadcast stars, and tunable lasers. The network design is for 100 GHz optical frequency spacing, hierarchical architecture, and slotted frame time-division synchronization.

• A second project is ONTC (Optical Networks Technology Consortium), consisting of six companies, two universities, and one government laboratory. The system uses a reconfigurable, scalable, modular architecture, cell-by-cell wavelength translation, integrated with ATM-SONET, and both analog and digital transport. The demonstration system uses a four-node reconfigurable WDM network, eight-wavelength laser arrays, WDM add-drop, and cross-connects.

• The third project is the MONET (Multiple Wavelength Optical Networking) consortium to determine the commercial feasibility of transparent optical networks, introducing a prototype network control and management system. There are testbeds for local exchange network and long distance, with a Washington, D.C., field experiment run by a consortium of telecommunications companies and government agencies. The goals are national-scale networking, reconfigurability, interoperability and transparency, network control and management, scalability, and economic studies. Milestones include a 2,000-km long-distance testbed and a local exchange testbed using cascades of erbium-doped fiber amplifiers, comparing WDM LAN ring, star, and mesh architectures.

• A fourth project is NTONC (National Transportation Optical Network Consortium), consisting of six companies and one university. There will be a demonstration network in the San Francisco Bay region, using a WDM SONET ring network. In this system, WDM is used for point-to-point long-haul or exhausted regional routes. It operates at 2.5 Gb/s, over four to eight wavelengths, and contains four WDM cross-connects and a WDM four-wavelength add-drop ring.

buffer, or regenerate the signal. Logical digital switches are typically better able to handle sophisticated control functions and can perform buffering, regeneration, and retiming of signals. In the logical switch, the control and signal do have the same form, but this switch must run at the bit rate and is not transparent to upgrades in capacity or format. Nonetheless, the change to sophisticated transmission formats that can carry mixed traffic types and to packetized traffic such as ATM requires even more logical complexity in the switching systems, making logical switching increasingly necessary. ATM switching will require switches at nodes that can handle terabits per second of information. The largest ATM switches commercially available today have a throughput of about 20 Gb/s.

High-density optical interconnects allow the electronics to do the logic and the optics to do the interconnect function (see Figure 1.11 for more details). Considerable research progress has been made in two-dimensional array interconnects and the associated smart pixel technology. These technologies appear capable of tracking the growth in switching capacity of large switches over the next 15 years, and they may be crucial in making possible such large logical switching machines. These technologies may also become important in high-performance computing.

Note that the size of the digital switching market makes it a significant challenge for optics to contribute to the technology. In 1995 the U.S. market for conventional central office switching equipment was $6.3 billion; worldwide, it was $24 billion and growing at 6% per year. This total does not include the markets for ATM, digital, cross-connects, and other broadband switching, some of which are growing much faster (20% to 60% per year) although still relatively small. It is too early to project photonics markets in switching since technology insertion is not fully in place.

Current results are encouraging, but to achieve optical interconnect insertion into logical switches, continued R&D is necessary on high-yield integration of optoelectronic and electronic devices and in general on the merger of optoelectronics and electronics technologies. All forms of photonic switching require continued development of devices and packaging to reduce costs and establish reliability.

All-Optical Switching

Will there be an architecture that relies on all-optical switching? It will be difficult to compete with silicon-based electronic logic. Electronic technology, including packaging, is mature and inexpensive. Issues that cause difficulties for all-optical switching systems are synchronization and clock recovery, buffering and memory, and logic. In addition, ultrafast optical switching devices require too much power, are too simple logically, and are not yet practical for large-scale use. Although some research systems at universities have demonstrated all-optical switching, a number of issues must be resolved before they can become practical. The future of ultrafast optical switching is not yet clear, and more research is needed. Because devices and concepts are in an early stage of development, there are still major opportunities for invention and innovation in optical devices and materials. Nonlinear materials, fibers, and switches are excellent topics for university research. In the future, telecommunication systems may experience a graceful transition to all-optical switching as more and more of the electronics functions are taken over by optics.

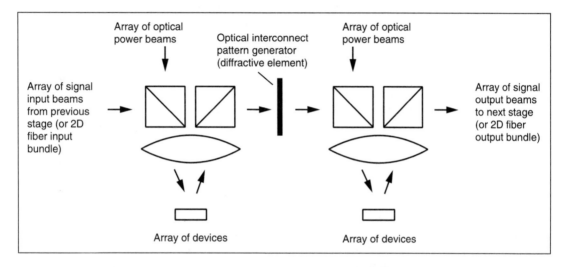

FIGURE 1.11 High-density optical interconnects that link electronic logic arrays solve a number of problems of very large digital switches. At the chip level, *using optics provides protection from electrostatic discharge on the inputs and from simultaneous switching noise, reduces the area of off-chip drivers and pads, and reduces the power dissipation of off-chip line drivers. Chip-to-board interconnects use optics to overcome pin inductance on chips and chip carriers, the limited number of pins, and the use of a large number of power and ground pins because of off-chip driver current requirements. On-board and board-to-backplane interconnects use optics to overcome the limited number of pins on board connectors, connector inductance, impedance matching, wave reflections, line termination, electrical isolation, cross talk between lines, and bandwidth limits of lines. These optically interconnected systems require integrated receivers, receiver arrays, and smart pixels.*

The existing technologies for high-density optical interconnects are in the form of optical data links, with multimode vertical cavity surface emitting lasers (VCSELs) or arrays. Technology for two-dimensional (2D) arrays is emerging: smart pixels, modulators, VCSELs, dense integration of modulators and detectors with silicon electronics, array optomechanics, and laser power sources.

Two-dimensional array interconnects are being researched now; experimental systems using "smart pixel" technology, for example, are demonstrating thousands of high-speed optical interconnects directly off the surface of silicon chips using free-space optics or fiber bundles.

An example of the use of 2D-array interconnects in an experimental free-space photonic switch is sketched above. In the smart-pixel version of such a system, electronics performs the logic functions and optics performs the interconnections.

(Courtesy of D.A.B. Miller, Lucent Technologies.)

Optical Image Processing and Computing

Historically, coherent optical information processing was analog, driven initially (30 years ago) by interest in synthetic aperture radar and image analysis, for which lasers and optics could provide correlations and spatial filtering. A separate stream of research developed from the invention of holography, which provided technology to enable optical associative memories and matched filters. More recently, reflection holograms have been proposed for interconnects between integrated circuits and for neurocomputers.

In the 1970s, researchers looked into using optics to perform computationally intensive operations such as vector-matrix and matrix-matrix multiplications. Special-purpose discrete optical processors were designed and demonstrated, using film as a mask for processing the input data set. The spatial light modulator (SLM), with transmission or reflection that varies across a plane, grew out of the need for a real-time programmable mask. Electrically activated liquid crystal SLMs operate at kilohertz frame rates and can accommodate more than 65,000 parallel channels. These remained expensive speciality items until the liquid crystal television, introduced in 1985, afforded researchers a low-cost alternative. However, their performance was poor.

Research on analog optical processing flourished during the 1980s, with the emphasis being on optical signal processing and neural networks for pattern recognition and image processing. Motivated by the introduction of optical bistable devices (optical transistors) in the late 1970s, researchers in the United States, Europe, and Japan also began investigating the use of optics in digital computing.

Both analog and digital technologies have been under development to replace (or enhance) electronic computers and specialty signal processors. As the optics research has been under way, however, the performance of electronic computers has continued to improve dramatically, presenting a formidable competitive challenge for optics. As a result, optical computing appears to be useful only in specialty applications in which there is a high degree of parallelism.

The generic geometry of an analog optical image processor begins with a two-dimensional coherent image, created in real time by reflecting laser light off (or transmitting it through) an SLM. For this step, traditional liquid crystal SLMs are seeing competition from magneto-optic and electro-optic semiconductor SLMs, which can be faster. Next, in an analog processor, the coherent light is sent through optical components (lenses, holograms, prisms, pinholes, gratings, or a programmable SLM) that do the optical processing. Finally, the processed signal is read out visually or on a CCD (charge-coupled device) array. In a digital processor, the programmable SLM is replaced by a two-dimensional smart pixel array or an array of nonlinear optical devices; digital optical processing uses logic gates and thresholding.

The applications envisioned for these technologies have typically been correlators or niche applications, but in many historical cases, these optical systems have lost their niche to high-speed silicon. Nonetheless, optical image recognition can be commercial; there is a fingerprint recognition product from Canada, for example (see Table 1.4).

The future of optical computing lies in exploiting the synergies between electronics and optics to approach problems that are difficult or impossible for all-electronic processors. Potential image processing applications lie particularly in rare-event problems, which electronic computers find difficult, such as fingerprint identification or recognition of a face in a crowd. Examples include electro-optic SLMs and smart-pixel arrays (based on semiconductor multiple quantum wells, liquid crystals, or magneto-optics) for use in microdisplays. Recently, quantum-well 600×600 arrays bump-bonded to silicon driver chips have been demonstrated that operate at megahertz frame rates.

Future applications of optical processing and computing lie in processing, storing, and displaying large space bandwidth product data. The need to perform these functions efficiently will grow rapidly in the information age of the next millennium. Furthermore, these functions exploit a primary advantage of optics (parallelism) in applications that have an optics-only solution (image display). It is in these computer peripherals, and in the interconnects between computers and peripheral devices, that optics fits best into computing. The size of the market is not yet defined, however, since the technologies are still under development.

For these optical technologies to have a future, they must be applied to practical problems. As for optical data links, processing information optically puts a premium on low cost. The action remains in developing practical hardware, which requires close cooperation between device and systems researchers. Devices developed without an understanding of the system are no more valuable than systems developed without an understanding of the devices. The technologies that will succeed are those that satisfy both systems and device needs.

TABLE 1.4 Examples of Optical Inspection and Machine Vision Systems That May Use Optical Processing

Security	Transportation	Manufacturing	Medicine and Health	Environment
Novelty filtering	Vehicle avoidance	Part recognition	Diagnostic cytology	Spectral imaging
Face recognition	Sign recognition	Integrated circuits	Cancer detection	Photogrammetry
Fingerprint identification	Augmented vision	Produce inspection	Flow cytometry	Monitoring change

Overall Issues

Because the competing technologies are electronics based, optical systems compete with a rapidly moving target as silicon chips become faster and faster and as parallel computing becomes more standard. Unlike other areas of information technology, in information processing, optics plays no unique role—so it must always be held to the electronics standard.

Several key technical issues remain open. Will optical computing find a niche where it is practical? Will optical image processing find a niche where it is practical? Will optical switching in telecommunication systems be centralized or distributed? What will optical cross-connects look like? The challenge is to continue the design of cost-effective devices that can be manufactured and packaged cost-effectively.

There are also important issues for optical information processing in university education and research. As noted in several places in this report, university programs rarely have a systems perspective. Because the advantages of optics for information processing lie in areas that require knowledge of both hardware and software, university education is rarely very relevant to successful engineering in this area. Similarly, in research, optical devices for information processing cannot be considered in isolation since the use of optics is determined by the system design, which is still a matter of debate. Because systems-related work in universities is relatively weak, companies involved in optical information processing typically train their employees on the job. If industry provided more state-of-the-art devices to universities, more useful systems studies could be carried out in academia. Intellectual property concerns may currently be a barrier to this in the United States.

Optical information processing faces some grand challenges: cost reduction of optical and optoelectronic components, packaged subsystems, and full systems; seamless merging of optics and electronics; optoelectronic device development driven by systems needs and system design driven by device realities; and full exploitation of wavelength, space, and time with optics.

Optical Storage

Market Size and Current Trends

The data storage market, fueled by an insatiable appetite for generating, collecting, and storing information, is growing exponentially. Multimedia applications such as video on demand, interactive games, advertising and home shopping, desktop video editing, video and data servers, document and medical imaging, and engineering and architectural drawings have made storage the fastest-growing segment of the computer hardware industry today. Increased storage requirements are distributed throughout the hierarchy, from high-end libraries to desktop personal computers in the office or home. Typical capacity requirements and the way they are distributed are shown in Table 1.5.

The amount of information stored electronically is already enormous (see Figure 1.12). Extrapolating the growth rate in recorded information over the past 50 years suggests that by the year 2000 the total will be 10^{20} bits, or 12 exabytes. This information will be stored on a combination of magnetic disks and tape, optical disks and tape, and a variety of optical formats (Table 1.6) based on the compact disk and the digital versatile disk (DVD).

CD- and DVD-based technology has accelerated the convergence of consumer and computer systems—a convergence that will put pressure on non-CD-format optical storage. This powerful impact of the compact optical disk is illustrated in Figures 1.13 and 1.14. The estimated worldwide market for data storage hardware in the year 2000 is expected to be in excess of $120 billion.

Trends in Storage Technologies

The storage industry is extremely cost-competitive, with price per megabyte being a key metric. Consumers want more capacity, cheaper systems, higher data rates, and products packaged in smaller form factors (i.e., more, cheaper, faster, smaller). Magnetic disk storage, the cornerstone of the industry, is characterized by its storage density. Packing more and more information onto each square inch of disk surface increases storage capacity, decreases the cost per megabyte, and increases the data rate because the linear density increases. At the same time, equal or greater capacity can be packaged in a smaller form factor. Since the first magnetic disk storage system was introduced in 1957, storage density has increased at a 30% compound annual growth rate. Everyone has predicted that such a growth rate can never be sustained, and they are correct: Since 1990 the annual increase has been 60%!

TABLE 1.5 Typical Storage Capacity Requirements for Selected Applications

	Medical X Rays	Movies on Demand	Document Imaging
Main server	Medical center	Regional distributor	Insurance company headquarters
	10,000 GB	70,000 GB	5,000 GB
Local server	Hospital department	Local vendor	Regional office
	20 GB	350 GB	100 GB
Client storage	Doctor's workstation	Television set-top box	Claims administrator
	2 GB	0.1 MB	2 GB

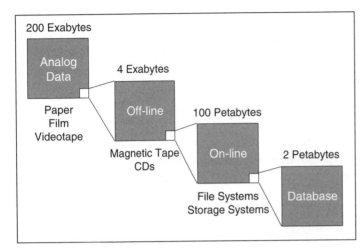

200 Exabytes

Analog Data

Paper
Film
Videotape

4 Exabytes

Off-line

Magnetic Tape
CDs

100 Petabytes

On-line

File Systems
Storage Systems

2 Petabytes

Database

FIGURE 1.12 A pictorial overview of the data stored in the world in 1995. The variety of storage methods is indicated, from computer RAM to paper. An exabyte (EB) is 10^{18}, or a billion billion, bytes. The 4 EB of electronic storage in 1995 is expected to increase to 12 EB by 2000. (Courtesy of A. Chandra, IBM Research.)

It is estimated that by the year 2000, magnetic storage density will be 10 gigabits per square inch (Gb/in.[2]). The fundamental limit is believed to be determined by the size of the smallest magnetized region that remains stable against thermal demagnetization—estimated by the most optimistic researchers to be between 100 and 1,000 Gb/in.[2]. At these density levels, gigabyte capacity drives can be packaged like semiconductor processors and plugged into boards as storage modules. The cost is projected to be 1 cent per megabyte in 2000 and less than 1/4 cent per megabyte in 2005. With the addition of removability, once the distinguishing feature only of tape and optical storage systems, magnetic storage will continue to be the dominant high-end storage technology for the foreseeable future.

Optical storage became an important part of the low-end storage market in 1982, with the introduction of the compact disk. The first success of CD technology was a consumer product capable of storing 75 minutes of high-quality music in digital form. The next application was the CD-ROM (read-only memory) used to store data for low-end computers such as PCs and workstations. This marked the beginning of the convergence of storage technologies for the entertainment and computer markets. Today, CD-ROMs have already replaced floppy disks as the preferred medium for mass distribution of programs, video games,

FIGURE 1.13 Dramatic cost reductions and performance improvements continue to bring the world of computers and the world of consumer electronics closer and closer. The optical compact disk has had a powerful impact on the merger of these two worlds. (Courtesy of A. Bell, IBM.)

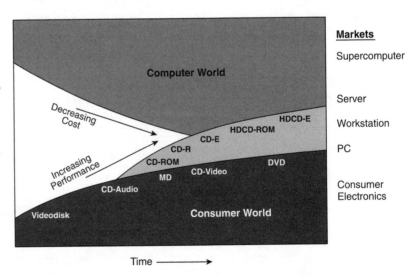

Markets

Supercomputer

Server

Workstation

PC

Consumer Electronics

Computer World

Decreasing Cost

Increasing Performance

Videodisk

CD-Audio

MD

CD-ROM

CD-R

CD-E

CD-Video

HDCD-ROM

HDCD-E

DVD

Consumer World

Time ⟶

and reference material. CD-ROM installations are growing rapidly, as indicated in Figure 1.15. Low-end CD-based technologies are propelling optical storage sales toward a figure in excess of $70 billion in the year 2000, including $20 billion for CD-based drives and $40 billion for CD-based media (including content).

The original CD technology offers permanent, non-rewritable storage that can be removed easily from the drive and can be mass produced for less than 0.1 cent per megabyte. Another advantage is that optical readout using laser light does not cause deterioration of the stored information, such as that caused by a needle in a gramophone pickup. A successor CD technology with higher capacity is already in production. Digital versatile disks with capacities ranging from 4.7 to 17 gigabytes (GB) will soon allow storage of a full-length movie on a single disk. Agreement has already been reached on a DVD standard by which a four-layer DVD-ROM will hold 17 GB on a double-sided disk (see Table 1.6).

Starting from permanent, nonrewritable storage, CD technology has moved to compatible writable and rewritable storage systems. In 1983-1984, 5.25-inch WORM drives were offered, followed by 14-inch WORM drives in 1987, and 3.5-inch and 5.25-inch rewritable drives in 1988. (WORM stands for write once, read many times. Rewritable means that previously stored information can be erased and new information stored.) This trend toward erasability and higher storage capacity would move optical storage into competition with high-end magnetic storage, but optical storage density is not increasing on the same steep slope as magnetic storage density.

FIGURE 1.14 Compact optical disks. They may not be aware of it, but hundreds of millions of people have lasers in their homes. Lasers are hidden inside their compact disk players and computer CD-ROM (read-only memory) drives, which they use to read the information stored on compact disks such as the one shown here. These disks are made of injection-molded plastic. The information is stored digitally in the form of a pattern of pits pressed in the plastic. It is recovered by sensing changes in the reflection of a focused laser beam. Digital storage ensures high-quality reproduction of sound or data, and because the readout beam passes through the substrate, reproduction is insensitive to dust or scratches on the plastic surface. In addition, noncontact readout with a laser beam does not degrade the stored information in the way that passing a needle across a gramophone record did.

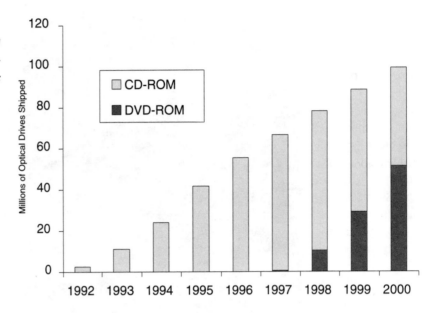

FIGURE 1.15 Installations of CD-ROM and DVD-ROM readers are growing rapidly. NOTE: 1997 = estimate; 1998-2000 = forecast. (Data courtesy of J. Porter, DISK/TREND.)

TABLE 1.6 Hierarchy of CD-Based Technologies That Has Evolved Since Their Introduction to the Market in 1982

Name	Technology	Capacity and/or Time	Media	Form Factor	Market Introduction
CD-audio	Read only	63-74 min	Injection-molded plastic	12 cm, ss, sl	1982
CD-ROM	Read only	682-778 MB	Injection-molded plastic	12 cm, ss, sl	1984
CD-R	Write once	682-778 MB	Dye polymer, phase change, ablative	12 cm, ss, sl	1990
CD-RW	Read-write	682-778 MB	Magneto-optic, phase change	12 cm, ss, sl	1997
DVD-video, DVD-ROM	Read only	4.7 GB, 133 min	Injection-molded	12 cm, ss, sl	1997
		8.5 GB, 236 min	Plastic, multilayered	12 cm, ss, dl	—
		9.4 GB, 259 min	Laminated	12 cm, ds, sl	—
		17 GB, 472 min		12 cm, ds, dl	—
DVD-R	Write once	3.9 GB	Dye polymer,	12 cm, ss, sl	—
		4.7 GB	phase change	12 cm, ss, sl	—
DVD-RAM	Read-write	2.35 GB	Magneto-optic, phase change	?	—

NOTE: CD = compact disk, DVD = digital versatile disk, R = recordable, RAM = random-access memory, ROM = read-only memory, RW = read-write; dl = double-layer, ds = double-sided, sl = single-layer, ss = single-sided.

The easy removability of optical storage media has made it possible to construct optical library systems for archiving large amounts of data. A good example of today's capabilities is the use of 5.25-inch erasable magneto-optic disks, with a capacity of 4.7 GB each, in libraries with mechanical pickers that can access any of 144 disks. This system provides a library capacity in excess of 600 GB.

It should be noted that fast access time is a very important attribute of a storage system, particularly as capacities increase. Users complain already about the relatively slow access times of CD-ROMs. The inherent parallelism of optics should offer an advantage in access speed, but this goal has not yet received enough attention. Array technologies such as those investigated for optical processing should be applicable.

Education Issues

With the exception of a few university centers, such as at Carnegie Mellon University and the University of Arizona, U.S. education in storage, either magnetic or optical, is nonexistent. In the case of optical storage, this is understandable because few domestic companies are developing products, and consequently there is a lack of employment opportunity. Research into the more advanced storage mechanisms (e.g., holographic and two-photon storage) is more widespread but is not well connected to or directly supported by the storage industry.

International Competitiveness

Most of the development and manufacture of optical storage systems for consumer and computer use does not reside in this country. In contrast to magnetic disk and tape storage, where 80% of the worldwide market is owned by U.S. companies, less than 10% of the optical storage revenue is attributed to U.S. companies. It does not appear that this trend will change.

In contrast, activity to develop advanced storage methods based on holography, two-photon recording, and so on, resides in this country with little activity offshore. This could change, however, as the technologies move closer to becoming products, especially if they have wide applicability and become attractive targets.

Key Unresolved Issues

What is unresolved, of course, is how far conventional optical storage density can be pushed and whether the cost-performance trade-offs will permit it to be more competitive with magnetic recording or relegate it to niche markets. The former question is a concern of optics and recording physics; the latter relates to manufacturing and the market. Current trends suggest that the convergence of consumer and computer storage may favor the upward extension of the low-end, entertainment type of storage, rather than the continued development of high-end technologies. A strong driving force would be the compatibility of all forms of information storage—music, games, video, program distribution, working files, backup files (i.e., true multimedia)—and the use of a common, universal drive. If this were to happen, high-end optical storage as we know it today would either disappear or be relegated to highly specialized niche markets. Similar considerations apply to more advanced storage technologies, with the further caveat that we must first determine and demonstrate their technical feasibility and performance before we set out to define markets.

Opportunities, Challenges, and Obstacles

Paths along which optical storage can move to higher densities are depicted in Figure 1.16. In combination, these approaches offer as much as a hundredfold improvement in capacity.

Higher storage densities require either making and reading smaller marks on the surface of a recording medium or storing information within its volume. To accomplish the former, resolution must approach the diffraction limit or, where possible, exceed it. Since the recorded spot size is directly proportional to the wavelength of the laser source and inversely proportional to the numerical aperture of the focusing objective lens, changing the wavelength and the numerical aperture are two high-leverage approaches to increasing capacity.

Because of its size, cost, and reliability, the laser diode has been a critical component in the success of all levels of optical storage, from CD-audio to massive optical library systems. The transition from early 820-nm laser diodes to today's 640-nm devices has been a slow, often painful process, primarily devoted to achieving the requisite reliability. This transition was helped by the CD-audio business, which developed reliable low-power lasers. Little help came from the enormous resources pumped into communication lasers, as they moved in the opposite direction, to longer wavelengths in the near infrared. For storage, transition to a blue laser would provide a fourfold leap in recording density. It does not appear, however, that the technology employed in 820- and 640-nm lasers can be extended to much shorter wavelengths, certainly not into the blue. Two options remain: (1) doubling the frequency of these longer-wavelength lasers through some nonlinear interaction, or (2) developing new direct-bandgap materials that produce blue light directly. The nonlinear route has shown promise with the generation of 40-50 mW of power at 435 nm, but the packaging is prohibitively large, and the cost is currently far too high for implementation in a competitive storage drive. More recent work on smaller, integrated devices producing about 10 mW of power is encouraging, but much work remains to be done. Direct-bandgap lasers in the blue

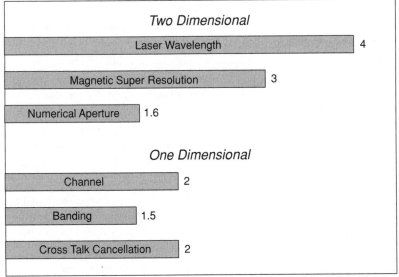

FIGURE 1.16 *Illustration of paths to higher optical storage density, including the use of shorter-wavelength sources (e.g., blue lasers), increasing the numerical aperture of the objective lens, the use of multiple data layers, and the use of the third dimension (via holography). These approaches are relatively independent of each other and can be implemented in combination to offer improvement in capacity by a factor of 100. (Courtesy of B. Schechtman, National Storage Industry Consortium.)*

have been sought for many years, but only recently has their feasibility been established with the demonstration of continuous-wave (cw) operation at room temperature in III-V materials such as gallium nitride, which have been reported with lifetimes of more than 100 hours, and in II-VI materials such as zinc sulfide and zinc selenide, which produce a few milliwatts for about 10 hours. With improved reliability, power levels of a few milliwatts may be attractive for read-only applications such as audio, video, and program distribution.

Increasing the numerical aperture of the objective lens significantly beyond the 0.6 value used today is certainly possible from a theoretical point of view, but it may not be practical in a storage system. The key problem is retaining the removability of the storage medium while keeping the sensitivity to contamination manageable. The usual solution is to place the recording layer on the underside of a thick (1.2-mm) transparent cover layer. However, if the disk is tilted or warped, as often happens, aberrations are introduced into the wavefront, and diffraction-limited imaging is compromised. As the numerical aperture increases, the tolerance to tilt decreases, and the thickness of the cover layer must be reduced; eventually contamination protection, and hence removability, is lost. Because of these problems, there seems to be little promise to developing low-cost, high numerical aperture lenses.

Numerous other techniques provide the ability to make smaller marks, including the solid immersion lens and near-field optical probes. Storage densities of 10 Gb/in.^{2} and 45 Gb/in.^{2} have been demonstrated with these two methods. To achieve these densities requires placing an element such as a glass hemisphere or a drawn fiber tip in close proximity to the recording layer, usually flying over it at distances less than

100 nm, much as in magnetic recording. In general, the removability of the storage medium becomes a problem, because great care must be taken to prevent contamination from getting between the optical head and the medium. This is certainly a viable option, but one must expect to compete head-to-head with magnetic recording, so a clear density advantage is mandatory. Unfortunately, demonstrated storage densities have yet to show this margin.

Another method of increasing effective storage density is to use multilayers—data layers separated by thin transparent spacers or air gaps. The high numerical aperture of the focusing objective ensures that there is minimum cross talk between layers as close as 10 or 20 microns. Read-only systems with 50 layers and WORM or erasable systems with 20 layers are feasible. DVD technology, for example, will use four layers, two on each side of a double-sided disk. Extending this technology to many more layers will require development of the requisite manufacturing capability, as well as innovative optical design to control the aberrations encountered in imaging layers at different depths.

To push CD and DVD technologies to higher effective storage densities and performance levels, U.S. industry should develop multilayer storage media; low-cost optical systems for writing and reading data; and efficient, low-cost techniques for mass replication and assembly of multilayer disks.

The use of a third dimension, such as the volume of the storage medium or the recording wavelength, appears to be the way to win the density battle (see Figure 1.17). The use of multilayer recording described above is one way of recording discrete bits throughout a volume. Others include two-photon recording, electron-trapping optical memory (ETOM), and persistent spectral hole burning. Holographic storage also uses the volume; however, it stores information not as discrete bits but as interference patterns (i.e., the information about an individual bit is distributed throughout the volume rather than localized at a point). These systems are capable of very high storage densities, theoretically as high as $1/\lambda^3 \approx 10^{12}$ bytes/cm^3, and capable of data rates in excess of 1 Gb/s. Key to the success of these technologies will be the development of an optimized organic or inorganic recording material and reduced complexity of the optical system required to write and read data. Near-infrared sensitivity would enable the use of high-power laser diodes, thereby enhancing reliability, lowering cost, and improving system packaging. The necessary component technologies, including spatial light modulators, detector arrays, lasers, and deflection optics, are here today but need to be configured for specific applications. The potential performance advantages of these technologies, such as high data rates and parallel optical access, make them particularly attractive for large image and multimedia databases and network servers.

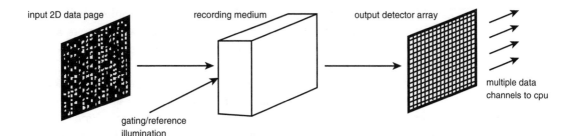

Features	Candidate Technologies	Issues
• Rapid access	• Two-photon	• Materials
• High data rates	• Electron trapping (ETOM)	• Low-cost components
• Large capacity	• Holography	• Defining markets
• Low S/MByte	• Photochemical hole burning	• Window of opportunity
• Removability		
• High reliability		

To retain the U.S. technological edge in three-dimensional recording, industry and universities should nurture and accelerate the development of advanced three-dimensional recording media, the design of low-cost optical systems, and the study of systems integration and architectures. It is imperative that these activities be coordinated among university and industrial researchers.

The challenge for U.S. industry and universities is to pursue aggressively a family of optical storage solutions that have clear cost or performance advantages over competing technologies in the 100 GB and larger storage range. Capacity growth from smaller consumer-based systems will be slowed by attempts to maintain backward compatibility and the need to adopt standards. Rather than compete in this game of incremental development, the United States has an opportunity to leap forward and establish a new storage paradigm. **The Defense Advanced Research Programs Agency (DARPA) should establish a program to seek new paradigms in optical storage that will reach toward the theoretical storage density limit of about 1.0 terabyte (TB) per cubic centimeter, with fast (>1 Gb/s) recording and retrieval.**

FIGURE 1.17 An overview of potential three-dimensional (3D) optical storage techniques. The sketch shows the generic layout of such a system: a two-dimensional (2D) data page serving as the input, the 3D recording medium, and the 2D array of detectors that serves as the output and the link to the user. Candidate technologies and the main features of these techniques are listed.

Displays

The growth of broadband networks and new information age services continues to increase the importance of text, graphics, image, and video information and, with it, the need for new and ubiquitous display devices. In many cases, displays are becoming a critical enabling technology for new information systems that could not exist without them.

This trend is creating exciting challenges and new opportunities for display technology. The technology is intrinsically optical and serves as the intermediary between electronic information systems and the eye of the human user.

Until 1970, the display market consisted of cathode-ray tubes (CRTs) that performed video-rate imaging, primarily for the consumer television market but also for scientific instrumentation and military cockpit systems. The CRT, which enjoyed its hundredth birthday in 1997, employs an electron beam to scan an analog signal across a phosphor that emits light in proportion to the energy of the beam. By the early 1970s, this emissive display technology had matured. The major challenge became achieving low-cost, high-volume manufacturing to meet the consumer needs of the mass television market. This challenge was met by transferring the technology and manufacturing these displays offshore.

Although the CRT was (and still is) the cheapest display technology, several drawbacks have become apparent. CRTs are heavy and occupy a large volume, require high voltages, and consume more power than their flat counterpart, the liquid crystal display (LCD). They are very hard to read in direct sunlight. Given these system constraints, it is unlikely that the CRT will achieve the futuristic vision of a "hang-on-the-wall" television. To realize this vision, research started in the late 1960s and early 1970s on flat-panel display (FPD) technologies, including flat electroluminescence displays, field emission displays, plasma displays, and LCDs. The most successful and prolific of these is the LCD.

The history of liquid crystal displays (see Box 1.6) reminds us that many early innovations in this field started in the United States and Europe. Invented in 1968 at RCA, the first liquid crystal switch or display element was a low-contrast scattering device. Such displays consist of tiny organic molecular rods that move under the influence of an electric field. They modulate (turn light on and off) in proportion to the

....

BOX 1.6 HISTORY OF LIQUID CRYSTAL DISPLAYS

- Liquid crystal dynamic scattering device is invented at Sarnoff Laboratories, 1968.

- Twisted-nematic (TN) LCD is discovered by Schadt and Helfrich, 1971.

- Active-matrix liquid crystal display (AMLCD) is invented by Brody at Westinghouse, 1972.

- Clark and Lagerwall invent ferroelectric LCD, 1980; licensed to Canon.

- Scheffer invents supertwisted nematic (STN) displays at Brown Boveri, 1983, and moves to Tektronix.

- World LCD market reaches $1 billion, 1991—mostly TN, STN, and AMLCD technologies.

- Buzak invents plasma-addressed LCDs, 1994; licensed to Sony, 1994.

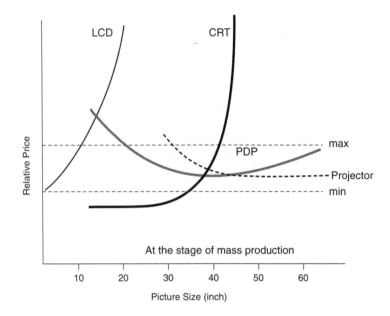

FIGURE 1.18 *Predictions made in 1993 of relative mass market prices for several key display technologies. Studies such as this are an important basis for the expectation that major market segments will be dominated by different technologies: LCDs for medium-sized displays, CRTs for large displays, and PDPs or projection displays for very large displays. NOTE: PDP = plasma display panel. (Courtesy of NHK.)*

applied electric field. In 1970, Rockwell demonstrated the first commercial production of an LCD.

The advent of the liquid crystal display has changed the CRT's dominance of the display market. There are now two dominant mass display market segments: medium-sized displays (about 10.4 to 12.1 inches along the diagonal) and large displays (greater than 15 inches). Medium-sized displays, mostly LCDs, principally serve the computer sector, such as laptops and PCs. Large displays, mostly CRTs, principally serve the television entertainment sector. In addition there are important niche markets for small displays, projection displays, very large displays, and displays for military and avionics needs.

Because of its low cost (about $50 for 10-inch displays in quantities of 100,000), the CRT still dominates the mass market for large displays. Despite the drawbacks mentioned above, no technology appears to be on the horizon to challenge the CRT in this market. However, this is a mature technology and no major CRT innovations are expected. Scaling the CRT to very large sizes (e.g., 40 inches) is difficult and costly, and it is doubtful whether CRTs will be able to provide the large high-resolution displays needed for future high-definition television (HDTV) systems at a reasonable cost (see Figure 1.18).

Medium-Sized Displays

The initial twisted-nematic (TN) LCDs were less than 2 inches (diagonal) and were used primarily in calculators, watches, radios, and later, miniature television sets. Almost all of these displays were manufactured in Japan. At the time, it was not clear that LCDs could be scaled to larger sizes. Japan continued to invest in the development of LCD technology, whereas the United States focused on such new technologies as electroluminescence and field emission FPDs. It is estimated that Japan has invested more than $3 billion in scaling up the LCD technology to produce medium-sized (10.4-inch) actively addressed LCDs, originally invented by Peter Brody at Westinghouse Corporation.

A fortuitous collision of technological innovations occurred with the advent of the personal computer and subsequently the portable, laptop computer. These systems required medium-sized displays. Without such a parallel development, Japan might have lost its huge R&D investment in LCDs. Instead, by 1991, Japan had the largest share of a billion-dollar LCD market. The subsequent development of active-matrix liquid crystal displays (AMLCDs) produced higher-contrast displays. By 1996, the annual LCD market had reached $7.5 billion; it is expected to grow to more than $20 billion by 2001 (Figure 1.19).

Although the early development of FPDs occurred in U.S. R&D laboratories, the United States has not invested in creating a capability for high-volume, low-cost display manufacturing. The United States currently has only 5% of the worldwide market, mostly concentrated in niche applications for military and avionics use. Japan enjoys a 50 to 60% market share, and Korea is expected to achieve a 25% share by 2001. Taiwan has also recently announced a billion-dollar FPD initiative.

Medium-sized displays are now produced in the tens of millions of units a year. Typically 10.4 to 12.1 inches in diagonal, AMLCDs for laptop computers have become a commodity, selling for less than $500; prices are still dropping by 20% or more per year. It appears unlikely that the United States can compete in the commercial and consumer markets that require displays in this format, at least in the near term.

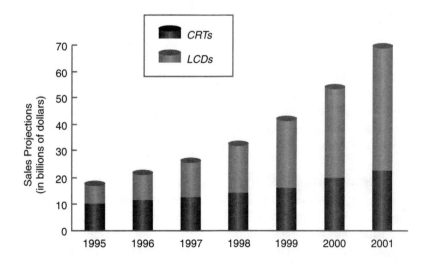

FIGURE 1.19 Worldwide market for electronic displays. Over the next several year, LCD sales are expected to grow rapidly.

Although the size of the FPD industry alone is considerable, it is an enabling technology for even larger collateral markets in display components, systems, and applications. The most obvious example is the laptop computer, which simply could not exist without the medium-sized flat-panel display. The market values of enabled technologies such as electronic display drivers ($20 billion), computer and peripheral systems ($200 billion total), and evolving two- and three-dimensional augmented reality systems and games (probably more than $100 billion by 2003) approach a total of $300 billion in leveraged commerce.

The success of the LCD has now made it the entrenched technology in the mass market for medium-sized displays. Experts doubt, however, that AMLCD technology can be economically scaled up to the large and very large displays (more than 16 inches). This means that there are opportunities for new technologies to meet the needs of current niche markets such as those for very large and very small displays.

Small Displays

The next potential mass display markets appear to be for very large displays and for applications that integrate small and miniature displays. With regard to small displays, the application areas include (1) portable and wearable computer displays; (2) electronic imaging engines for projection displays; and (3) handheld displays in industrial tablets, personal digital assistants, and games (Figure 1.20). Displays for wearable computers and Web browsers include head-mounted eyeglass-style displays in the format of a 35-mm slide (1.3 inches diagonal) or one-quarter of that (0.7 inch).

Handheld displays will probably have a footprint smaller than a 3 × 5 inch notecard. Although miniature AMLCDs (less than 0.7 inch) can be used in these applications, cost limitations may necessitate the use of passively addressed twisted- or supertwisted-nematic (STN) LCDs. Emissive technologies such as field emission and electroluminescence offer wider viewing angles than LCDs, an advantage for handheld devices. It is estimated that by 2001 the industrial tablet market will reach $2.6 billion, with personal digital assistants reaching $500 million and the game market growing exponentially.

A group of U.S. companies is currently leading the exploration of new approaches for miniature and micro displays. A consortium formed to develop field emission displays includes such companies such as Pix Tech, FED Corporation, Raytheon, SI Diamond, and Motorola. There is also a strong thrust to develop electroluminescent displays and other potential breakthrough technologies, including liquid crystals on silicon reflection-mode displays, digital micromirrors, and mirror-grating displays. There is no clear winner yet in the race to produce a high-quality miniature display for the kinds of applications listed above.

FIGURE 1.20 One future market expansion for displays is in flat, light, low-volume visual interfaces for personal digital assistants, Web browsers, augmented and virtual reality engines, and portable and wearable computers. The display shown here is a full-color VGA-sized device that carries nearly the information content of a personal computer monitor. (Courtesy of Planar America.)

What is needed for the potential mass market is low-cost manufacturing, low power dissipation, high contrast ratio (an advantage of emissive displays), low-complexity imaging optics (for head-mounted and projection displays), low-voltage materials and devices, and hardware or software systems for three-dimensional imagery and interactive displays. In addition, there are specific needs for some technologies, such as faster switching for nematic liquid crystals with 8 bits of gray scale and electroluminescent materials with better white phosphor (especially in the blue).

Projection Displays

Projection displays can be considered a close relative of the slide projector. A key difference is that the "slide" is changeable in real time, under the control of a computer or an electronic terminal. The system projects large images onto a screen or the wall, but the technology for "changing the slide" is closely related to that of miniature LCDs.

For boardroom, classroom, or family room wall-sized displays, electronic projectors using LCD or micromirror technology are being commercialized by several U.S. companies. These systems use 1.3-inch active-matrix displays made on glass for transmission displays or on a silicon substrate when used in reflection. The latter presents an opportunity for the U.S. semiconductor industry. According to the Semiconductor Industry Association's *National Technology Roadmap for Semiconductors,* growth in this industry will be produced by the following:

- Achieving finer lithographic features;
- Improving interconnects through on-chip optics; and
- Creating new applications for integrated circuits (such as silicon-based displays).

Electronic projector sales exceeded $1 billion in 1996 and are expected to double every year for the next few years. The potential market may exceed that of laptop computers by 2001. In addition to the technology needs described above for miniature displays, projection technology will require extremely bright illumination sources, with small spot sizes and long service lives.

Very Large Displays

Very large (nonprojection) displays are designed to deliver 20- to 40-inch images for such applications as high-end graphics workstations and future HDTV. Currently this is a niche market, but it has the potential to grow into a mass market. Conventional LCDs and CRTs are not expected to be scalable to these large sizes at a reasonable cost. Two flat-panel technologies now under development promise to reach 40-inch size, plasma displays and plasma-addressed LCDs. The United States has a technology edge in large plasma displays, and the development of a 21-inch full-color display with 1280×1024 resolution was announced in 1995. Plasma-addressed LCDs were invented in the United States (at Tektronix) but have already been licensed to Sony. As for other display technologies, a challenge for U.S. industry is to develop a high-volume, low-cost manufacturing capability before the mass market develops.

Military and Avionics Displays

Military and avionics displays are an important niche market for U.S. industry. With the increasing emphasis on graphics, images, and video information, the U.S. military finds displays increasingly critical. Typical applications are medium-sized displays for the digital battlefield and cockpit displays, as well as miniature devices for simulators and tank and gunsight displays. Although medium-sized displays are produced for the mass consumer market, it appears impractical to insert this technology directly into military applications, which have to work in a wide range of light levels and whose construction must be more rugged than that of civilian products.

Government support of the flat-panel display industry has provided strong innovative impulses and has been a critical element in establishing a strong technology position in the current niche markets for small displays and very large displays. This could provide the catalyst and basis for U.S. reentry into the commercial and consumer markets, particularly if niche markets show fast growth and become mass markets. It should be stressed, however, that successful reentry will depend critically on the development of a high-volume, low-cost manufacturing capability. Note that the market for military displays is no more than 5% of the worldwide mass display markets. These mass markets can generate $3 billion to $4 billion annually in R&D resources that can be used to place their low-cost manufacturing on a fast learning curve. U.S. efforts toward reentry have to be measured against these vast resources.

Educational and R&D Issues

Display technology is a major outlet for the U.S. research community in optics and condensed-matter and materials physics. This community continues to create potential breakthrough technologies, but there is only relatively weak coupling of materials and device work to programs on display systems and applications. In fact, this study has not identified any course at a U.S. university in which display technology is taught in grand perspective, including materials, device, systems, and applications issues. In view of the growing importance of displays, there is a need to develop such cross-disciplinary educational curricula.

Academic research on displays is limited to less than a dozen U.S. universities. The most notable academic program is at Kent State University, which has focused on building a center of excellence in liquid crystal technology. Other noteworthy university programs include Princeton University (organic luminescent displays), the University of Michigan (AMLCDs), the University of Colorado at Boulder (ferroelectric LCDs and liquid crystal on silicon), Georgia Tech (liquid crystal on silicon), and the Massachusetts Institute of Technology (liquid crystal on silicon). It is often claimed that U.S. universities find it difficult to obtain research contracts for display work.

Most U.S. industrial display R&D is pursued by small and medium-sized companies, with a focus on military and avionics applications. These programs are concentrated on CRT, plasma, electroluminescent, field emission, and liquid crystal technologies.

There is widespread consensus that the United States should try to reenter the mass display markets. However, there is no well-understood, broadly supported, well-coordinated strategy involving industry, universities, and the government that has the goal of accomplishing this task.

Summary and Recommendations

Information Transport

Optical technology overwhelmingly dominates the long-distance transmission of information. More than 100 million kilometers of optical fiber have already been deployed worldwide, and deployment continues at the rapid rate of greater than 20 million kilometers per year (more than 2,000 km per hour, which is faster than Mach 2). Moreover, the transmission capacity of a single fiber has increased exponentially by about a factor of 100 per decade. A capacity of 1 Tb/s has been demonstrated in the research laboratory, and 20-Gb/s and faster systems are being deployed commercially. The introduction of optical amplifiers and wavelength division multiplexing has been essential to recent advances, and these two key enablers have considerable potential for further innovation and application to WDM networking.

Various optics-enabled systems are being evaluated worldwide for broadband applications in local exchange access. These include analog lightwave transmission, fiber to the curb, and fiber to the home. Over the next 20 years, the installation of a broadband local exchange access infrastructure will require an estimated $150 billion in capital investment in the United States alone and probably three to five times as much worldwide. This is a large-scale challenge for optics and associated technologies. The local access market is considerably larger than that for long-distance technology. Fiber to the home (FTTH) is generally considered the ultimately desired broadband technology. It offers a future-proof installation that can be adapted for most conceivable broadband services once demand for them develops. To make FTTH viable, cost reduction is needed in several key enabling technologies: low-cost lasers, low-cost packaging, low-cost manufacturing, and passive network topologies that allow cost sharing and reduce operational costs. FTTH will make it possible to avoid outside plant powering, which will result in significant savings in cost and operations. Lifeline telephony provided by FTTH will not function at home during power failures unless special provisions are made for powering.

Increasing access bandwidth is a global challenge for the future. It will enable society to enjoy a variety of future information services. The rapid emergence of broadband telecommunications networks and broadband information services will have an enormously stimulating impact on commerce, industry, and defense worldwide. **Congress should challenge industry and its regulatory agencies to ensure the rapid development and deployment of a cost-effective broadband fiber-to-the-home information infrastructure.** Only by beginning this task now can the United States position itself to be world leader in both the broadband technology itself and its use in the service of society.

The technology for analog lightwave transmission is improving rapidly and is enabling systems of lower cost and improved performance. This progress is broadening the range of applicability from cable television distribution to remote links for mobile radio cellular systems, as well as microwave and potentially millimeter-wave systems. Information transmission via laser beams promises lower cost and higher capacity for future links between satellites. The required know-how for this optical technology does exist at U.S. universities and federally funded laboratories; however, unlike their counterparts in Europe and Japan, U.S. industry and the U.S. government have not invested in the development of experimental optical payloads that would allow realistic economic assessment and timely market entry.

Although U.S. universities are quite strong in the physics, materials, and component areas of information transport technology, they are relatively weak in applications and systems. There is a need for better linkage of university device research to systems research, in order to motivate and guide the former and speed the development of the latter. There are considerable research opportunities in component integration to make WDM technology more cost-effective and broaden its applicability. Low-cost optoelectronic packaging is essential for reducing system costs and should be considered an integral part of device research. Funding agencies should support better linkages between university researchers and industry and between research on devices and on multidisciplinary systems.

Processing

Data Links

Low-cost optical data links are on the verge of becoming practical in local area networks that link together personal computers and workstations, file servers, printers, and data storage systems within computing clusters. To date, however, optical data links have achieved only modest market penetration, since copper is usually cheaper. The Optoelectronics Industry Development Association (OIDA) expects annual revenues for data communications equipment to reach $25 billion by 2003, with $7 billion of the total being for optoelectronic equipment. By 2013, OIDA expects the total market to double and the optoelectronic component to reach $30 billion.

This market is global, and low cost is crucial. Further market penetration requires cost reductions, especially in manufacturing connectors and cables; packaging and alignment; improved molding or plastic packaging; and modeling and simulation tools for high-frequency, low-cost packages. Many optical data links employ parallel technology, using arrays to achieve economies of scale. The challenge is to enable research in the new R&D environment, in which low-cost, high-volume global markets mean low marginal profits. Strong university research must continue and should be focused on longer-term issues that are unlikely to be addressed by industry. Collaboration between industry and universities is vital.

Optical Networking and Switching

As telecommunications bandwidth increases, the cost of switching is becoming an ever-larger fraction of the cost per bit, and low-cost switching technology is becoming increasingly important. The increasing importance of low-cost switching will motivate the exploration of WDM and the use of optics in high-density logical routing switches. Switching system devices and concepts are in an early stage of development, so there are still major opportunities in materials and devices. There is strong competition from Europe and Japan in both WDM technology and high-density switching fabric technology.

Experimental switching systems using WDM technology are being tested in Europe, Japan, and the United States. WDM systems use integrated-optics crossbar switches to route different colors of light in different directions. These switches route the entire optical signal and thus are independent of the modulation format or the number of multiplexed signals. Network management and control systems are the major need for commercial acceptance of WDM optical networking and switching. Standards are necessary. Issues to be resolved include simplicity versus complexity, components, and architectures. Practical wavelength conversion devices are needed to allow full flexibility in WDM systems.

High-density optical interconnect switching is a systems approach based on the current technology of electronic logical switches. Optics is introduced for routing within electrical logical switches to overcome the electrical limitations in high-density switches. The technology requires two-dimensional arrays of integrated receivers, transmitters, and smart pixels. The use of optics in high-density switching fabrics is still in the research stage. Component development and systems integration studies are under way. Significant deployment of optics in switching may start in about 5 years, be important in 10, and be crucial for successful systems in 15. Market size projections cannot yet be made, however, since the technologies are too immature.

Purely optical switching has not yet become a reality because of difficulties in buffering, memory, and logic. In addition, ultrafast optical switching devices require too much power, are too simple logically, and are not yet practical for large-scale use. All-optical switching systems are unlikely to be practical unless there is a breakthrough in technology. Nonlinear materials, fibers, and devices that could lead to breakthrough technologies are excellent university research topics. Additional opportunities lie in exploiting ultrafast optical technology, such as short-pulse sources for WDM. Continued strong investment is recommended, with research carried out in both universities and industry.

Optical Image Processing and Computing

Optical logic has great difficulty in competing with electronics for computing since electrons perform logic much more easily than light. Breakthrough technologies will be required before general-purpose optical computing is likely to be important. Applications for optical processing are in correlators and niche applications for machine vision and pattern recognition. High-speed silicon provides strong competition for optical processing in many cases. Optical image recognition can be commercially viable, however, particularly in rare-event problems.

The future of research on optical processing and computing lies in the application of technology to practical problems, with a premium on low cost. The development of practical hardware can come only by close interaction between device research and systems research.

General

University education and research must do a better job of interrelating devices and systems. Better interrelation of university and industry research is also necessary. Research should focus on cost reduction of optical and optoelectronic components, packaged subsystems, and full systems. It should strive for a seamless merger of optics and electronics via improved systems integration and device integration. Optoelectronic device development should be driven by systems needs, and system design by device realities. Designers of optical information processing systems should more fully exploit wavelength, space, and time.

University research should be carried out with an understanding of industrial needs. Collaboration between universities and industry should be encouraged, and issues that stand in the way, such as intellectual property, should be addressed on a national level. It would be helpful for students and faculty to do part of their research on-site in industry. Device research and development should be carried out in a systems context. Although this usually occurs in industry, it must happen more in universities. Industry can assist universities by donating or lending components, enabling university researchers to carry out systems studies.

Storage

The requirement for digital information storage is growing at an enormous rate; it is expected to exceed 10^{20} bits in the year 2000, with an estimated annual market for data storage in excess of $100 billion. Of this market, optical storage will be about $12 billion. Each year, the number of bytes of shipped magnetic disk storage doubles; 125 petabytes shipped in 1996 will become 2 exabytes in the year 2000. These numbers do not include consumer-based electronics and entertainment storage (i.e., CD and DVD technology), which will add another $20 billion in drive sales alone with media sales exceeding $40 billion. Much of today's information is distributed and stored on paper, perhaps as much as 90% of it in the form of newspapers, magazines, and books—15 million tons of paper will be used annually by the year 2000. An increasing amount of the original information for these publications, however, is being captured electronically. Multimedia, parallel, and network-centric computing is increasing the need for larger, faster, cheaper storage.

Magnetic disk storage density is currently at 2.6 Gb/in.^{2} and is growing at an annual rate of 60%. This translates into lower cost and higher performance. It is estimated that the cost of storage in the year 2000 will be 1 cent per megabyte and in the year 2005 less than 1/4 cent per megabyte. The limit for magnetic storage density is unknown at present but is estimated to be in excess of 100 Gb/in.^{2}. Optical storage densities have increased by a factor of seven, from 0.39 Gb/in.^{2} (CD-ROM) to 2.7 Gb/in.^{2} with the recent introduction of DVD-video, and will continue to increase to 9.8 Gb/in.^{2} when double-sided, double-layer DVD-ROM disks are available in a few years.

CD-ROMs have already replaced floppy disks as the preferred medium for the distribution of programs, video games, and reference material. DVD technology with 4.7 to 17 GB capacity per disk is available to allow storage of a full-length movie on a single disk. Recordable CD technology is already available, with the DVD version not far behind. Extensions to CD and DVD technologies share many of the same pathways to higher density and have the added value of commonality with a large installed base. The inherent removability of optical storage media has made optical library systems an important part of the storage hierarchy, providing reasonably rapid on-line access to large databases. These systems are often used for archiving data. A large fraction of the data stored in optical libraries is on WORM media that use ablative, phase change, or dye polymer materials. Erasable libraries generally use magnetooptical materials. Read-only CD libraries are available that will transition to WORM as the use of CD-recordable technology become more widespread. DVD is expected to follow a similar product path. Erasable CD and DVD systems are on the horizon.

The market for magnetic disk storage is dominated by U.S. industry, which has an 80% share of worldwide revenue. The market for CD-ROM drives, conventional optical drives, and CD libraries is dominated by foreign companies, with greater than 90% of the market. There is essentially no CD-audio player industry in the United States. It should be noted that even though U.S.-based companies dominate the magnetic disk market, all major manufacturers build their drives offshore.

Three-dimensional storage, such as holographic, ETOM, and two-photon, has the potential to provide inexpensive, rapid access to large databases. The key to success is identifying and optimizing the proper materials systems. Storage and retrieval of information as large blocks (pages) of data can provide very high transfer rates in excess of 1 Gb/s. These storage technologies are ideally suited for the storage of digital images such as medical images, satellite images, geophysical data, and movies. The enabling technologies for information input (e.g., spatial light modulators) and information output (e.g., CCD arrays) are generally available. Most R&D on three-dimensional storage is performed in the United States; although many of the programs are exploratory and some have stopped short of fruition, two major industry-university consortia are in place to commercialize holographic recording. The unique requirement of specific market segments for systems with special performance characteristics is a driving force that will allow optical storage products to penetrate niche markets.

Information storage of all formats, including electronic, optical, film, and paper, will continue to grow unabated, and storage products have enormous market potential. Magnetic recording will be the mid- to high-end, high-performance technology of choice for the foreseeable future. Extensions of the optical CD and DVD technology will capture a substantial share of the low- to midrange storage market. CD and DVD technologies will start to penetrate the high-end library storage market. Low drive cost and media standards that promote interchangeability and backward compatibility will be key to the evolving success of CD and DVD. U.S. industry is not playing a leadership role in the development of conventional, CD, and DVD optical storage systems. Although there is a window of opportunity between 2000 and 2005 for the deployment of three-dimensional storage techniques, they must compete on a cost-per-megabyte basis with the aggressive evolution of magnetic and optical (CD and DVD) technologies in the same time frame. The United States has a lead of about 2 to 3 years in the development of three-dimensional storage. To be successful, an optical storage product need not displace all other forms of storage but can coexist with them by providing cost-effective solutions to a subset of the market. Optics is a key enabling technology in the information storage marketplace.

To push CD and DVD technologies to higher effective storage densities and performance levels, U.S. industry should develop multilayer storage media; low-cost optical systems for writing and reading data; and efficient, low-cost techniques for mass replication and assembly of multilayer disks.

To retain the U.S. technological edge in three-dimensional recording, industry and universities should nurture and accelerate the development of advanced three-dimensional recording media, the design of low-cost optical systems, and the study of systems integration and architectures. It is imperative that these activities be coordinated among university and industrial researchers.

DARPA should establish a program to seek new paradigms in optical storage that will reach toward the theoretical storage density limit of about 1.0 TB/cm^3, with fast (>1 Gb/s) recording and retrieval.

Displays

Display technology is critical for the development of new information systems and services. Major change is transforming the display market and creating new opportunities for innovation. Once dominated by the CRT, the market is now split into two mass markets of about $20 billion each, plus a few niche markets. The mass market for large displays (greater than 15 inches, mostly televisions) is still dominated by the CRT, while the mass market for medium-sized displays (less than 15 inches, mostly computers) is dominated by LCDs. The U.S. military finds displays critical and has to maintain core competence in this field; however, military needs support only a small niche of the display market.

No major innovations in CRT technology are expected, and scaling CRTs to very large sizes (greater than 40 inches) is expected to be difficult and costly. There are two approaches to overcoming this barrier for large workstation, simulator, and television applications: plasma displays and plasma-addressed LCDs.

In medium-sized displays, there is a convergence of the computer and television formats. The industry standard is the active-matrix LCD, manufactured for 10.4-inch laptop computer applications. More than 10 million such displays were sold in 1996. The future trend for AMLCDs will be to increase their size to 12.4 inches and eventually to 14.1 inches.

There is no clear winner yet in the race to develop a high-quality miniature display for head-mounted, handheld, and projection applications. Candidates include emissive technologies such as active-matrix electroluminescence, field emission, and liquid crystal on silicon; reflection-mode technologies include digital micromirrors and micromirror or grating displays. For head-mounted and handheld applications, R&D on small displays should focus on ergonomics. Before head-mounted displays are accepted by the consumer, they will have to be much lighter and more comfortable, and they should not cause nausea when used. Improved optical system designs have to be developed to make reflection displays as cost effective and high in contrast as their transmissive and transmission-mode counterparts.

Improvements are needed in many display technologies, including better white phosphors, faster-switching nematic liquid crystals, better lamps, and brighter blue LEDs.

Although many currently used display technologies were invented in the United States, their development is for the most part carried on overseas, as is 95% of display manufacturing. The United States still has a lead in the development of very large and very small emissive and reflective display technology. Translating U.S. inventions into a share of the global market requires further R&D on display systems and applications. This competence is currently scattered and limited to a handful of U.S. universities and less than a dozen small and medium-sized companies. Because of the rapid learning curve and the large R&D resources (about $2 billion) generated by the mass market—heavy Japanese and Korean investment in low-cost manufacturing technology is expected to result in improved performance while lowering prices by at least 20% per year—it will be extremely difficult to displace liquid crystals from the mass market for medium-sized flat-panel displays. Major opportunities exist for new technologies to enter the niche markets for small displays, projection displays, and very large displays, and in the long term, the U.S. lead in these niche technologies, leveraged by investment in military displays, may establish a base for U.S. reentry into the mass consumer and commercial markets.

The United States should devise and develop a broadly based and well-understood strategy to capture a significant share of the future global display market. This will require a well-coordinated effort that brings together partners from government, universities, and industry. Important components of such a strategy include timely development of a low-cost manufacturing capability, ensuring a core competence in displays for military needs, and developing multidisciplinary university courses and curricula on display systems and applications. Other key elements are building a broad consensus on the evaluation, roadmapping, and prioritization of promising display technologies and improving the coupling of the device, systems, and graphics software communities. The strategy should consider, among other options, U.S. mass market reentry via current market niches (e.g., very small or very large displays) in which the United States has a strong technical position.

2

Optics in Health Care and
the Life Sciences

Optics has affected the lives of most Americans by changing the practice of medicine and offering new approaches to major health problems, such as the treatment of heart disease, cancer, kidney stones, knee injuries, and eye diseases. The use of optics and fiber optics has led to less invasive ways of treating disease by replacing open surgery with minimally invasive therapies. The basic research in biology that leads to new insights into the treatment of disease has benefited from technical advances ranging from optical methods of gene sequencing to new and more precise microscopies.

This broad use of optical techniques has led to new approaches to biological research problems, new methods of medical diagnosis, and new ways to treat diseases. Tools developed for use in research have evolved into tools for patient treatment, and new and increasingly sophisticated research apparatus continues to emerge, improving our ability to study and control basic biological processes.

It is the intent of this chapter to show how optics and lasers have changed the practice of medicine in ways that most readers have experienced, either directly or through a family member, and to give some view of how optical science may affect the health care of the future. In addition, the reader will have a better sense of how optics is involved in health care technologies used for applications as diverse as the determination of viral load in HIV and, potentially, the monitoring of blood glucose levels in diabetics.

The material in this chapter is organized into three main topics: (1) surgery and medicine, (2) biology, and (3) biotechnology. The chapter concentrates on revolutionary developments, ones that have led to new techniques for research, diagnosis, or treatment or that could do so in the future. It concludes with some general remarks on health care and the life sciences as a whole, highlighting the key challenges and opportunities that the field faces and making some recommendations to government, academia, and the private sector.

Surgery and Medicine

Optics has enabled laser surgery, optical diagnostic techniques, and visualization of the body's interior (see Figures 2.1 and 2.2 and Boxes 2.1 and 2.2).

Although the applications of optics to surgery and medicine have increased rapidly since the invention of the laser in 1960, a number of optical techniques were used before that time. The development of rigid and flexible endoscopes—devices that allow the inside of canals (e.g., blood vessels) and hollow organs (e.g., the colon) to be viewed— is discussed in some detail elsewhere (Katzir, 1993). A number of rigid endoscopes were used in the nineteenth century, and the first flexible medical endoscope using optical fibers was demonstrated in 1959.

It is worth noting that the use of microscopes by pathologists to examine tissue in order to diagnose disease was a well-established medical application of optics long before the era of the laser. The microscope is still the essential tool of the modern pathologist, although it has been made optically more advanced by the advent of computer-designed lenses and high-quality antireflective coatings. Some clinical specialties use specially modified microscopes. Ophthalmologists use a modified microscope, called a slit lamp, to project a slit-like beam of light into the eye to detect scattering objects within the cornea and lens. Advances in microscopy continue and include efforts to automate microscopy to allow initial screening for disease and infection.

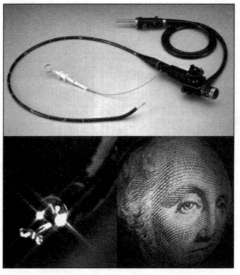

FIGURE 2.1 A flexible gastroscope, used to examine the inner surface of the stomach, together with a view of the distal (insertion) end showing working channels for tools. The image of part of a dollar bill, taken using a flexible colonoscope with a charge-coupled device (CCD) camera at the distal end, illustrates the excellent resolution available. (Courtesy of Olympus America, Inc., and N. Nishioka, Massachusetts General Hospital.)

Arguably the most extensive use of optics in health care is in the fabrication of eyeglass frames, lenses, and contact lenses. This market was estimated at $13.2 billion in 1994 and consists of the 145 million people—55% of the total population—who wear corrective lenses (American Optometric Association, 1996). The ophthalmic market has evolved, with a variety of safe and light plastic lenses increasingly replacing glass and with continued growth in the use of antireflection and ultraviolet (UV) blocking coatings. An additional change is the move from bifocal and trifocals with discrete zones to progressive lenses where the refractive correction varies smoothly from the bottom of the lens. Many manufacturing changes for ophthalmic optics are being implemented to enable these advanced features to be delivered to the consumer on demand (in an hour). These evolutionary developments are important but are outside the main interest of this report and are not discussed in more detail.

Introduction of Lasers

The medical potential of the laser has been explored almost from the invention of the ruby laser in 1960. These initial experiments were often of the "point-and-shoot" variety, unguided by an understanding of the mechanisms by which the laser interacted with tissue or of ways to optimize these interactions. Ophthalmology was the specialty that adapted and incorporated laser techniques into clinical practice most rapidly, in large part because the interior of the eye was optically accessible (Krauss and Puliafito, 1995). By the end of the 1960s, some understanding of the mechanisms by which the laser interacts with the retina had been obtained, with both thermal and mechanical effects identified.

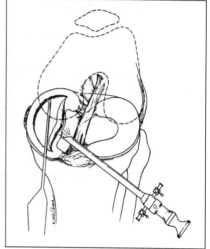

FIGURE 2.2 Schematic diagram of an arthroscope, a rigid viewing scope commonly used for knee surgery. A variety of surgical tools can be passed through the working channels of the scope. (Courtesy of T. Narashima, Scientist/Imagemakers.)

· ·

BOX 2.1 TELEMEDICINE

Telemedicine has the potential of bringing access to medical specialists to remote communities in the United States such as Indian reservations, to underserved communities in the United States, and to the entire world. The use of high-speed communications systems to transfer medical images, such as x-ray radiographs and optical micrographs of histology specimens, has been demonstrated at a number of sites. One major East Coast hospital regularly receives and reads radiographs from Saudi Arabia, returning reports within the same day. The use of teleconferencing systems to allow medical consultations involving patients who may be thousands of miles from physician consultants is also being studied in pilot projects. The technology underlying these systems is discussed in more detail in Chapter 1; it includes the development of fiber-optic communications networks and image processing and computational schemes allowing image compression. However numerous financial and legal issues must be clarified, including the malpractice aspects of teleconsultations.

Several groups are developing CCD arrays for the detection of x rays used in medical imaging. When commercialized, such devices will provide x-ray image information directly in digital form, avoiding the need to scan and digitize conventional x-ray film, and will facilitate the transport and storage of radiographs.

C h a p t e r   2

BOX 2.2 "GETTING SCOPED"

Flexible and rigid viewing scopes have changed medicine in ways many Americans have encountered. The repair of a torn meniscus in the knee is usually performed using a rigid arthroscope, through which a number of surgical tools are passed. This technique has changed knee surgery from an inpatient procedure to an outpatient one with reduced pain and convalescence. The colonoscope is used routinely to examine patients for possible colon cancer. The resulting early detection of colon cancer is often life-saving. Many gynecological procedures have become less invasive through the use of a laparoscope, which passes through the abdomen to allow access to the uterus. Laparoscopic techniques also enable numerous other procedures, such as gall bladder removal, which is discussed later in the section on minimally invasive therapy.

....

Medical applications spread from ophthalmology into the general area of surgery, with these applications generally developing around the most readily available lasers. It is important to note that lasers can emit either short pulses of light (pulsed lasers) or a beam of light that is always on (continuous-wave, or cw, lasers) because the effects of pulsed and cw laser light can be quite different. These were primarily the pulsed ruby laser; the cw argon ion and carbon dioxide (CO_2) lasers; the Nd:YAG (neodymium-doped yttrium-aluminum-garnet) laser, primarily in the cw mode; and the cw dye laser. The ability of the cw CO_2 and Nd:YAG lasers to cut tissue while producing coagulation led to their use as general surgical lasers. Many companies entered the medical laser marketplace, often without a strong scientific understanding of the effects of lasers on tissue. In addition, the role of the Food and Drug Administration (FDA) in the regulation of new laser devices was not as well established as it is today, allowing the introduction of medical laser systems with unproven efficacy.

Since the early 1980s a number of changes in the nature of medical laser research have occurred. There was an increasing interest in the mechanisms of laser-tissue interactions, and new clinical applications based on these interactions came into use. One of the driving forces behind this change was the initiation of the Medical Free Electron Laser (FEL) Program by the Department of Defense (DOD) in 1985. Although the program was specifically aimed at developing FEL applications, the novel pulse structure of the FEL led to an increased interest in pulsed laser effects, which in turn led to an increased understanding of laser-tissue interactions based on conventional lasers.

Today, the use of optics in surgery and medicine is large and growing. For example, worldwide sales of medical laser systems reached $890 million in 1994, $1,070 million in 1995, and $1,295 million in 1996 (estimated), and they were forecast to reach $1,460 million in 1997

(Arons, 1997). The corresponding figures for U.S. sales were $535 million in 1994, $695 million in 1995, $830 million in 1996, and $960 million forecast for 1997.

Understanding the Interaction of Light with Tissue

The optical properties of tissue were studied, leading to the awareness that most tissue is an inhomogeneous substance with multiple absorbers such as melanin (the primary pigment in skin), oxyhemoglobin (a constituent of blood), and proteins. The significance of these absorbers varies with the wavelength of interest; for wavelengths greater than 1 μm, for example, water is the primary absorber. For reference, the wavelength range of visible light is about 0.4 to 0.7 μm; the wavelengths of lasers used in medicine extend to both the short (ultraviolet) and the long (infrared) side of the visible spectrum. New clinical treatments grew from increased insight into light-tissue interactions. With an understanding of the different absorption properties of various tissue components and of the depth that light penetrated into tissue came the insight that thermal effects could be confined to the optical penetration depth by using laser pulses short enough that no thermal diffusion occurred during the pulse. This led to the concept of "selective photothermolysis" in which particular sites in tissue, such as blood vessels, are targeted with minimal effect on surrounding tissue. This concept is exploited in dermatology, where the treatment of skin lesions characterized by abnormal blood vessels, such as port wine stains, is often required.

New kinds of laser effects were discovered. There was increased awareness and utilization of the fact that lasers could be used to produce tissue effects other than the purely thermal ones involved in early laser surgery. The ability of pulsed lasers to cause a number of mechanical effects was recognized, studied, and used. Some of these photomechanical effects relied in turn on the ability of pulsed lasers to initiate nonlinear effects; specifically, the ability of pulsed lasers to produce optical breakdown in water was used to generate cavitation bubbles and launch stress waves. These mechanical effects found clinical use in ophthalmology, where they are employed in a procedure referred to as "photodisruption." This procedure is used to treat a side effect of cataract surgery, the formation of an opacification on the membrane that holds the opaque lens, by rupturing or tearing a portion of that membrane. Here, a simple laser procedure now avoids the need for a second, invasive surgery.

The use of optical breakdown made it possible to deposit laser energy in biological media that had no linear absorption, a conceptual change. Subsequently, laser-induced mechanical effects found a second clinical application, the fragmentation of urinary tract calculi (stones) in patients, a procedure known as laser lithotripsy. This technique complemented the existing method of treating urinary tract stones, which involved the use of acoustic pulses generated by a machine (the shock-wave lithotripter), that was several times more expensive than the laser system. The laser technique, which used an optical fiber to deliver light to the stone, allowed the fragmentation of stones at locations that could not be accessed by the shock-wave lithotripter because pelvic bones blocked the acoustic pulses. The mechanisms involved in this application were investigated after the effect was demonstrated and clinical trials initiated. In these studies, optical techniques from the physical sciences, such as pump-probe measurements and high-speed flash photography, played a significant role in clarifying the mechanisms of stone fragmentation.

An additional nonthermal use of lasers in medicine is using light, primarily from laser sources, for cancer treatment. Drugs injected into a patient can be selectively activated by illuminating the area of interest; this can lead to the photochemical destruction of tumors. This treatment, known as photodynamic therapy (PDT), is being investigated for the treatment of a number of cancers and has recently been approved for palliation of esophageal cancer. PDT is discussed in more detail below.

Today many different lasers are being used to irradiate a variety of tissue targets. Table 2.1 lists the most commonly used lasers, their wavelengths, the tissue targets, and the therapeutic interaction desired.

TABLE 2.1 Common Medical Lasers and Some of Their Applications

Laser	Wavelength (nm)	Target(s)	Applications
ArF Excimer	193	Tissue protein	Refractive surgery
Argon ion	488, 514	Hemoglobin	Retinal photocoagulation
Nd:YAG, frequency doubled	532	Hemoglobin, tattoo pigments	Tissue cutting and coagulation, tattoo removal
Pulsed dye	577	Hemoglobin	Removal of vascular lesions
Continuous dye	630-690	Photosensitizers	Photodynamic therapy
Visible diode	650-690	Photosensitizers	Photodynamic therapy
Pulsed ruby	694	Tattoo pigments	Tattoo removal
Infrared diode	800 (nominal)	Hemoglobin, absorbing dyes	Retinal photocoagulation, tissue welding
Nd:YAG	1,016	Water	Tissue cutting and coagulation, many surgical applications, tattoo removal
Ho:YAG	2,100	Water	Tissue cutting and shrinkage
Er:YAG	2,940	Water	Skin resurfacing, hard and soft tissue cutting (experimental)
CO_2	10,600 (nominal)	Water	Skin resurfacing, tissue cutting and coagulation, surgery

Table 2.1 includes standard surgical lasers, such as Nd:YAG and CO_2, as well as some lasers whose uses are still experimental. The Er:YAG laser, for example, has been studied as a tool for dentistry (Wigdor et al., 1995) but has only recently been approved for dental use.

Finally, the potential of optics and lasers for obtaining information about tissue for use either in clinical diagnosis or in providing feedback control of surgical laser systems has received increasing attention. The need for feedback control arose as situations were encountered in which the tissue response to laser irradiation depended critically on the flux and the light dose. Tissue welding, the use of lasers to join tissue by localized heating, is optimal over only a small temperature range, which makes it difficult to obtain reproducible results. A number of feedback systems, based either on tissue temperature or on changes in tissue optical properties, have been studied in an attempt to obtain reliable laser-based tissue welding. If feedback control can enable tissue welding to be performed by most surgeons, it may complement sutures for applications, such as plastic surgery, where minimal scarring is desired. A number of studies have investigated the use of laser-induced fluorescence, both from substances naturally occurring in tissue and from externally administered fluorescent marker dyes, to delineate tumors and potentially to aid in the early detection of cancer. Optical radar techniques have been applied to biological tissue, starting with the skin and soon thereafter the eye; more details are given in the discussion of optical diagnostic techniques.

Minimally Invasive Therapy

The growth of optical and laser techniques in medicine was in large part due to the fact that devices for delivering light to the inside of the body became available. In the 1990s, advances in such areas as CCD (charge-coupled device) camera technology and innovative new approaches by surgeons led to the development of what is now referred to as "minimally invasive therapy" (MIT; see Box 2.3). In addition to optics, the development of specialized surgical tools to allow traditional surgical manipulations such as cutting, suturing, and stapling to be performed through tiny incisions was another technology that enabled MIT. The concept of MIT is the replacement of traditional "open" surgery— with its large incisions and direct viewing of the surgical field by the physician—by several small incisions, typically punctures on the order of 5 to 10 mm in diameter, through which viewing devices and surgical tools can be passed.

Optics is critical to MIT since the main concept is to use video rather than direct viewing to minimize the surgical invasiveness of the procedure. Quartz optical fibers, developed initially for fiber-optic communications, are capable of transmitting many of the laser

BOX 2.3 MINIMALLY INVASIVE THERAPY

Advantages

- Shorter hospital stays
- Reduced patient trauma, morbidity
- Shorter convalescence; faster return to work
- Decreased expenditure on pain medication

Basic Components

- Imaging (primarily optical; ultrasound, magnetic resonance, computerized tomography may also be used)
- Tissue manipulation tools
- Source of directed energy (electrocautery, laser, focused ultrasound)

Role of Optics

- Video cameras
- Flexible endoscopes
- Rigid laparoscopes
- Laser sources
- Tissue characterization

••

wavelengths of interest for therapeutic applications. An optical fiber can usually be added to an endoscope by passing it through one of the already available "channels" designed for irrigation and the passage of tools, resulting in an instrument that allows both viewing and laser irradiation. The main exception to this approach has been the CO_2 laser, whose infrared wavelength could not be delivered readily using available fiber optics; in this case, rigid or flexible metallic waveguides were used as delivery channels. Despite efforts to develop an infrared fiber capable of both transmitting CO_2 laser radiation and surviving in the wet environment of the human body, such fibers have not reached the point of clinical use.

In addition, although the initial applications used only optical imaging to obtain information about the surgical field, the combination of x-ray, magnetic resonance, and other imaging technologies to produce fused images is envisioned by workers in MIT today. The fusion of imaging modalities may be necessary to allow internal access to solid organs such as the liver.

The first extensive application of MIT was to a new procedure for gall bladder removal, the laparoscopic cholecystectomy. In this procedure, four incisions admit a viewing device, a gas infusion device to inflate the abdomen, and two surgical tools. The surgeon operates by viewing a TV monitor while manipulating tools for cutting, stapling, or suturing tissue. Although a laser was initially used to stop bleeding during surgery, it was soon found that the use of a much less expensive electrocautery device was equally satisfactory, and use of the cauterizing laser in this specific MIT procedure has become minimal.

A major impact of laparoscopic cholecystectomy is a drastic reduction in recovery time for the patient, with attendant savings in lost wages and lost time to employers. Hospitalization time is decreased from 4 days to 1, leading to a major cost savings. The acceptance of the procedure can be most directly appreciated by comparing the number of conventional and laparoscopic cholecystectomies between 1988 and 1994, tabulated in Table 2.2. Today, the laparoscopic procedure is the method of choice.

Other examples of MIT abound; as surgeons become more skilled in the techniques involved, more complex procedures have been performed, including hernia repair and colon surgery. In orthopedics, knee and shoulder surgery is routinely performed using dedicated rigid or flexible fiber-optic viewing instruments called arthroscopes, together with dedicated miniature surgical tools for specialized operations.

A word of caution is needed in considering the future of MIT techniques in the present health care environment. Since most surgery is paid for by health care providers, the acceptance of a particular MIT technique is determined by whether it reduces direct cost to the provider, not by overall societal benefits such as a decrease in time lost from work. In some cases, the direct costs of an MIT procedure can be higher than those of the older, more-invasive technique because additional tools and more sophisticated equipment are required. Thus, the introduction of new MIT techniques will require that direct costs do not increase substantially or that patient demand is such that the minimally

TABLE 2.2 Growth Patterns in Minimally Invasive Surgery—Traditional and Laparoscopic Cholecystectomies

Year	Traditional Procedures	Laparoscopic Procedures
1988	537,000	0
1989	545,000	1,000
1990	535,000	25,000
1991	410,000	125,000
1992	150,000	480,000
1993	75,000	525,000
1994	85,000	575,000

Source: W. Grundfest, Cedars Sinai Hospital.

invasive technique will obtain reimbursement regardless of direct cost. Some studies have pointed out that evaluations of the economic benefits of MIT have produced wildly different conclusions (Cuschieri, 1995), indicating the need for care in making quantitative statements about its economic benefits.

Advanced Therapeutic Applications of Lasers

Currently, numerous advanced therapeutic applications of lasers are being investigated. This report illustrates the types of new clinical approaches being investigated by using a few examples that show the diversity of approaches being studied and their relation to the enhanced understanding of basic mechanisms obtained from fundamental studies. In addition, the specific optical issues involved in each of these examples are illustrated.

Laser Refractive Surgery

For the correction of visual defects such as nearsightedness, astigmatism, and farsightedness, laser-refractive surgery has attracted intense clinical and public interest. The basic concept is simple. Since most of the refractive power of the eye comes from the cornea, the outer surface of the eye, relatively small changes in the curvature of the cornea can correct a large number of visual defects that currently require eyeglasses or contact lenses. In a generic sense, the concept is to perform corneal "sculpting" using a laser (McDonnell, 1995; Seiler and McDonnell, 1995).

Although a number of approaches to corneal sculpting have been used, the basic concept relies on the observation, made in the materials science community in the early 1980s, that UV laser radiation from pulsed excimer lasers can be used to ablate both polymers and tissue with minimal damage (typically less than 1 μm) and with high precision and control. This basic observation served to guide the development of a number of different excimer laser systems for refractive surgery. All of these systems were designed to ablate tissue from the cornea in a controlled and predetermined manner to produce a change in the refractive power of the eye, but they differed in engineering details. In the course of development of these systems, numerous problems involving optical engineering, the safety and efficacy of the procedure, wound healing, pharmacology, and regulatory issues required solution. A powerful driving force for solving these problems was the perceived size of the market. Approximately 25% of the population of the United States suffers from myopia (nearsightedness) and constitutes potential customers for excimer laser photorefractive keratectomy (PRK).

Intense effort went into developing a system that could produce the desired correction. Clinical trials were needed to determine how severe a myopia could be successfully treated, as well as to find the limitations on treating other refractive defects such as hyperopia (far-sightedness) and astigmatism. Long-term follow-up was necessary to determine the stability of these laser-induced changes in the cornea. After extensive clinical trials and experimentation, the FDA approved a commercial excimer laser system for PRK in October 1995.

During the course of development of PRK, one of the initial problems encountered was the formation of haze in the treated cornea. Years of experimentation showed that this could be controlled by control of the laser beam profile, combined with the pharmacological treatment of some patients. A variation on PRK that shows promise in early clinical trails is LASIK (laser-assisted in situ keratomileusis) in which the anterior surface of the cornea (corneal cap) is first microtomed (cut off) to reveal the central stroma of the cornea. The stroma is appropriately ablated with an excimer laser, and the corneal cap is then replaced. The thickness profile of the cornea has been changed without affecting the anterior (front) surface of the cornea. Minimal haze is associated with this procedure, and no sutures are required.

Other laser technologies are being explored for corneal sculpting. One approach relies on the use of an Ho:YAG infrared laser to heat and controllably shrink portions of the cornea outside the central visual field, avoiding the haze problem. A central issue here is biological: Will the reshaped cornea retain the new shape or relax to the original one?

Cardiovascular Applications

Heart disease is the leading cause of death in the United States, and the search for alternatives to expensive coronary bypass surgery has been active, with laser systems initially offering a promising approach. Cardiovascular applications provide an example of both the potential of laser techniques and the potential pitfalls in applying technological solutions to complex biological systems (Deckelbaum, 1994). The attractive feature of the laser in cardiovascular applications is its ability to deliver energy via an optical fiber to sites in very small vessels.

Laser angioplasty, the use of lasers to remove blockages in arteries, is a well-known concept. Although many techniques used for angioplasty, such as laser ablation, inflatable balloons, and high-speed rotating cutters, are effective at removing blockages, a major problem shared by all of these approaches is restenosis—the vessel reclosure that occurs within 6 months in about 40% of all angioplasties. Consequently, emphasis has shifted from the development of new angioplasty techniques to the development of methods of controlling restenosis, including the insertion of metal stents and the use of photo-chemical therapies (see Box 2.4).

BOX 2.4 LESSONS FROM LASER ANGIOPLASTY

The story of application of the laser to angioplasty serves as an example of some of the pitfalls in applying technology to complex biological problems. The problem was first perceived as one of removing plaque that reduced the inside diameter of blood vessels. The first approach was a purely thermal one. The "hot tip" was a laser-heated metal tip at the end of an optical fiber. This proved a poorly controlled way of removing plaque that had the additional risk of perforating the artery. With the discovery that excimer laser ablation of tissue led to minimal thermal damage, intense commercial activity was focused on developing excimer laser angioplasty systems. The success of these systems was limited by the restenosis problem common to all angioplasty procedures. The applicability of these systems was also limited to situations in which the cheaper and more conventional balloon angioplasty approach could not be applied directly. All in all, the complexity of the biological problem was not understood, and the lure of a large commercial market led to overselling the capabilities of the laser. Today, the emphasis has shifted to understanding and controlling restenosis, which causes failure of nearly half of all angioplasties within 6 months. In addition, a new technology, the use of metal stents that are inserted into the artery to prevent restenosis, was successfully introduced. Laser angioplasty is an example of an application in which the biological response of the tissue, rather than the sophistication of the optical tools, was the critical issue in the clinical acceptability of a technique. It also illustrates how rapidly new and sometimes inexpensive technologies that compete with lasers are introduced.

Use of the pulsed dye laser for thrombolysis, the destruction of blood clots provides an example of the application of selective absorption of laser energy to achieve a desired clinical effect (see Figure 2.3). The ideal system for thrombolysis would deposit laser energy into the clot, but not into the wall of the vessel, and would lead to ablation of the thrombus. Studies of the absorption of light by blood clots showed a strong absorption in the blue-green region of the spectrum, far greater than that of normal vessel walls or vascular grafts or sutures. The much higher concentration of blood in the clot compared with normal vessels underlies this contrast and provides a method for selective targeting. The use of a pulsed laser, rather than a continuous one, allows the energy to be deposited before thermal diffusion occurs. The laser thrombolysis approach has moved from animal experiments to clinical trials at several centers. In an interesting aside, in coronary arteries, light has been delivered by a "fluid catheter" consisting of a flowing radiographic contrast medium, which acts as a light guide—an effect first demonstrated by Tyndall to the Royal Institution in London in 1854.

Photodynamic Therapy
Cancer therapy was the initial application for photodynamic therapy (PDT), a photochemical approach to the selective destruction of tissue (Figure 2.4), and the treatment of a variety of cancers at sites ranging

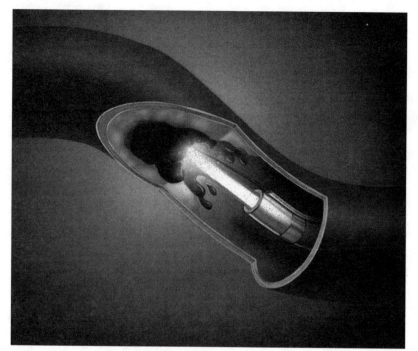

FIGURE 2.3 Schematic diagram illustrating laser thrombolysis—use of a laser to destroy a blood clot. (Courtesy of K. Gregory, Oregon Medical Laser Center.)

from the skin to the bladder is being studied. Current research has expanded to include some exciting noncancer applications such as the destruction of abnormally growing blood vessels in the eye, which can lead to blindness; the treatment of psoriasis, a skin disease; and the treatment of rheumatoid arthritis. The basic concept of PDT is activation of a drug by light. This activated drug in turn transfers its energy to molecular oxygen, which can then destroy tissue. The concept of selectivity is central to PDT since one does not want to destroy normal tissue. The photosensitizer used is chosen because it accumulates preferentially in the tissue of interest (e.g., within a tumor). Further selectivity can be achieved by delivering light only to the tissue of interest. The mechanism of action of PDT varies with the photosensitizer used; originally, it was thought that the photosensitizer accumulated in tumor cells and activation by light resulted in tumor cell destruction. More recent research has shown that destruction of the tumor vasculature (blood vessels) is often a significant additional tumor-killing mechanism.

The development of PDT as a viable cancer therapy has been more complex than the schema given here. The need to deliver a controlled dose of light throughout a tumor led to the study of light propagation in tissue, which scatters light strongly, and to the development of delivery devices, usually modified optical fibers, for bringing light to the tissue of interest. In early experiments, one photosensitizer, hematoporphyrin derivative (HpD), which was activated with 630-nm light, was used

FIGURE 2.4 Schematic
diagram illustrating the
process of photodynamic
therapy. (Courtesy of M.
Hamblin, Massachusetts
General Hospital.)

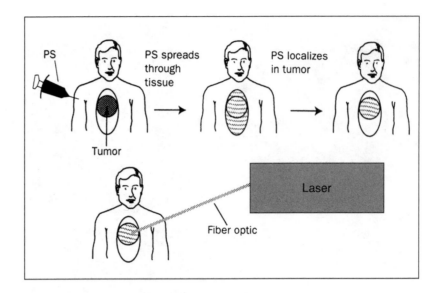

FIGURE 2.4 Schematic diagram illustrating the process of photodynamic therapy. (Courtesy of M. Hamblin, Massachusetts General Hospital.)

almost exclusively. As PDT developed, new sensitizers were studied to obtain better selectivity, reduce the long-lasting skin phototoxicity associated with HpD, and take advantage of the potential of compact, efficient diode lasers as PDT sources.

Photosensitizers used for PDT typically have absorption spectra in the range of 630 to 700 nm, with some of the newer species having absorptions at wavelengths as long as 800 nm. For many years, the standard light source for PDT was a dye laser pumped by an argon ion laser; more recently, the experimental use of diode lasers that emit red light has begun. Typical power requirements are of the order of 1 to 10 W and are currently available from commercial diodes at wavelengths of 660 nm and longer. Laser diodes offer significant advantages over the existing dye laser systems for PDT: They are far more compact, efficient, and reliable. Whereas the efficiency of an argon ion laser is about 0.01%, typical diode laser efficiencies can range from 20 to 50%. Light-emitting diodes (LEDs) are also starting to be used as sources for PDT in situations where optical fiber delivery is not required, such as the irradiation of cells in culture and the irradiation of skin. When fiber delivery is needed, lasers are the preferred source because their output can be readily coupled to fibers.

Models of optical propagation in scattering media are finding increasing use in the development of PDT applications. The implementation of PDT techniques frequently requires the development of new light delivery systems for irregularly shaped sites in the human body, as well as devices to measure the light dose delivered to tissue. Such dosimetry is often critical for safe and effective treatment since underdosing can result in untreated regions and overdosing can lead to the destruction of normal tissue. Initial animal studies indicate that PDT employing fiber-optic delivery can be used to treat rheumatoid arthritis of the knee. Computer modeling of light propagation is being combined with kinetic models of photodestruction of the dye molecules to obtain a better understanding of the required dosimetry.

Optical Diagnostic Techniques

Although the initial applications of lasers to medicine focused on therapeutic applications, in the early 1980s a number of groups began to explore tissue diagnosis as well. The promise of noncontact and, in many cases, noninvasive acquisition of diagnostic information was one of the driving forces for these studies, as was the hope that the well-developed spectroscopic techniques used in the physical sciences would have application in medicine. The availability of optical fibers to deliver light to the inside of the body via endoscopes would allow examination of sites such as the bladder, colon, and lung.

Some optical techniques are well established in clinical practice, such as laser Doppler velocimetry to measure blood flow and the pulsed oximeter used in all hospitals today. Ophthalmologists now use fundus cameras to obtain pictures of the retina and, with the use of fluorescent dyes, images of retinal blood flow. However, a number of potentially powerful techniques are still in the laboratory stage.

Blood Monitoring

Recent progress in biotechnology has led to the development of a whole new generation of optically based instruments that provide a cost-effective means for doctors to monitor blood chemistry (Box 2.5), immune system function, and cardiopulmonary efficiency at or near the patient's bedside.

One excellent example of such a system that is already in widespread use is the pulsed oximeter, which allows monitoring of O_2 saturation levels in blood. The measurement is totally noninvasive, requiring only that a disposable probe be attached to the patient's fingertip. The probe incorporates inexpensive LEDs and photodiodes; it determines O_2 saturation levels by measuring the ratio of the absorption of hemoglobin at two colors in the orange and near-infrared (IR) spectral regions. The commercial success of this technology is due in large part to the development of inexpensive, disposable optical assemblies that allow patient monitoring during anesthesia as well as after an operation. This optically based measurement has now become part of the accepted "standard of care" and is as common for tracking patient postoperative recovery as measuring blood pressure and heart rate.

Optical technology is now being combined with state-of-the-art fluidic processing to make new bench-top devices for blood analysis.

These devices provide blood chemistry tests at the patient's bedside or in the doctor's office, immediately supplying information that can allow timely intervention in critical situations and reduce pain and suffering. One example is an instrument that measures the level of the drug theophylline in the blood of patients suffering respiratory distress brought on by a severe asthmatic attack. Theophylline gives rapid relief to the patient but must be carefully administered since the therapeutic dose

···

BOX 2.5 GLUCOSE MONITORING IN DIABETES

Diabetes affects more than 16 million people in the United States. Treatment of diabetes and its complications represents one of the largest single portions of health care costs in the United States. Careful monitoring of glucose levels can significantly reduce complications due to diabetes, which can lead to retinal disease and blindness or to kidney disease and failure. Glucose monitoring results in major improvements in the quality of life and in medical cost savings, but it requires periodic blood testing—often several tests per day. Current glucose monitors are optically based instruments that require only a small blood sample, usually obtained by using a lancet to prick a patient's finger. The blood sample is applied to a reagent strip that has specific, carefully controlled optical properties. Enzymatic reagents embedded in the strip react with the blood sample and change color in proportion to the amount of glucose in the sample. A small optical reader incorporating visible and infrared LEDs and a solid-state detector performs a two-color reflectance measurement on the reagent strip. The ratio of the reflected powers is used to calculate glucose concentration. The reader is about the size of a deck of cards and costs less than $100. It is battery powered and can easily fit into a pocket or purse. The cost of a single test is roughly $0.25, and millions of tests are performed each day in the United States alone. This type of portable, easy-to-use, diagnostic instrument has revolutionized the monitoring of glucose levels in diabetics.

Careful monitoring of glucose levels significantly reduces the onset of complications that can lead to retinal disease and blindness or to kidney disease and failure. The major cause of discomfort to the user in these systems is the need for constant lancing of the finger to provide a fresh blood sample. The inconvenience and discomfort of this procedure can lead to poor patient compliance, resulting in inadequate monitoring. Hopefully, the next generation of instruments will provide comparable accuracy in a totally noninvasive manner, eliminating the need for a blood sample. Several dozen groups are working on a variety of approaches to develop this type of instrument. Most of these approaches are based on in vivo, noninvasive, spectroscopic measurements of glucose via its absorption properties in the near IR or other modifications of optical properties that track glucose levels. There are many blood components with interfering absorption spectra that complicate making these types of noninvasive measurements. No FDA-approved methods for noninvasive measuring of glucose currently exist. Moreover, to be of greatest benefit a noninvasive instrument must supply the convenience, portability, and affordability of the current method. Development of this type of external, noninvasive glucose monitors is hindered by our limited understanding of the in vivo spectroscopy of blood components. Moreover, this type of instrument may not be possible using existing technology.

range for this drug is narrow and an overdose can cause seizures. The theophylline concentration in a blood sample is measured by using an optical scattering technique that combines immunochemistry with the light-scattering properties of latex beads. Several of the optical elements in this instrument are integrated into an inexpensive, disposable, injection-molded plastic cartridge. This instrument allows measurement of the theophylline level in less than 3 minutes. Previously, blood samples had to be sent out to be analyzed at a blood chemistry laboratory, causing delays of up to several hours.

Optical Tumor Detection

Initial studies of tumor detection demonstrated that fluorescence-based techniques using either exogenous marker dyes or endogenous (natural) fluorophores could be used to mark gross, visually detectable tumors. Subsequent work has emphasized the ability to determine whether small, visually undetectable lesions can be identified by spectroscopic techniques. Such optical methods might help guide conventional tissue biopsy to the most suspicious regions or might in some cases alleviate the need for a biopsy; the terms "optically guided biopsy" and "optical biopsy" have been used to describe this approach generically.

There are numerous variations on the optical biopsy concept. Different spectroscopic techniques such as fluorescence, reflectance, and Raman scattering have been employed. These techniques have been used both to make measurements at single points with an optical fiber and to obtain images using either conventional or intensified CCD video cameras. Although the exact implementation of these concepts varies, the basic idea is always to find a spectral signature of the abnormal tissue that differentiates it from normal tissue and to develop algorithms for utilizing these signatures. This approach has been used with laser-induced fluorescence studies of a number of organs, including the colon, bladder, and cervix. Although a number of encouraging results have been obtained, large-scale in vivo studies are generally needed before these approaches gain clinical acceptance. Such studies have already been performed in the lungs using autofluorescence imaging and in the bladder using fluorescence imaging of a marker dye. A major engineering challenge will be to make optical systems that yield new and accurate information but are inexpensive enough to be accepted even in today's cost-conscious health care environment.

Another approach to optical biopsies uses the fluorescence lifetime of a molecule, rather than its spectrum, as a source of information. The lifetime is the time for which a fluorescent molecule emits light after a rapid optical excitation pulse. This has the advantage that compounds whose fluorescence spectra overlap can be monitored by using differences in lifetime. In addition, lifetime changes can be used to monitor processes, such as binding of a fluorophore to tumor tissue, that cannot as easily be detected spectrally. As in fluorescence spectroscopy, both point measurements and imaging of lifetimes have been demonstrated. In addition, lifetime measurements have been obtained using both time-domain and frequency-domain techniques, discussed in more detail below.

Imaging and Spectroscopy in Scattering Media

X-ray mammography is the standard screening technique for breast cancer. However mammograms require highly-trained radiologists for interpretation, and even at its best, mammography fails to detect a significant number of breast cancers, especially in younger women. An optically based mammography system could complement the existing technology if it were able to find the cancers that x-ray mammography misses. Today, a number of optical techniques aimed at this goal are being explored (see Figure 2.5).

The use of light to create images of the interior of tissue is an attractive idea whose roots can be traced to studies of tissue transillumination in the 1920s. However, these early studies failed to overcome the effect of tissue opacity. Whereas some materials are opaque because they strongly absorb visible light, others such as tissue may be opaque because photons traveling within these media are highly scattered. A small number of photons travel straight through such substances and can be used to make shadowgraphs of internal structures in a manner similar to x rays. However most of the light is transported through these materials in a process similar to heat diffusion (Gratton and Fishkin, 1995; Yodh and Chance, 1995).

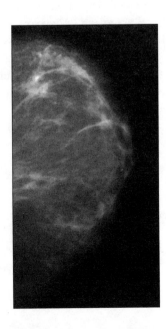 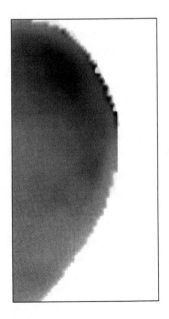

FIGURE 2.5 X-ray mediolateral mammogram (left) and corresponding optical mammogram (right) of a 0.5-cm diameter tumor in a 72-year-old woman. The optical mammogram was obtained using two laser diodes operating at 690 and 810 nm in a frequency-domain imaging system. (Courtesy of S. Fantini, University of Illinois at Urbana-Champaign.)

In the biophysics and medical communities there are extensive research efforts to overcome the effects of scattering and use diffusing photons to view body function and structure. These efforts are based on the existence of a spectral window between 700 and 900 nm, in which photon transport within tissue is dominated by scattering rather than absorption. Thus, to a very good approximation, near-infrared photons diffuse through human tissues and can be used for a variety of biomedical applications.

In a typical measurement, the researcher uses an optical fiber to inject near-infrared photons into tissue or a tissue-like medium and a second optical fiber to detect photons at other locations. Microscopically, the injected photons experience thousands of elastic scattering events while traveling from one fiber to the other. Occasionally, the photons are absorbed in this process and are unde-tected. Microscopically, individual photons undergo a "random walk" within the medium, but collectively, a spherical wave of photon density is produced and propagates outward from the source. Typically, quanti-ties such as the photon energy density within the sample are measured to verify light transport models.

The patterns of light energy density or photon density waves are dis-torted as they traverse scattering media. Recent experiments and simu-lations using short-pulse (time-domain), amplitude-modulated (frequen-cy-domain), and cw sources have utilized these distortions for spec-troscopy and imaging of deep tissues. It is feasible to use these waves as probes of biological samples whose extent is of the order of 1 cm, or about 100 transport mean free path lengths.

Since tissues are often quite heterogeneous, it is natural to contem-plate making images with the diffusive waves. Resolutions comparable to those of PET (positron-emission tomography) and MRI (magnetic res-onance imaging)—several millimeters—are highly desirable, but a range of problems exists for which resolutions of ~1 cm are useful. A simple example of the utility of imaging is the early localization of a head injury that causes brain bleeding or hematomas. Here prototype devices already detect the presence of small brain bleeds at the limit of detection by x-ray computed tomography. Such sensitivities in detecting hematomas suggest it may be possible to localize small blood vessel expansion (aneurysms), which must be detected at levels below 1 cm to avoid danger of rupture.

The medical utility of near-infrared spectroscopy and imaging approaches ultimately depends on whether the tissue has enough opti-cal contrast to differentiate normal from abnormal tissue or body func-tion. Spectroscopy is useful for the measurement of time-dependent variations in the absorption and scattering of large tissue volumes, such as might occur following a head injury. Imaging is important when a localized heterogeneity of tissue is involved, for example, an early breast or brain tumor, a small brain bleed, or an early aneurysm. Here images enable experts to identify the site and extent of the trauma and to differentiate it from background tissue.

The sources of image contrast when using a light probe are different from those of other imaging techniques such as those based on x rays, magnetic resonance, skin temperature measurements (thermography), and ultrasound. Spectroscopic information is available as a result of the intrinsic absorption of tissue or as a result of the absorption of contrast agents (optically absorbing chemicals) that may be introduced into the body. The fluorescence spectra and lifetimes of some fluorescent dyes, which may be intrinsic or extrinsic, are sensitive to the local environments within tissue and may be useful for marking tumors. Variation in light scattering affords a novel source of contrast that is demonstrably related to intracellular organelles such as mitochondria. It also depends on fat, water, and perhaps even glucose concentration in tissue.

Tumors are another type of structural anomaly that optics may be able to detect, localize, and classify. A whole subfield of MRI has developed based on the permeability of blood vessels in rapidly growing tumors. These blood vessels will leak paramagnetic contrast agents (small molecules) into the tumor space at a faster rate than into the adjacent normal tissue. The optical approach would utilize this enhanced permeability in a different way, by tagging small tumors with an optical contrast agent, for example, Indocyanine Green (ICG), which is strongly absorbing in the region near 800 nm. In the long term it may be possible to design contrast agents for specific properties of tumors, such as the membrane potential of their organelles. The optical method, in addition, has other criteria by which tumor growth may be observed: larger blood volume resulting from a larger number density and volume fraction of blood vessels residing within the tumor; blood deoxygenation arising from relatively high metabolic activity within the tumor; and increased concentration of cell organelles involved in metabolism, such as mitochondria. Recently, the first images of human breasts obtained using near-infrared light have been obtained; while these are early results, they do demonstrate the ability to detect relatively large tumors.

There are many other examples of potential medical applications in areas as diverse as the neonatal brain and lungs and the adult breast. Optical tools may make possible a range of physiological studies of hemodynamics in relation to the oxygen demand of various body organs. Among the most important are the changes of oxygen delivery that occur in the brain, especially during mental activity.

Many developments make the possibility of light-based images of the interior of tissue realistic. On the technological side, the development of small and efficient light sources, detectors, and control electronics, along with continual improvements in computational capabilities, make possible sensitive noninvasive optical instruments capable of rapid and repeatable measurements. On the fundamental side, advances in the understanding of photon transport make it possible to address the problem of light transport in scattering media, such as tissue, with unprecedented theoretical sophistication and clarity.

An alternative to using models of diffusive transport to extract optical information from tissue is offered by the fact that multiply scattered photons take longer to travel between two points. By selecting photons that arrive at a particular time after an optical pulse is launched, the scattered photons can be rejected. Unscattered photons (often called ballistic photons), or photons that maintain approximately straight-line trajectories (snake photons), can be detected and used to produce shadowgraphs of objects or structures within tissue. Typically, femtosecond or picosecond light pulses are used for illumination. Such an approach can be used to image the bones of the hand or to detect an opaque object within tissue. The limitation of this approach is that the number of snake or ballistic photons available for imaging decreases rapidly with distance in tissue and limits the path length to a few millimeters. The uses contemplated for these techniques are similar to those mentioned above: identification of internal bleeding and detection of breast tumors, among others.

Optical Coherence Tomography

Optical coherence tomography (OCT) is another near-infrared imaging technique under intensive development by groups in the United States and abroad (Huang et al., 1991). It is essentially an optical ranging technique that produces images similar to those obtained with ultrasound, but with much higher resolution.

Technically, OCT uses low-coherence interferometry to produce a two-dimensional image of optical reflection and backscattering from tissue microstructures in a manner analogous to ultrasonic pulse-echo imaging. Coherent detection is used to reject multiply scattered photons, and the interferometer system is used to select only photons that have traveled a specified distance into the tissue. The beam is scanned to turn the one-dimensional depth profile into a two-dimensional image. The resolution of OCT is a few micrometers both laterally and axially; by comparison, the resolution of conventional low-frequency ultrasound is of the order of 100 μm.

OCT was initially applied to imaging of the retina and has progressed to the point where a commercial system for ophthalmology is available. The resolution of the system is such that otherwise undetectable subsurface retinal changes can be seen, allowing the monitoring of injury to the optic nerve from glaucoma. More recent research has focused on the development of beam scanning systems compatible with endoscopes to allow application of the technique within the rest of the body. The arteries and the colon are target organs that are being actively explored by OCT. Applications to biology, such as in vivo imaging of the beating heart of a tadpole, have been demonstrated.

Laser-Hyperpolarized Gases for Magnetic Resonance Imaging

A novel and elegant optical technique, recently developed, is the use of circularly polarized laser light to align the nuclear spins of xenon (Xe) or helium-3 (^{3}He) atoms, which enhances their usefulness as MRI contrast agents (Middleton et al., 1995). The resulting hyperpolarized xenon or ^{3}He has a magnetization that is about 10^5 times that obtained from the protons of water molecules in tissue, resulting in much stronger magnetic resonance signals per atom. Xenon is soluble in lipids and has been used as a brain probe. Helium-3 gas serves as contrast medium for use in regions that contain little water, such as the lungs. Images of ^{3}He-filled human lungs have recently been obtained (see Figure 2.6). In addition, the transport of ^{3}He from the lungs can be monitored, raising the possibility of its use as a functional imaging agent. The use of this technique with commercial MRI machines requires some relatively expensive modification to tune the xenon or helium resonances; low-cost optical pumping systems are needed to make the total system a cost-effective modification of an MRI facility.

Feedback Control of Therapeutic Lasers

Optical techniques for feedback control of therapeutic lasers are being applied to systems in which the laser effect occurs too rapidly to allow manual control or in which more precise control than manually possible is necessary to obtain the desired effect. The ablation of burnt skin (eschar) by a scanned high-power CO_2 laser beam in order to prepare a bed for skin grafting, as well as successful grafting, has been demonstrated in animals and, more recently, in humans. A major advantage of the laser technique is that it avoids the bleeding that accompanies surgical removal of eschar, thereby saving large amounts of blood. Feedback control may be necessary to limit ablation to burnt skin in regions

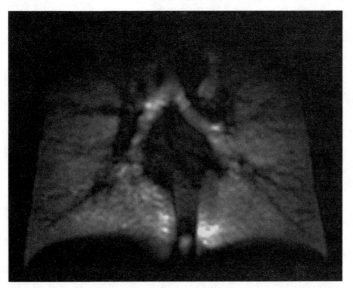

FIGURE 2.6 Magnetic resonance image of human lungs obtained using nuclear spin-polarized helium-3 (^{3}He) gas to provide contrast. The spin-polarized ^{3}He is produced using laser optical pumping. (Courtesy of W. Happer, Princeton University.)

where the depth of the burn varies substantially. Feedback control is also being studied as a means of better controlling a number of clinical laser treatments, such as retinal photocoagulation, tissue welding, and the removal of pigmented skin lesions. Optically based feedback control requires a tissue optical property that changes as the desired biological end point is approached.

Nontechnical Considerations

The current health care environment requires that new technologies show clear cost-effectiveness compared to existing methods. This cost-effectiveness has been measured in terms of direct cost to health care providers, rather than total societal costs and savings. Changes in the health care environment, specifically the trend to managed care, in which the total (direct plus lost time) cost to the employer providing managed health care is considered are changing this accounting and encourage minimally invasive therapies. Nevertheless, cost considerations affect the introduction of new therapies and diagnostics.

The process of obtaining FDA approval for new medical devices and procedures is of continuing concern to companies that develop new therapeutic and diagnostic techniques based on optical science and engineering. The approval process is perceived to be slow and expensive, limiting commercial interest to large-volume, highly profitable applications. However the FDA plays an important role in ensuring that new laser therapies are safe and effective. Although they are important, regulatory matters lie outside the scope of this study.

A final concern is the availability of funding for the development of medical applications of optical science and engineering. The National Institutes of Health (NIH) supports some technology development under the basic research (RO1) grant program; for example, there is a specific study section devoted to radiology technology development. The National Center for Research Resources also supports biotechnology development, which includes modeling and simulation, imaging technology, and imaging computation. There is not a specific study section devoted to optical technology, and there is a perception that the disease-oriented structure of NIH makes it difficult to obtain support for novel technologies that cut across traditional NIH boundaries. NIH's Small Business Innovative Research (SBIR) program, which has a special study section for lasers and optics, does cut across such boundaries. A similar, regular (RO1) study section for optical technology may be warranted.

NIH should establish a study section for RO1 grants devoted to biomedical applications of light and optical technology. An initiative to identify the human optical properties suitable for noninvasive monitoring should also be established.

Tools for Biology

In addition to their medical and surgical applications, optical instrumentation and methods contribute enormously to fundamental research and discovery in biology. Modern methods now go far beyond the capabilities of the conventional microscope and other traditional instruments, providing new insights and enabling the discovery of hitherto unknown biological processes and new drugs. Many of these new approaches are strongly dependent on advances in laser technology, the development of sensitive detectors, improvements in optical components, and the development of advanced image processing hardware and software. Equally important for progress has been the creative development of optically based contrast-enhancing molecular probes, typically fluorescent molecules. Some exciting techniques are at very early stages of development; an example is x-ray microscopy, discussed in Chapter 7 of this report.

Most of optics' contributions to biology fall into three categories: visualization, measurement and analysis, and manipulation.

Visualization Techniques

Advances in biological visualization result both from new, more powerful imaging devices and from new scientific insights into the systems being imaged. Technological advances have changed the way instruments such as the conventional optical microscope are used. CCD cameras and computers with image processing software now complement traditional film cameras for recording and analyzing data. Fluorescence microscopy is enhanced by highly sensitive CCD cameras that detect weak signals. Highly sensitive point detectors, such as the silicon avalanche photodiode, have enabled development of the scanning confocal microscope. Lasers are coupled into both conventional and scanning confocal microscopes to serve as intense monochromatic sources for fluorescence excitation. The optical quality of microscope objectives is enhanced by the use of antireflection coatings and computer-aided lens design. These advances in microscopy and image processing are being combined with automation to improve clinical laboratory techniques. A major example is the application of automated microscopy to the analysis of Pap smears, used to detect cervical precancer, in an effort to reduce errors involved with manual reading.

At the same time, new scientific insights are leading to new uses for biological visualization. Microscopes have long been used to examine cells and tissue under static conditions. They are now being applied to examine the dynamics of cellular processes. New techniques make it possible to track molecules as they enter and leave cells and to determine how different chemical reactions are spatially distributed within cells. Both conventional and new microscopy techniques are being made more powerful by the development of fluorescent molecular probes, which can tag specific cell types or cell surface markers and monitor specific molecular signaling processes.

Confocal Scanning Laser Microscopy

Confocal microscopy selectively images specific layers within a sample, enabling three-dimensional visualization of individual cells in thick samples and even of elements within living cells. Confocal microscopy uses a variety of techniques to reject light from other than the layer of interest (Pawley, 1995; Wang et al., 1992). Both structural and functional information can be obtained. Structural information is obtained either from light-scattering features or from fluorescent probes (labels). Measurements of ion concentrations and other indicators of biological function within cells can be obtained using fluorescent indicators.

The history of the commercialization of the confocal microscope illustrates some of the technologies that enabled its development. At the time of its invention in 1957, the epifluorescence technique now common in biological microscopy was not well developed. There were no lasers to excite fluorescence, there were relatively few dye-labeled antibodies, and even dichroic (beam-splitting) mirrors were not available. There were no personal computers capable of storing images and manipulating them to quickly form three-dimensional displays. A basic practical flaw in the first confocal microscope was the lack of a normal, low-magnification mode of operation, which made it difficult to find the specimen. Commercial confocal microscopes finally began to appear in the 1980s. Successful commercial manufacturers worked closely with biologists, helping to generate images that made the covers of influential biological journals.

The scanning confocal microscope obtains three-dimensional information by collecting reflected light or fluorescence from individual pixels in a sample plane and rejecting light that originates above or below that plane, allowing views from deep within a sample to be obtained. The illuminating light is scanned to obtain information from many pixels within a plane, and computer processing is used to generate images. Although the lateral resolution is only slightly enhanced over that of the conventional microscope, the depth resolution improvement is significant and is the source of much excitement in the field.

Confocal microscopes are expensive ($150,000-$300,000) and complex to operate. As a result they are not yet in significant use in clinical settings, as opposed to research laboratories. A number of potential clinical applications are being investigated, however. Recently, several groups have developed scanning confocal microscopes designed

TABLE 2.3 Issues in the Improvement of Confocal Microscopy

Problem	Short-Term Solution	Long-Term Solution
Expensive ($150,000-$300,000)	Standardize, optimize	Increased production Cheaper cw lasers
Complex to operate	Automate operation (better sensors)	Education
Photodamage limits use	Use longer wavelengths (requires better sensors) Two-photon operation	Inexpensive 50-fs lasers
Limited clinical use	Education	More education Better sensors

Source: After J. Pawley, University of Wisconsin at Madison.

to examine tissue in vivo, with the ultimate goal of replacing skin biopsy in some dermatology applications. The increased production of standardized and optimized confocal microscopy systems, the availability of cheaper continuous wave (cw) laser sources, and more automated operation are among the steps necessary to increase clinical usage. Table 2.3 summarizes some of the current limitations of confocal microscopy and suggests some short- and long-term solutions.

An alternative approach for obtaining similar three-dimensional imaging capabilities relies on computation rather than optics to eliminate the effect of light coming from regions above and below the sample plane of interest. A normal wide-field microscope is used, and computational techniques, referred to as deconvolution, remove the contribution of light originating outside the desired regions. This technique has been implemented by at least one commercial vendor. At present, relatively slow data acquisition resulting from the computationally intensive approach is a drawback of this method. However with the continually decreasing cost of computational power, this approach is likely to find wider use in the future.

Two-Photon Microscopy

Recently a nonlinear absorption phenomenon has been used to obtain depth-resolved fluorescence microscopy images within biological samples without the necessity of confocal microscopy optics. The simultaneous absorption of two photons at sufficiently high light intensities was first predicted in 1931, but observation of these effects required the availability of intense pulsed laser sources. Since the advent of the laser, numerous experiments demonstrating two-photon absorption have been reported, but its application to imaging is relatively new. The depth resolution obtained is improved compared to confocal microscopy, but several more significant advantages are also gained.

Many biologically interesting molecules require ultraviolet light to excite the fluorescence; unfortunately, UV light is strongly absorbed by biological molecules of the host material, preventing imaging at depth. In addition, UV light can rapidly destroy the molecule of interest. In the two-photon approach, red light, typically from a femtosecond-domain (about 10^{-13} s) pulsed laser source, is used for the excitation of fluorescence; two-photon absorption occurs in only a small region around the focal spot of the objective. Since two-photon absorption is proportional to the square of the illumination intensity, signals are obtained from a small spatial region around the diffraction-limited spot of the beam focus, without the need for a spatial filter as in conventional confocal fluorescence microscopy that uses linear absorption. Outside the focal region, the intensity of the red excitation light is low enough that negligible two-photon absorption occurs; consequently, in this region, photodamage by the excitation light is minimized. The use of red light for excitation not only eliminates the need for expensive UV optics but makes it possible to excite molecules in tissue hosts that are opaque to UV.

The possibilities of the two-photon approach are still being explored; for example, "chemical cages" containing molecules of interest can be opened instantaneously by light and the ensuing chemistry studied on a microscopic basis. The technique is applicable both to intrinsic tissue fluorophores and to extrinsic dyes chosen for their marking properties. Other biological applications being explored include imaging DNA stains in developing cells and embryos; imaging the metabolic activity of cells; and measuring the concentration of intracellular calcium, an important indicator of cell function (Williams et al., 1994).

This practical application of two-photon absorption provides an excellent example of how advances in the fields of optics, physics, chemistry, and engineering can lead to a development with great importance for biology. The theory of two-photon absorption came directly from the discovery of quantum mechanics. Its experimental demonstration required the discovery of the laser, and its application to biology has been enormously facilitated by the development of reliable solid-state femtosecond pulse lasers. Present tunable femtosecond lasers, often based on $Ti:Al_2O_3$ (sapphire) laser technology, appear to be adequate and would be expected to drop in price as the technology develops. The development of new applications of the technique will probably drive the field.

Nonimaging Surface Microscopies

Both conventional and confocal microscopes have a spatial resolution that is limited by the diffraction of light; a visible-light microscope can resolve objects no smaller than about 0.5 μm. This limitation can be circumvented by no longer forming images but using light emanating from a scanning nanometer-scale optical tip to sense the local optical environment by measurements of the reflection, transmission, or fluorescence within some tens of nanometers of the tip. The technique is called near-field scanning optical microscopy, NSOM. The concept was first suggested in 1928 in a paper that discussed the possibility of fabricating an optical aperture much smaller than the wavelength of light and positioning the aperture a distance much less than the wavelength of light from the sample. The spatial resolution is thus determined by aperture dimension rather than by diffraction, which becomes operative only in the far field. The optical tip can take a number of forms, including specially narrowed optical fibers and hand-crafted hollow metal guides. The fabrication of tips is being aided by the use of photolithographic techniques developed in part for microelectronics applications. The tips can be spatially manipulated with atomic-scale accuracy using techniques developed for scanning tunneling microscopy (STM) and force microscopy.

The fluorescence intensity or absorption of the sample can be measured as the tip is scanned and the signal used to generate an image in a manner similar to other scanning microscopies. Resolutions of about 20 nm are regularly obtained with this method, which has been employed to observe single protein molecules and is being tested as a means of locating pieces of cells, such as the ribosome, that have eluded structure determination by x-ray diffraction. Localized measurements of fluorescence lifetime, described below, have also been performed using NSOM, raising the possibility of highly localized environmental probing within cells. It should be noted, however, that the wet environment of biology makes NSOM technically more difficult to apply than with dry, solid samples. Commercialization of NSOM has begun, with at least one source of research instruments.

Nonimaging microscopy is also being combined with spectroscopy. In this approach, researchers measure the phase shifts of an optical beam reflected from a sample in contact with a vibrating STM tip. The phase shift depends directly on the absorption spectrum of the molecule in the neighborhood of the tip. Current spatial resolution is about 1 nm. This technique is at an early stage of development but demonstrates an approach that could bring optical spectroscopy truly to the level of atomic-scale spatial resolution. The future success of this endeavor will require progress in strategies for labeling specific biological sites by probe molecules and advances in nanopositioning technology.

Development of Molecular Probes

Optical microscopy studies of tissue have long relied on absorbing stains to reveal specific cellular structures of interest; without such stains, most medical histopathology would not be possible. Traditional stains are a way of generating contrast where there is normally none; today antibody-linked fluorescent dyes are used to mark specific cell surfaces. The general strategy of many of the new optical technologies is also to tag the structure of interest with a dye molecule. Much of the progress in fluorescence microscopy is linked to the development of ever more specific fluorescent probes, which may be chosen for their ability to intercalate into DNA or to label specific ions such as calcium. These fluorescent probes may be used to locate and examine specific sites by direct visualization with a microscope, or they may be sensed by a variety of optical techniques such as flow cytometry (discussed below) (Tsien, 1994). Quantitative fluorescence-based measurement techniques are being introduced; an example is measurement of the length of long DNA molecules with fluorescence instead of pulse field electrophoresis.

Lifetime imaging is an important way to take advantage of the properties of fluorescent probes. Probe molecules can be quite sensitive to changes in their local environment; the fluorescent lifetime, fluorescence quantum efficiency, and emission wavelength of dyes can all change with environment (e.g., the local pH in a cell or the particular binding site on a protein). The most commonly measured property is fluorescence lifetime, whose variation with the above environmental factors must be determined for each dye. Lifetimes can be measured in the time domain using pulsed laser sources and fast detection schemes; alternatively, frequency-domain techniques can be employed to obtain equivalent information using modulated cw lasers and phase-sensitive detection. The frequency-domain approach has a significant advantage in being able to measure nanosecond-domain lifetimes using relatively simple equipment. Both techniques have been used to obtain microscopic and macroscopic images that show regions having specific lifetimes. Lifetime methods, although in their infancy, have already been applied to some important research questions concerning intracellular signal transduction involving protein kinase C as well as measurements of calcium ion (Ca^{2+}) concentrations.

New dye molecules with novel fluorescence properties are increasingly needed to meet the requirements of emerging technologies for visualization and other biological applications. Ideally the dye chemist would like to control the photostability, two-photon cross section, fluorescence yield, nontoxicity, fluorescence wavelength, and lifetime of a dye. In the case of marker molecules for use with tissue in vivo, fluorescence in the wavelength region longer than about 650 nm is desirable to minimize interference from tissue autofluorescence, which occurs at shorter wavelengths. Probes that are highly sensitive to specific chemical or physical conditions, subtle differences in pH, or the presence of chemical structures of different types are needed. For example, the systematic discovery of probes that are specific to calcium ions has allowed investigators to optically record cells responding to a variety of stimuli that induce the appearance of this ion. Probes to target specific regions of a cell or tissue, such as the DNA of the cell nucleus or the cellular cytoplasm, are also needed.

In addition to the advantage of having probes that are sensitive to chemical or physical conditions of the target, another strong driver in the development of fluorescent probes is the desire to replace traditional radioactive markers, given the rapidly rising costs of handling and disposing of radioactive tracer materials.

Every molecule undergoes some degradation when it absorbs a photon. The goal of detecting single molecules optically demands probes of even higher photostability than previously required. If a probe molecule photochemically degrades with an efficiency as low as 10^{-4}, a single molecule of this probe can generate only 10^4 fluorescent photons in toto before it becomes inactive. Combining photostability with the optical properties mentioned above provides a significant challenge.

Measurement and Analysis Techniques

The same sophisticated new optical probes that are so useful for biological visualization also make possible the application of optical measurement and analysis methods to such biological problems as gene sorting, mapping the human genome, and investigating cellular control and communication.

Flow Cytometry

Flow cytometry (Figure 2.7) is a technique for rapidly (tens of microseconds) analyzing individual particles that range in size from large plankton (1.3 mm long) to individual molecules (Shapiro, 1995). From the development of the technique in the late 1960s to today's sophisticated research and clinical instruments, this technology has continued to make a major impact in modern-day biological research and clinical medicine. Particles to be analyzed are suspended in a liquid medium, and stains or dyes that bind to specific parts of the particle are added. Single cell suspensions can be stained for DNA content, RNA content, cell surface molecules that identify different cell types, and physiological parameters such as pH or calcium concentration. The particles are then introduced into a fluid flow and passed through a nozzle that produces a stream of droplets containing individual particles. These particles pass through a region in which focused visible

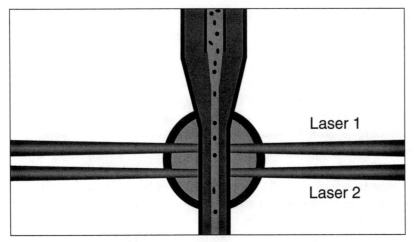

Laser 1

Laser 2

FIGURE 2.7 Schematic diagram illustrating the principles of flow cytometry. (Courtesy of J. Jett and B.L. Marrone, Los Alamos National Laboratory.)

laser beams can excite fluorescence. Multiple sensors are used to detect fluorescence signals, which are recorded. The sensors may also be used to detect light scattering by the particles. In some cases, signals from the sensors are used to activate an additional cell sorting process based on deflecting previously charged droplets by charged deflection plates.

The development of optically based measurement techniques and new probes occurred in parallel with the application of this technology to basic biological studies and routine clinical assays. Routine clinical applications of flow cytometry fall predominantly into two categories: immunophenotyping and DNA content measurement. Immunophenotyping, the identification and enumeration of white blood cells by analysis of surface molecules with fluorescent-dye-labeled antibodies, is used in various medical applications including monitoring AIDS progression (Box 2.6) and leukemia or lymphoma diagnosis. DNA content measurements provide clinicians with information about the number of proliferating cells in a population and the normality of the cellular DNA content. Such measurements have application in grading cancer cells and determining disease prognosis. These applications of flow cytometry in the clinical arena are used worldwide.

Sorting capabilities of flow cytometers are used to physically separate large numbers of human chromosomes. The sorted chromosomes provide template DNA for the construction of recombinant DNA libraries. Chromosome-specific libraries have been generated for each of the human chromosomes. Such libraries are an important component of genetic engineering, a technique that allows specific genes to be inserted into cells and organisms. The availability of these materials has also played an important role in the establishment and rapid progress of the Human Genome Project, which promises new understanding of the genetic basis of disease.

C h a p t e r   2

BOX 2.6 IMMUNE SYSTEM MONITORING FOR HIV

The AIDS epidemic is an excellent example of a critical medical problem that is being studied using optical biomedical instrumentation. AIDS currently affects more than 100 million people worldwide and is the leading cause of death among young adult males in the United States. Our understanding of this terrible disease has grown out of intense scientific research that has occurred over the past 15 years. A large portion of the research has focused on the impact of the AIDS virus on the human immune system. The primary tool used in this research has been the flow cytometer. For example, using flow cytometry, immunologists were able to determine the precise subgroup of white blood cells, the CD4 cell, that is attacked by the virus. The flow cytometer has evolved from the primary scientific tool used to understand the impact of the AIDS virus on the immune system into the principal clinical diagnostic instrument that is now the standard of care for monitoring CD4 levels in infected individuals. Flow cytometry data on CD4 concentrations in peripheral blood are used to guide physicians in choosing the antiviral and antibiotic drug therapies appropriate at various stages of the disease.

Another class of optical instrumentation that is of critical importance in the battle against AIDS is the automated genetic sequencer. Using this instrument, which typically incorporates a scanning laser fluorimeter, scientists have been able to sequence the complete genome of the AIDS virus. This information has provided insight into the structure of the surface proteins of the virus and has helped lead to effective methods for sensitive detection of viral proteins in peripheral blood. Detecting viral protein in a peripheral blood sample is currently the accepted diagnostic method for verifying HIV infection. Gene sequencing instruments are also used to monitor genetic changes in the virus that signal the evolution of viral mutants resistant to drug therapies and mutants that might elude the current generation of tests used to ensure the safety of the U.S. blood supply.

It is interesting to note that flow cytometers and automated gene sequencing instruments were developed in the late 1970s and early 1980s, precisely the time when the AIDS epidemic began. This timing was quite fortunate since without these instruments, our knowledge of the AIDS virus, its common modes of transmission, and possible strategies for combating it would have been severely affected and the epidemic would most definitely be significantly worse.

The next generation of AIDS diagnostic techniques will focus on determining the concentration of free HIV in peripheral blood, the viral load. This diagnostic measurement has proven to be of great importance for developing promising new anti-HIV drugs, the protease inhibitors, and for determining effective therapies involving combinations of these antiviral drugs. Several different techniques have been developed using DNA chemistry for viral recognition and optical detection for quantification, for example, quantitative competitive polymerase chain reaction (PCR) and branch DNA. Both of these techniques are usually performed in sophisticated molecular biology laboratories and are not yet suitable for a typical hospital clinical laboratory.

The impact of flow cytometry on modern biomedical research is large. The total annual market for flow cytometry instruments and reagents is estimated to be $300 million worldwide, of which U.S. manufacturers control on the order of 90%. Presently, clinical applications account for two-thirds of the total market, or about $200 million annually. One measure of the impact of flow cytometry is that, on the average, three out of four issues of *Science* contain an article with flow cytometric data. The technology is used worldwide, even in developing countries with limited funds for high-technology instrumentation. In these countries the major application is the analysis of white blood cell subpopulations in AIDS patients.

In the future, flow cytometers will be easier to use, more compact, and located in smaller hospitals or even doctor's offices. An integrated system with a flow cytometer on a chip that contains excitation source, detection, and fluidics is a realistic goal. On the research side, sensitive flow cytometry techniques orders of magnitude faster and more sensitive than currently used methods are being developed to analyze the size of DNA fragments and to sequence DNA. A variety of technological improvements are needed for this to occur. New compact light sources that emit light in the blue and ultraviolet will be needed to match the dyes currently in routine use. Detection and light filtration systems that are compact, efficient, and easy to use are also necessary.

One of the unique aspects of flow cytometry, whether in a clinical or a research laboratory, is that competent cytometrists must be well founded in a variety of disciplines from computer science to biology to optical sciences. Currently, there is no interdisciplinary degree program that adequately prepares either users or developers of the technology for the breadth of information and understanding that they need.

Bioengineered Fluorescent Indicators

A number of novel fluorescent indicators based on molecular biology have become available that serve as indicators of processes going on within living cells. For example, the green fluorescent protein (GFP) from a luminescent jellyfish is a protein that spontaneously modifies itself to generate a strongly fluorescent internal chromophore. Two mutants of different colors can engage in fluorescence resonance energy transfer, which can then be spectroscopically studied to monitor the presence or absence of protein-protein interaction inside living cells. Optical readouts of membrane potential, protein phosphorylation, and proteolysis are also under development. Even more recently, techniques have been developed to incorporate the gene for a bioluminescent molecule into bacteria and other molecules (Contag et al., 1995). This has allowed tracking of the spread of bacteria, as well as the action of antibiotics, throughout the body of small animals. The same approach may be useful in signaling successful gene therapy. More broadly, it appears we can now alter the optical properties of living organisms in order to monitor the spread and control of disease in living animals and eventually humans.

Micromanipulation Techniques

A new application of optics in biology is the use of light to actively manipulate the molecules, mechanisms, and structures that determine biological function. Laser beams can be used, with proper handling, to create optical traps or "tweezers" that capture and manipulate cells and even subcellular organelles. Optical tweezers are even being used to determine the forces involved in the locomotion of single biological molecules.

The force that light can exert was predicted by James Clerk Maxwell in his theory of electromagnetism of 1873 but was not demonstrated experimentally until the turn of the century. One reason for the delay is that radiation pressure is extraordinarily feeble. Milliwatts of power (corresponding to very bright light) impinging on an object produce piconewtons of force (1 pN = 10^{-12} N). The advent of lasers in the 1960s finally enabled researchers to study radiation pressure through the use of intense, collimated sources of light. By focusing laser light into narrow beams, researchers demonstrated that tiny particles, such as polystyrene spheres a few micrometers in diameter, could be displaced and even levitated against gravity using the force of radiation pressure. Under the right conditions, the intense light gradient near the focal region can achieve stable three-dimensional trapping of dielectric objects. Optical traps can be used to capture and remotely manipulate a wide range of larger particles, varying in size from several nanometers to tens of micrometers (Svoboda and Block, 1994). Subsequently, it was shown that these "optical tweezers" could manipulate living things such as viruses, yeasts, bacteria, and protozoa. Experiments during the past few years have begun to explore the rich possibilities afforded by optical trapping in biology.

Although still in their infancy, laser-based optical traps have already had significant impact. Tweezers afford an unprecedented means for manipulation on the microscopic scale. Optical forces are minuscule on the scale of larger organisms, but they can be significant on the scale of macromolecules, organelles, and even whole cells. A force of 10 piconewtons, equal to 1 microdyne, can tow a bacterium through water faster than it can swim, halt a swimming sperm cell in its track, or arrest the transport of an intracellular vesicle. A force of this magnitude can also stretch, bend, or otherwise distort single macromolecules, such as DNA and RNA, or macromolecular assemblies, including cytoskeletal components such as microtubules and actin filaments. Proteins such as myosin, kinesin, and dynein produce forces in the piconewton range. Optical traps are therefore especially well suited to studying mechanics or dynamics at the cellular and subcellular levels.

The possibilities for further development and use of optical tweezers in biology and medicine are extraordinary. There are many areas in which optical tweezers can be expected to provide visual images or better understanding of biological processes that involve motion. For example, the micromechanics of DNA-modifying enzymes (such as DNA and RNA polymerases) can be observed and protein synthesis manipulated at the most basic level; receptor-ligand interactions can be manipulated by physically constraining the reactants; small structures such as biosensors and microtubules could be constructed; mechanical properties of filaments can be measured directly; and forces allowing cells to crawl or chromosomes to move from place to place can be determined.

The National Science Foundation (NSF) should increase its efforts in biomedical optics and pursue opportunities in this area aggressively. This will require a broader interpretation of the NSF charter regarding health care in order to support promising technologies that bridge the NIH and NSF missions.

Biotechnology

Just as optics is playing an important enabling role in the development of new research techniques for fundamental biology, it is also becoming increasingly important in the biotechnology industry. Many of the devices and techniques discussed above in the context of biological research, such as flow cytometry and fluorescent molecular probes, play similarly important roles in biotechnology applications. In a general sense, biotechnology involves measurement, manipulation, and manufacture of large biologically significant molecules such as proteins and DNA. Among the applications for which optical methods are most important are genetic sequencing and pharmaceutical development.

DNA Analysis

The development of new instrumentation for DNA sequencing has been driven by the Human Genome Project, which is the largest government-funded project in the health sciences. The general strategy of all such instruments involves tagging the four distinct bases that occur in DNA with fluorescent dyes that have different emission wavelengths. Currently an argon ion laser is used to excite fluorescence. Sequence information is obtained by monitoring the multicolored fluorescent emission from large (50 cm × 70 cm) electrophoretic gels.

High-efficiency confocal laser scanning systems, which are commercially available, currently provide the fastest method for gene sequencing. Although they represent a major improvement over first-generation instruments, these devices are still considered approximately 100 times too slow to meet the goals of the Human Genome Project. The next generation of instruments, currently under development, incorporates integrated optics, hollow fibers for capillary electrophoresis, and red and infrared dyes for better spectral separation of the fluorescent indicators.

The polymerase chain reaction (PCR) used for DNA amplification is pervasive in biology today, being used for detection of viruses in blood, monitoring of viral loads in AIDS patients, detection of inherited disease tendencies, and forensics. Although current PCR systems are of laboratory bench-top size, the availability of miniaturized optics allows the development of miniaturized versions. These micro-PCR systems will allow quantitative detection of the nucleic acids formed and will use microspectrometers to monitor fluorescent tags in real time. The ultimate goal is to combine these optical monitors with control and analysis software that will determine the thermal cycling used in the PCR process. It is interesting to note that the problem of miniaturizing the liquid handling aspects of such systems presents formidable technical challenges whose solutions have yet to be found.

Oligonucleotide probe arrays, sometimes referred to as DNA chips (Figure 2.8), combine both optical and chemical techniques to obtain genetic information. Oligonucleotides are small polymers made up of nucleotides, which are subunits of DNA (Lipshutz et al., 1995). The basic goal of these chips is to make possible the performance of a large number of operations probing the sequence of DNA in parallel. The chips are made by light-directed chemical synthesis, which is in turn based on photolithographic techniques developed for the semiconductor industry and on solid-phase chemical synthesis. The photolithographic techniques are used to "deprotect" or activate small synthesis sites consisting of hydroxyls on a solid substrate. The sites are selected using photolithographic masks. The activated region can then be reacted with a chemical building block to produce a new compound. By combining many of these activation steps with multiple cycles of photoprotection and chemical reaction, a chip with a high-density checkerboard array of oligonucleotides can be produced. For example, if the resolution of the chemical process is 100 μm, 10^4 sites can be produced per square centimeter.

These sites are essentially probes for specific DNA sequences. The target or unknown sequence is labeled with a fluorescent dye and exposed to the chip. It binds most strongly to sites that match a portion of its DNA sequence, resulting in localized patches of high fluorescence. Laser scanning confocal microscopy, described previously, is used to produce a map of fluorescence intensity versus site on the chip. Since the chemical composition at each site is known from the synthesis procedure, the unknown sequence can be deduced.

Applications envisioned for these probe arrays include rapid sequencing of DNA as well as the detection of mutations associated with resistance to antiviral drugs used in the treatment of AIDS. Although the commercial success of the DNA chip will depend on many factors, including the development of competing technologies, it illustrates the way sophisticated optical techniques, developed in part for the semiconductor industry, are being used for biotechnology.

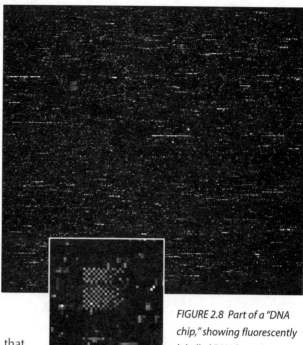

FIGURE 2.8 Part of a "DNA chip," showing fluorescently labelled DNA bound to an 8,000-site GeneChip® probe array. (Courtesy of Affymetrix, Inc., Santa Clara, Calif. Copyright © Affymetrix, Inc. All rights reserved. Affymetrix and GeneChip are registered trademarks used by Affymetrix, Inc.)

Pharmaceutical Screening

Pharmaceutical screening to find drugs that have optimal biological activity for a particular clinical application is a good example of the potential impact of advanced fluorescent indicators on biotechnology. These applications, now in the early stages of development, would allow the screening of very large numbers of potential pharmaceuticals using only minute quantities of the candidate drug and small groups of cells.

The pharmaceutical industry has developed very large libraries of semirandomly generated candidate compounds for drug discovery. The libraries contain thousands to millions of different chemicals, usually synthesized by combinatorial sequences of reaction steps. The libraries now encompasses a wide variety of chemical families, including many that could be suitable for orally active drugs to treat major diseases. However, screening these huge libraries to find which members possess optimal biological activity is a tremendous challenge. Only picomole quantities of each candidate are available, so most traditional pharmaceutical assays are too insensitive. Thus, there is a great need for bioassays that can be miniaturized to microliter or smaller assay volumes and performed at the rate of thousands to millions per day. Such bioassays have to be easily adaptable both to known drug receptors and to the thousands of new potential macromolecular targets being found by human genome sequencing.

Optically based methods to accomplish this are being investigated. The basic concept is to combine recent improvements in microscopic imaging with new fluorescent indicators of intracellular signaling to allow bioassays on single cells or small groups of cells. Cells can now be genetically engineered to be responsive to signaling pathways of interest or to mimic target disease processes. They are then grown by tissue culture in billions or trillions as required. Zeptomole (1 zmol = 10^{-21} mol) to attomole (1 amol = 10^{-18} mol) quantities of compound suffice to activate or inhibit individual cells, which can be imaged in microscopic volumes.

The best known intracellular fluorescent indicators report calcium signals and are already in use for drug screening at the cellular level. However, gene expression is a more universal and stable readout, which can be monitored by introducing an optically easy-to-detect enzyme for the protein that the cell would normally express. For example, reporter enzymes, such as β-lactamase, together with carefully designed, membrane-permeant fluorogenic substrates can disrupt fluorescence energy transfer in the substrate and change the emission color from green to blue. This color change is so dramatic that it can easily be seen by the unaided eye and is precisely quantifiable by two-color flow cytometry or standard ratio image processing. Flow cytometry should enable selection and cloning of cell lines whose β-lactamase expression is optimally sensitive to known drugs, hormones, or disease-mimicking alterations. The same enzyme system provides a nondisruptive optical readout to measure the effect of novel drug candidates on single cells or small clusters of cells. In this way the cumulative activity of nearly any specific signal transduction pathway of choice may be monitored optically.

The practical challenge is now to integrate the techniques of molecular biology, cell culture, optical signal transduction, organic synthesis, microscale liquid handling, high-performance optical imaging, and automated data analysis into a coherent, robust, and economically viable system.

Summary and Recommendations

Surgery and Medicine

Optics has enabled the development of rigid and flexible viewing scopes that allow minimally invasive diagnosis and treatment of numerous sites inside the body, such as the colon, the knee, and the uterus. Lasers have become accepted and commonly used tools for a variety of surgical applications. These include the CO_2 laser, the high-repetition-rate, frequency-doubled Nd:YAG laser (KTP 532), and the Nd:YAG laser. Lasers and optics have made possible noninvasive treatment of many diseases of the eye and have become essential to the practice of

ophthalmology. Inpatient procedures have often become outpatient
ones as a result. Lasers are now used extensively in dermatology for
the treatment of pigmented lesions, tattoos, wrinkles, and other problems. This use has become widespread because research has led to an
understanding of how to target specific tissue sites by the proper choice
of laser wavelength and pulse width. Biological response, rather than
the sophistication of a particular optical technique, is often the critical
issue in clinical applications. Close cooperation between physical scientists and physicians is necessary to successfully address clinical problems. One example is laser angioplasty. New infrared solid-state lasers
are being used to complement the more established CO_2 and YAG surgical lasers. The Ho:YAG laser offers compatibility with existing quartz
fiber optics and may replace CO_2 in some cases. The Er:YAG laser
is unique in its ability to cut bone with minimal thermal damage.
Photomechanical effects have been recognized as clinically significant
and often useful; they are used commonly in ophthalmology and urol-
ogy. Light-activated drugs are being used to treat both cancer and noncancer diseases by photodynamic therapy. These photochemical treatments are able to affect not only cells and tissue, but also specific
growth factors and signaling processes in tissue. Noninvasive monitoring of basic body chemistries, such as glucose concentration, remains
a major challenge for optics. The basic science required for the development of such monitoring techniques is often missing or incomplete.
As laser medicine and surgery have moved from being almost entirely
empirical arts to having a solid basis in the underlying physics and
chemistry of laser-tissue interaction, new and less painful laser treatments for numerous diseases have been developed. The disease-oriented structure of NIH does not encourage the funding of biomedical optical technology programs.

Lasers and fiber-based instrumentation have enabled many new minimally invasive therapies that reduce total (direct plus lost time) health
care costs. Optically based diagnostic methods are less developed than
therapeutic ones, but they offer potentially improved techniques for the
medical laboratory (more accurate blood tests), the clinic (techniques to
complement x-ray mammography), and home care (noninvasive glucose
monitoring). New laser technologies and effects are now quickly assimi-
lated by the medical care community. However, the FDA regulatory
process makes commercialization of new technologies costly. Close
cooperation among optical scientists, physicians, and FDA personnel
may improve the process. Optics and lasers will continue to facilitate
the development of new medical systems. Visible diode lasers, diode-
pumped solid-state lasers, light-emitting diodes, and compact optical
parametric oscillators are some of the devices on which such systems
will be built. Feedback control will attract increasing attention as opti-

cal and magnetic resonance imaging systems are coupled with laser-based treatment systems. Mechanisms should be developed for encouraging increased public and private investment in noninvasive optical monitoring of basic body chemistries. Clearer separation of the roles of the public sector—basic science and proof of principle—and the private sector—device development—is needed. Better understanding of how light interacts with tissues will continue to be important for the development of optical techniques for treatment and diagnosis.

NIH should establish a study section for RO1 grants devoted to biomedical applications of light and optical technology. An initiative to identify the human optical properties suitable for noninvasive monitoring should also be established.

Tools for Biology

Confocal laser scanning microscopy and computed microscopy have enabled depth-resolved microscopic imaging that allows three-dimensional information to be acquired. Two-photon techniques have not only enhanced the capabilities of fluorescence microscopy but also opened up new possibilities for performing spatially localized photo-chemistry within cells. The potential of these techniques is relatively unexplored. Near-field microscopy, a nonimaging technique, allows microscopy with resolutions of tens of nanometers, far less than the diffraction limit for light. Fluorescent markers have replaced many of the radioactive tags used to mark the presence of specific molecules, such as proteins, and in DNA sequencing, thus eliminating the complications associated with handling and disposing of radioactive materials. Flow cytometry, which is based on laser and optical technology, has become both a standard clinical assay and a frequently used research tool. Optical micromanipulation techniques (optical tweezers) have found uses in the study of the forces involved in molecular locomotion and in the manipulation of cells and molecules within them. The use of fluorescence techniques as quantitative assays will grow as more quantitative measurement techniques are introduced.

New microscopies (confocal, two photon, near field) are extending the capabilities of traditional microscopy by enhanced resolution and the ability to image in depth. Lasers and optical methods have become an integral tool for many essential biological technologies and methods. The continual development of new, specific, and inexpensive molecular probes is necessary for optimal utilization of fluorescence-based techniques. The development of instrumentation that solves significant biological problems requires interdisciplinary teams that are aware of both available technology and biological questions. The advances in technology that are now being applied build upon long-term investments in basic research. Examples are the understanding of two-photon absorption, which builds on basic quantum mechanical calculations that are more than 60 years old, and the development of optical tweezers, which grew out of studies of optical levitation.

NSF should increase its efforts in biomedical optics and pursue opportunities in this area aggressively. This will require a broader interpretation of the NSF charter regarding health care in order to support promising technologies that bridge the NIH and NSF missions.

Biotechnology

Lasers have become essential parts of all systems used for DNA sequencing, ranging from those that are commercially available to more experimental capillary electrophoresis systems. Optics is being employed in a number of biotechnology applications, from sophisticated systems using DNA chips to simpler systems using transmission probes. Scientists, engineers, and technicians with cross-disciplinary training will enhance the transfer of optical science into biology and medicine.

References

American Optometric Association. 1996. *Caring for the Eyes of America.* St. Louis, Mo.: American Optometric Association.

Arons, I. 1997. Medical laser market hits new high. *Med. Laser Rep.* 11:1-2.

Contag, C.H., P.R. Contag, J.I. Mullins, S. Spilman, D.K. Stevenson, and D.A. Benaron. 1995. Photonic detection of bacterial pathogens in living hosts. *Mol. Microbiol.* 18:593.

Cuschieri, A. 1995. Whither minimal access surgery: Tribulations and expectations. *Am. J. Surg.* 169:9-19.

Deckelbaum, L.I. 1994. Cardiovascular applications of laser technology. *Lasers Surg. Med.* 15:315-341.

Gratton, E., and J.B. Fishkin. 1995. Optical spectroscopy of tissue-like phantoms using photon density waves. *Comments Mol. Cell Biophys.* 8:307-357.

Huang, D., E.A. Swanson, C.P. Lin, J.S. Schuman, W.G. Stinson, W. Chang, M.R. Hee, T. Flotte, K. Gregory, C.A. Puliafito, and J.G. Fujimoto. 1991. Optical coherence tomography. *Science* 254:1178-1181.

Katzir, A. 1993. *Lasers and Optical Fibers in Medicine.* San Diego, Calif.: Academic Press.

Krauss, J.M., and C.A. Puliafito. 1995. Lasers in ophthalmology. *Lasers Surg. Med.* 17:102-159.

Lipshutz, R.J., D. Morris, M. Chee, E. Hubbell, M.J. Kozal, N. Shah, N. Shen, R. Yang, and S.P.A. Fodor. 1995. Using oligonucleotide probe arrays to access genetic diversity. *BioTechniques* 19:442-447.

McDonnell, P.J. 1995. Excimer laser corneal surgery: New strategies and old enemies. *Invest. Ophthalmol. Vis. Sci.* 36:4-8.

Middleton, H., R.D. Black, B. Saam, G.D. Cates, G.P. Cofer, R. Guenther, W. Happer, L.W. Hedlund, G.A. Johnson, K. Juvan, and J. Swartz. 1995. MR imaging with hyperpolarized ^{3}He gas. *Magn. Reson. Med.* 33:271-275.

Pawley, J.B. 1995. *Handbook of Biological Confocal Microscopy,* 2nd ed. New York: Plenum Press.

Seiler, T., and P.J. McDonnell. 1995. Excimer laser photorefractive keratectomy. *Surv. Opthalmol.* 40:89-118.

Shapiro, H.M. 1995. *Practical Flow Cytometry,* 3rd ed. New York: Wiley-Liss.

Svoboda, K., and S.M. Block. 1994. Biological applications of optical forces. *Ann. Rev. Biophys. Biomol. Struct.* 23:247-285.

Tsien, R.Y. 1994. Fluorescent imaging: Technique tracks messenger molecules in living cells. *Chem. Eng. News,* July 18,: pp. 34-44.

Wang, X.F., A. Periasamy, and B. Herman. 1992. Fluorescence lifetime imaging microscopy (FLIM): Instrumentation and applications. *Crit. Rev. Anal. Chem.* 23:1-26.

Wigdor, H.A., J.T. Walsh, S.R. Visuri, D. Fried, and J.L. Waldvogel. 1995. Lasers in dentistry. *Lasers Surg. Med.* 16:103-133.

Williams, R.M., D.W. Piston, and W.W. Webb. 1994. Two-photon molecular excitation provides intrinsic 3-dimensional resolution for laser-based microscopy and microphotochemistry. *FASEB J.* 8:804-813.

Yodh, A., and B. Chance. 1995. Spectroscopy and imaging with diffusing light. *Phys. Today* 48:34-40.

3

··

Optical Sensing, Lighting, and Energy

A major fraction of all information received and analyzed by humans is received through the eyes, whether from reading a newspaper, watching television, or just observing our environment. The ability to optically sense and obtain information in this way is fundamental to our human existence and involves the traditional optical science and technology of the human eye, the vision process, corrective eyeglasses or contact lenses, and the use of lighting to permit the surroundings to be illuminated. Although advances have been made in some of these areas, for the most part the fundamental way we observe and see our immediate surroundings has not changed significantly over the past hundred years, with the exception that now artificial lenses and surgical techniques can improve some vision problems and better eyeglasses and corrective procedures are available.

What is significant, however, is the tremendous advance that has occurred recently in the development and use of new optical and infrared sensors and instruments that can detect and analyze our surroundings and present this information to us visually, thus greatly augmenting our normal visual process and in some cases showing details and information never previously seen. For example, a broad range of newly developed optical sensors and instruments are already used in everyday life, such as those that provide satellite pictures of clouds and weather patterns on TV evening news, infrared night vision scopes used by law enforcement, spaceborne probes to Jupiter that use optical instruments to measure and image the surface temperature of the planet, home security infrared motion sensors, and optical or laser probes to detect and display gas emissions from automobile highway traffic. Related to these advances in optical sensing and imaging technology are associated advances in the development of new, high-efficiency sources of light to illuminate our surroundings and in the use of optics and lasers for development of new energy sources. For example, new lighting sources are being developed that may reduce U.S. energy consumption by tens of billions of dollars per year, and new laser-based nuclear fusion power plants and mass-produced photovoltaic solar cells are being studied for long-range potential as cheap power in the next century.

This chapter presents a synopsis of recent advances in optical sensing instruments and techniques, lighting, and energy. The emphasis is on new or revolutionary optical technologies that are expected to significantly impact the future growth and well-being of our society. As such, technical areas of lighting, energy, and optical sensors that either are mature or are not expected to grow dramatically are not covered in as much depth. Although the topics include a rather broad range of optical fields, they are centered primarily on the generation of light (new lighting sources), the conversion of light to energy (solar cells and laser fusion research), and the use of optical and imaging sensors for the measurement and detection of a wide range of physical and chemical parameters (night vision scopes, video cameras, gas vapor sensors, traffic laser radars, bar-code scanners). The topics covered have been divided into four subsections: (1) optical sensors and imaging systems, with application in the environment, global imaging, astronomy, industrial/chemical sensing, video cameras, law enforcement and security, common optical sensors, and scanners; (2) lighting, including new light sources, light-emitting diodes (LEDs), and the use of lasers in entertainment; (3) applications of optics and lighting in transportation, including autos and aircraft; and (4) energy applications, including laser fusion, laser isotope separation, and solar cells. The role that advances in materials have played in many of these fields is also addressed, because the development of new optical materials is often the key factor enabling progress (Box 3.1).

Overall, this study finds that the areas of optical sensing, lighting, and energy account for sales, research, and development of about $19 billion per year in the United States. This figure includes about $3.5 billion for optical sensors and imaging instruments, $12 billion for lighting fixtures and lamps, $400 million for light-related energy research and solar cell production, and $2 billion for the use of optics in cars and airplanes. The total world market is estimated to be two to three times as large. Some of these applications have a great impact on other markets and represent key or enabling technologies. For example, the efficiency of lamps has a direct impact on the $40 billion that is spent each year in the United States on electricity for lighting. As such, a 50% change in lighting efficiency can have a $20 billion impact on the U.S. economy and an even larger impact on worldwide energy demand,

• •

especially in developing countries where lighting and energy production are still expanding. Another example is the real-time global mapping supplied by space-based optical imaging weather satellites. These maps affect a much larger market for weather and crop forecasts and help authorities develop forecasts and emergency plans for storms and hurricanes whose impact in dollars and lives saved is often incalculable. Where practical, these secondary impacts of optics applications are also covered in this chapter.

The discussion of each subtopic is based on the results of a workshop held by the committee, as well as on additional written inputs obtained by the committee. The main findings and conclusions, which cover key highlights and challenges, are collected at the end of each major section. Finally, recommendations based on the findings and conclusions are made to the government, academia, and industry, where appropriate.

Optical Sensors and Imaging Systems

Light reflected from objects has been used by humans for thousands of years as a way to see or remotely sense the presence and composition of the surrounding environment. In most cases, the reflected or transmitted light is seen directly by the eye, and differences in color or intensity over the visible wavelength spectrum are used to detect and differentiate objects and images. Although outside the portion of the spectrum that is visible to the eye, light at ultraviolet (UV) and infrared (IR) wavelengths contains additional information. For instance, absorption and possibly fluorescence at UV and IR wavelengths can be used to detect certain chemicals and pollutant gases, to see objects at night by using IR thermal radiation, and to measure the temperature and composition of a distant object. It is the spectroscopic or wavelength (color) dependent nature of the reflected or transmitted light that allows one to detect a particular feature or the presence of a particular chemical.

The use of optical sensors and imaging systems has been enhanced recently with the advent of small, inexpensive video cameras and detectors that operate in both the visible and the infrared; the development of new compact tunable laser sources; and the manufacture of compact optical spectrometer instruments. Although spectroscopic optical instruments have been used for the past hundred years, recent advances in these optical techniques and the reduction in their costs have led to the recent surge in their use in a wide variety of fields. The following sections outline the current use and projected growth of optical sensors in environmental and atmospheric monitoring; Earth and global surface monitoring; astronomy and planetary probes; industrial chemical sensors; imaging detectors and video cameras; law enforcement and security; and common everyday optical sensors, printers, and scanners.

Environmental and Atmospheric Monitoring

Optical systems can be used for the detection of a number of important gases or pollutants in the atmosphere. In many cases, each chemical has a distinct absorption spectrum in which different wavelengths (or colors) of a transmitted optical beam are preferentially absorbed according to the concentration and presence of the chemical or gas in the atmosphere. Several different optical techniques are used, depending on the substance of interest, its concentration, and the detection range expected from the instrument. An important point is that optical sensing can often be accomplished remotely, because the optical beam can be directed at a distant object and information about the composition and gases surrounding the distant scene can be deduced from backscattered light. In fact, optical remote sensing can be used to detect chemicals (or physical parameters such as speed and dust cloud density) at ranges from a few meters to several hundred kilometers in some cases. This capability has significantly changed the way we measure our environment. For instance, 30 years ago weather balloons were used to carry instruments aloft to sample the upper atmosphere; now, we use laser beams from the ground to make the same measurements. Similarly, where once we measured the severity of air pollution in Los Angeles by measuring the time it took a stretched rubber band to rot, we now use chemical and optical absorption instruments to obtain round-the-clock coverage of the concentration of ozone and other environmental gases. The advances in these areas are covered in the following sections.

Open-Path Gas Monitoring

Optical gas monitoring uses a beam of light that is transmitted through the open air or through a sample chamber (cell). The beams of open-air systems can cover paths of several hundred meters to several kilometers. Selective absorption of the light allows for detection of the compounds present and quantification of their concentrations. This is usually done by using a conventional optical spectrograph or a Fourier-transform infrared (FTIR) optical spectrometer that directs an optical beam through the atmosphere by means of a telescope. These optical instruments can be used as sensitive real-time monitors of the composition and concentration of environmental gases in the atmosphere or in a plume from a smokestack. They have been used to detect the concentration of organics, refrigerants, carbon monoxide (CO), nitrogen oxides (NO_x), ozone, and other gases in the environment and from industrial sources; to sense emission gases from automobiles over a highway; and to detect evidence of the manufacturing of chemical, biological, or nuclear materials. For example, Figure 3.1 shows an FTIR instrument used to monitor the perimeter of an industrial chemical plant to detect the accidental release of a hazardous gas by the plant. Although conventional analytical chemical techniques such as wet chemical analysis or gas chromatography are often used for this purpose, they do not offer real-time remote sensing or on-site capability as easily as optical monitoring does. The advantage of conventional chemical measurements is the longer historical use of these techniques and their lower capital cost, although their operational costs can be higher. Conventional analytical chemical sensors are still dominant, but optical methods now claim about 40% of the market and this fraction is rapidly growing. The current annual U.S. market for optical instruments used in this area is about $500 million (systems cost). The demand is driven by regulatory laws for source ambient air quality usage, although industrial process control is beginning to incorporate these techniques as well. The recent increased acceptance of such optical instruments by the Environmental Protection Agency (EPA) will certainly stimulate their more widespread use. The main technical challenge is for smaller and cheaper laser or optical spectroscopy devices. At present, there is a significant U.S. market, but the market in Europe is somewhat more

FIGURE 3.1 An optical-beam FTIR instrument used to measure gas emissions along the perimeter of an industrial chemical plant. (Courtesy of D.N. Hommrich, Essential Technologies, Inc.)

advanced. The slower U.S. development is due to the U.S. regulatory agencies' longer acceptance times for new environmental monitoring technology (currently about 5 years or more).

Lidar Remote Sensing

Laser radar (lidar) has been used for more than 25 years to detect from afar a wide range of atmospheric or environmental characteristics, such as temperature, gas concentration, and wind velocity. Lidar uses a laser beam to probe a remote target, aerosol layers, or gas clouds at ranges from 10 m to several kilometers and deduces the range and composition of the cloud or target from the detection of backscattered light. Combined with spectroscopic wavelength control, tunable lidars have detected and mapped ozone, water vapor, methane, and other pollutant gases in the atmosphere or in smokestack plumes. In the effort to understand global climate change, lidars have been used to monitor gas concentrations and temperatures in the upper atmosphere and the concentration of ozone, water vapor, and methane over the Amazon jungle. If their sensitivity is high enough, range-resolved lidar returns can be used to map in three dimensions the physical extent of a plume or haze region; this was done to map the global movement of volcanic ash clouds from the eruptions of Mount St. Helens and Mount Pinatubo. Airborne lidar systems have been used to make range-resolved maps of the density of haze over the Los Angeles basin. Figure 3.2 shows a plume of ozone detected and mapped using a differential-absorption lidar; this ozone plume was found over the mid-Pacific near Tahiti and was part of a smoke plume produced by biomass (trees) burning in Africa and transported thousands of miles by global winds. Also of

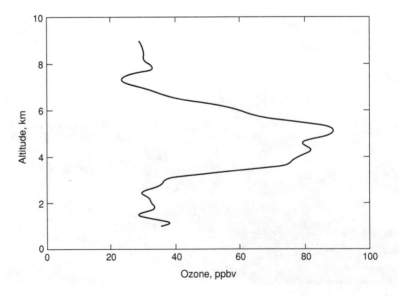

FIGURE 3.2 A plume (layer) of ozone detected by a differential-absorption lidar near the mid-Pacific (Tahiti) that had been transported by global winds from biomass (tree) burning in Africa. (Courtesy of E.V. Browell, NASA-Langley Research Center.)

importance is the recent use of a airborne range-resolved precision lidar by the National Aeronautics and Space Administration (NASA) to measure the canopy height and depth of foliage to determine the biomass coverage of Earth. This latter use will significantly increase our knowledge of the density of forests and jungle growth in remote sites, which is crucial for accurate predictions and understanding of the production of oxygen and uptake of carbon dioxide (CO_2) by plants on Earth.

A potentially significant new application for lidar will be its joint use with an open-path optical spectrometer instrument, since lidar can measure and map cloud or aerosol movements in three dimensions while the open-path instrument can determine the integrated gas concentration. Such measurements would yield gas flux values, which are most vital for environmental and gas emission regulatory detection.

Lidar instruments are still rather expensive and one of a kind; they are used more for research than commercial applications, with government funding for lidar still greater than private or commercial funding. The total U.S. market is on the order of $10 million to $20 million per year, but significant growth is expected as the required laser sources become available in more compact, less expensive forms, especially diode-pumped solid-state lasers and optical parametric oscillator (OPO) lasers. Thus, the growth potential for lidar remains dependent on developments in lasers. One important use now being evaluated is for aircraft wind shear and wake vortex detection at airports. Such a device would be an important enhancement to aircraft safety. Airlines, air cargo companies, and the U.S. Air Force are also interested in the use of on-board lidars for measurement of wind profiles below, above, and ahead of aircraft. Significant fuel savings could result from the use of such data. At present, significant work in this area is being done in Japan and Europe, as well as in the United States.

Another growing lidar market is for police laser radars to detect traffic speeds. The current annual world market for this application is on the order of $10 million to $40 million. Laser devices have the advantage over radar that the small laser beam can select a single automobile from a group of vehicles and can measure the range to the vehicle with an accuracy of better than a few centimeters. The type of traffic lidar shown in Figure 3.3 costs about $4,000, compared with about $1,500 for conventional microwave radar. Several U.S. states now each have several thousands of these lidars in use by local police agencies.

A related instrument, the laser range finder, is used by land surveyors to map distances to an accuracy better than 1 cm and for the detection of Earth's crustal movement (earthquakes) over fault lines or volcanic sites. Since a laser beam is very directional and small, it can be used for the precision determination of angular and distance measurements in both construction and land surveying, where the traditional

FIGURE 3.3 Laser radar used to measure traffic speed. The narrow laser beam can select one car from a group of vehicles, unlike conventional microwave radar. (Courtesy of the Institute of Police Technology and Management, University of North Florida.)

transit for survey work essentially has been displaced by an invisible, infrared laser transmitter and a retroreflecting mirror on a pole.

The question of eye safety is always of concern with lidar since the laser beam can be directed toward an urban population in some cases. Usually, this concern is handled by increasing the size of the laser beam transmitted into the atmosphere so that its intensity falls below the allowable eye safety value for direct ocular viewing (i.e., for a beam aimed directly into the eye). This value is about 10 mJ/cm^2 per pulse for an IR wavelength greater than about 1.4 microns. For lasers at visible wavelengths, the limit is several orders of magnitude lower, since the eye focuses visible light onto the retina, whereas infrared wavelengths are not focused but absorbed in the cornea and interior portion of the eye.

Optical Environmental Biosensors

A new type of optical biosensor, developed during the past decade, uses the combination of an optically active bioreceptor and a photodetector as an ultrasensitive sensor (Rogers and Gerlach, 1996; Vo Dinh et al., 1994). Biosensor materials change color or other optical properties in the presence of trace amounts of a known chemical or biological substance; they provide excellent specificity and sensitivity for a wide variety of environmental chemicals and biological agents. Most of these optical biosensor materials use an enzyme, DNA, and an antibody-based fluorescence label or other bioreceptor or an optically active bioagent that changes color or fluoresces in the presence of a specific substance. Most place the bioreceptor on the end of a fiber-optic probe or waveguide as the sensing end of the instrument, although some techniques use an optical microcavity with a chemically permeable membrane. These techniques are already being employed in the pharmaceutical and medical laboratory industries in the form of test kits and are used in polymerase chain reaction (PCR) applications as DNA probes. The market for such optical instruments was about $400 million in 1991 and is growing rapidly. Future markets are predicted to be about $1 billion annually for monitoring and bioremediation (making harmless) of hazardous waste dumps and $300 million annually for environmental sensing of water and air quality. Table 3.1 shows a list of some bacteria and viruses that are being detected using DNA-sensitive optical biosensors.

TABLE 3.1 Bacteria and Viruses Detected Using DNA-Sensitive Optical Biosensors

Disease	Causative Agent	Sample Source
Food poisoning	*Salmonella* bacteria	Food processing
Pneumonia	*Legionella* bacteria	Water samples
Diarrhea	*Giardia lamblia* bacteria	Water samples
Hepatitis	Hepatitis virus	Shellfish

Work is progressing to make optical biosensors more rugged and cheaper over the next 2 to 3 years, which would greatly expand their utility and commercial use for applications in medicine and public health (e.g., glucose sensors; see Chapter 2). In addition, current research is directed toward producing a complete optical biosensor on a chip, using techniques similar to those used to manufacture silicon integrated circuits.

Earth and Global Surface Monitoring

Optical sensors and television imaging systems based on high-altitude aircraft, balloons, and satellites have been used for more than three decades to detect and map weather patterns, mineral resources, ocean currents, and land topography on Earth's surface. Recently, more sophisticated optical instruments have been used that can detect the concentration of important greenhouse gases related to the study of global climate change. As such, there is both a commercial and a scientific use for high-altitude aircraft (U2), balloon, and satellite-based optical instruments.

Atmospheric and Global Climate Change

Remote sensing from satellites or high-altitude aircraft or balloons is a cost-effective way to obtain homogeneous, global measurements of critically important weather and climate variables such as atmospheric temperature and humidity profiles, cloud properties, stratospheric and tropospheric aerosol amounts, sea surface temperature, ocean color, sea ice coverage, stratospheric and tropospheric ozone, and other important trace gas concentrations. A wide variety of techniques are used, including passive microwave, infrared and visible spectroscopic imaging, solar and lunar occultation, and radar and laser ranging. Of these systems, about half are optics based, including those used for cloud properties, ocean temperature, trace gas measurements, and humidity profiles. For climate monitoring, long-term precision in the measurement of properties is required. Such measurements have been used to monitor the Antarctic ozone hole and to measure trends in cloudiness, Earth's radiation balance, and air temperature.

Since the first weather satellite was flown in 1960 there has been a continuous program of improvement along with the introduction of techniques to measure new physical variables from space. In the United States, NASA has its Mission to Planet Earth—Earth Observing System (EOS) program to collect a benchmark series of important visible and infrared global climate observations in the late 1990s and the early twenty-first century. The Upper Atmosphere Research Satellite (UARS), placed in Earth orbit earlier this decade, incorporates a wide range of optical spectroscopic instruments to measure freon and ozone-related chemicals in the upper atmosphere.

A vigorous program of innovation is under way to develop smaller, more capable, less expensive instruments. The National Oceanic and Atmospheric Administration (NOAA), the Department of Defense (DOD), and NASA are cooperating in developing the National Polar Orbiting Environmental Satellite System, a more efficient and capable system for operational weather and climate observations from satellites. The European Space Agency (ESA) and the Japanese space agency (NASDA) also have active Earth remote sensing programs for weather and climate research and are developing new tunable-laser lidar and long-path absorption technologies for this purpose.

Earth's Resources and Weather

Earth remote sensing satellites typically have optical or infrared instrumentation included in the satellite sensor package, sometimes in addition to microwave or radar sensors. The optical sensing instruments provide extensive knowledge of the global weather, agricultural resources, and land topography of Earth's surface (Office of Technology Assessment, 1990). Since the 1960s, satellite systems such as the Geostationary Orbital Environmental Satellite (GOES) system have provided near real-time photographs and digital images of clouds and weather patterns. For example, Figure 3.4 shows the image of a hurricane and its associated weather pattern off the coast of Florida in 1996. Since 1972, spectroscopic (or multispectral) wavelength bands in the visible to near-infrared region have also been used to detect agricultural parameters such as plant stress, plant density, and growth rates and to produce resource maps showing the location of minerals and sediment flow in rivers. For example, satellite-based multiwavelength optical imaging sensors have been used to map Earth's green biomass, i.e., plant density.

An estimated $28 billion has been invested in remote sensing satellites that include optical imaging systems (see Figure 3.5). Of this total, a relatively smaller amount, on the order of several hundred million

FIGURE 3.4 GOES-8 weather satellite image of Hurricane Fran in the Atlantic Ocean off the coast of Florida. (Courtesy of NASA Goddard Space Flight Center.)

dollars, represents the actual optical components and instruments used in the satellites. There are currently five operational optical imaging satellites in orbit: SPOT 3, Lansat 5, JERS-1, OFEQ 3, and IRS-2C. The majority of such systems are funded by governments including costs from satellite and downlink station design through deployment. Eight commercial remote sensing systems are scheduled to be launched by the year 2000: EarthWatch, Inc.'s Earlybirds (2) and Quickbirds (2); Space Imaging (1); and Orbital Science Corporation's SeaStar (1) and Resource 21 (2). The U.S. government plans to launch a series of Earth observing sensors starting in 1999. Other countries are planning to deploy an additional 10-15 satellites by 2000. Anticipated (and demanded) lower future launch costs are driving the systems toward lighter and cheaper packages. The trend is therefore toward an increase in the number of commercial satellites at lower costs ($50 million to $100 million each) for all applications, including agricultural and forest remote sensing. Optical systems capable of generating detailed digital elevation models (DEMs) have been identified as one of the biggest markets for this type of data and will be used for the generation of precision land contour and elevation maps for precision farming, watershed flow prediction, and land surveys. U.S. government policy regarding remote sensing data is contained in the Land Remote Sensing Act of 1992 and relates to the market for monitoring information, which is estimated to be on the order of $300 million per year for environmental uses. Approximately $2 billion per year for all applications (including

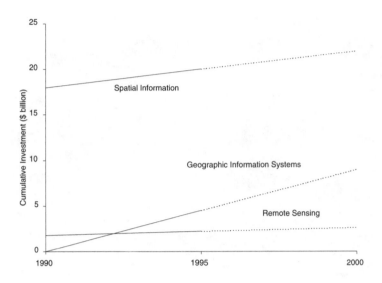

FIGURE 3.5 Cumulative investment in satellite remote sensing systems, 1990-1995, with projections to 2000. (Data for remote sensing include digital aerial photography.)

agricultural uses), has been forecast as potential value in expenditures by investment brokerage (commodities) groups and commercial businesses.

Astronomy and Planetary Probes

Optical sensors, telescopes, and instruments play a vital role in the field of ground- and space-based astronomy and studies of the solar system using spacecraft probes. During the past few decades, significant advances have been made in each of these areas. Although they have, at present, little commercial potential, this important field has significant scientific and societal benefit.

New Astronomical Telescopes

Optical telescopes have been used since the time of Galileo to study and map the heavens. Large telescopes are essentially "time machines" that peer far back in time toward the early universe. The field of astronomy is being rapidly changed by stunning advances in (1) optical engineering of larger telescopes; (2) techniques that compensate for optical distortions of Earth's atmosphere using "adaptive optics" and laser-excited artificial guide stars in the upper atmosphere; (3) coherent addition of two or more separate interferometric images from separate telescopes to increase the resultant resolution; and (4) the use of new ultrasensitive UV and IR detector arrays and computer-enhanced images. These advances are enabling researchers to actively engage the fundamental questions of astronomy and astrophysics: Are there planets around nearby stars? What is the origin of structure in the universe? What powers the galaxies? What is the nature of cosmic dark matter? Significant progress in answering some of these questions can be expected in the next few years as a result of recent developments in optical science and engineering.

Several advances have recently been made in the engineering of large telescopes. These include the development of large (up to 10 m in diameter) segmented mirrors based on new polishing techniques, ion milling for final figuring to a precise optical shape, and advanced computer-controlled alignment of the segments. Spin casting of large blanks has simplified the production of mirrors up to 8 m in size with near-final shape.

A particularly exciting development is the use of "adaptive" telescope mirrors to compensate for the distortion of stellar images produced by atmospheric turbulence. This technique makes use of a deformable mirror calibrated with the image of either a natural star or an artificial star produced by a laser. (An "artificial star" is formed by using a laser to create a tiny glowing spot high in the atmosphere.) Using adaptive optics, ground-based telescopes are now demonstrating diffraction-limited performance, albeit over relatively small fields of view. It can be expected that large ground-based telescopes will have higher resolution and light-gathering power than space-based telescopes, since both of these performance metrics depend on aperture size and ground-based telescopes can be larger than space-based ones. As an example of the performance gains associated with adaptive optics, Figure 3.6 shows an image without and with adaptive optics turned on at the 2.5-m Mount Wilson telescope. As can be seen, the increased resolution is dramatic. The adaptive optical system displays 0.07-arcsecond resolution, which is almost a hundred times better than past ground-based telescope systems but uses a telescope built approximately 80 years ago! The system on Mount Wilson uses a natural guide star. A related experiment at the 1.5-m Starfire Optical Range telescope at Kirtland Air Force Base, New Mexico, uses a yellow laser guide star

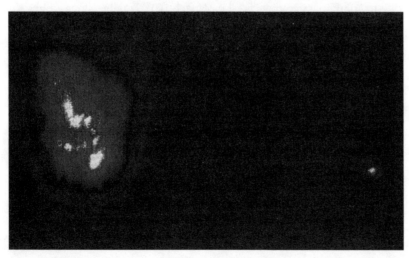

FIGURE 3.6 Telescope images with and without an atmospheric compensation adaptive optics system. The large, blurred patch at left is the star tau Cygni as seen without adaptive optics. The sharp, point-like image at right is the same star seen after correction by the atmospheric compensation system. (Courtesy of Mount Wilson Observatory.)

Chapter 3

as a reference for its adaptive optical system. Since the field of view of a state-of-the-art adaptive optical system is relatively small, it is necessary to have a guide star near the object to be studied. The laser guide star concept permits the (artificial) guide star to be placed anywhere in the sky that the telescope is aimed. A closed-loop laser guide star adaptive optics system has recently been developed at the Lick Observatory, and one is now under development for the 10-m telescope at the W.M. Keck Observatory on Mauna Kea, Hawaii. It should be added that similar adaptive optics or "rubber mirror" techniques are also being used for compensation of distortions within laser cavities (commercial lasers and the National Ignition Facility laser fusion system) and in manufacturing processes to correct for aberrations.

Recently, the Keck Observatory incorporated two individual optical telescopes to be linked as interferometric arrays so that individual images can be coherently added, thus increasing the spatial resolution and sensitivity compared to a single telescope. Such techniques have been used for years in radio astronomy (employing radio frequencies) and during the past few years for specialized optical studies, but this is the first time the technique has been used in the optical portion of the electromagnetic spectrum for general telescopic observations.

Over the past few years, the development of ultrasensitive electronic detector arrays, first in the UV and visible and more recently in the IR, has helped to automate the data collection process. These electronic arrays produce image data in electronic form, which can be processed and accessed over the Internet. This provides rapid data access to a much larger group of scientists than earlier imaging modalities. It also facilitates the remote operation of telescopes. The development of sensitive large-area charge-coupled device (CCD) detectors and associated data reduction techniques has also had a major impact on astronomy. Such arrays have led to the recent observation of gravitational lensing due to "dark massive objects" (i.e., "machos"). Finally, either ground-based or space-station-based imaging sensors will be used to detect debris in low Earth orbit.

Planetary and Space Probes

Unique optical instruments have been used for several decades for planetary, astrophysics, and Earth remote sensing from orbital and interplanetary platforms, including the detection and measurement of the temperature and atmospheric composition of planets (Venus, Mars, Jupiter), comets (Halley), and other celestial bodies. Optical spectroscopic cameras or imagers have obtained close-up spectrometer images of planets, asteroids, and satellites for discerning atmospheric gases and surface composition and the dynamics and evolution of planetary surfaces and atmospheres. Low scattered light imaging spectrometers have been used to yield information on mass exchange between binary stars, planetary system evolution, and stellar atmosphere surface science. The optical cameras and optical spectrometers on the Voyager and Galileo missions to the outer solar system, and those on-board the Viking and Pathfinder Mars lander-orbiter, provided images that have changed our fundamental view of the solar system and of mankind's role in the universe. Figure 3.7 shows an example of such an instrument used in the Galileo space mission.

The Hubble Space Telescope (HST) provides data for frequent new astronomical discoveries. A second-generation telescope, called Hubble II, is now on the drawing boards and will significantly exceed even HST's excellent performance.

A new NASA mission, the Origins Program, will search for life in the universe. One part of this program is to design and build a space optical instrument to detect and characterize planets around stars. Of recent importance was the Clementine mission, which detected water (H_2O) on the moon using a polarimeter.

The Origins Program and many planned NASA science missions depend on the development of new lightweight, compact, advanced optical instruments. Toward this end, expected improvements in image detector array complexity and resolution, and more sophisticated spectroscopic instrumentation, will greatly enhance the opportunities in space science. Among recent technological advances that will enable new space science are ambient temperature IR detector materials for the spectral range from 2 to 5 μm, agile spectroscopic optical filters, lens optimization modeling, and integrated optics spectrometers. These particular developments are examples within an extensive optics program technology structure that can be categorized into several generic components: optical testing, wavefront sensing and control, and spectroscopic sensor optical systems.

All of these instruments and new technologies must operate in the demanding environment of space: very wide temperature excursions; zero gravity, which may cause misalignment of Earth-built instruments; possible exposure to UV light, x rays, and ionizing particles; and the ability to operate unattended for several years with high reliability. These harsh environmental conditions drive requirements for special materials, devices, and designs.

FIGURE 3.7 The photopolarimeter-radiometer of the Galileo space probe, which mapped the atmospheric composition of Jupiter using optical and infrared spectroscopic instruments. (Courtesy of Raytheon Santa Barbara Remote Sensing.)

Industrial Chemical Sensors

A wide range of optical sensors are used in industry, and their application is growing rapidly in certain selected cases (Janata et al., 1994; Warner et al., 1996). One example is the use of submersible fiber-optic probes to control the flow and level of liquid chemicals and distillation processes. Other important applications include absorption spectroscopy optical fiber sensors to monitor the concentration of liquid and gas products; optical fiber-routed Raman spectra instruments used in distillation columns for control of chemical reagents and products; and specific chemically coated fiber tips that react only with a particular set of ions, enzymes, antibodies, or sugars. Many of these sensors use glass, silica, or hollow fiber-optic pipes to route the optical beam to a remote location for analysis. As a result, one of the main advantages of optical chemical sensors is that the chemical or substance can be measured in real time and in situ, with no need to extract a chemical sample and take it back to the laboratory for analysis. This is a significant advantage in monitoring a chemical reagent or substance in a hot reaction stack, in a flow chamber, or underground within a radioactive waste site.

The current annual U.S. market for optical chemical sensors is several hundred million dollars. It represents a niche market at present but is growing rapidly. Optical chemical sensors are still about 5 to 10 times more expensive than conventional chemical-related sensors, so current work is being conducted to make them cheaper, smaller, and more competitive. It is anticipated that future advances in semiconductor laser materials and manufacturing techniques will assist in this goal and produce a chemical sensor on a chip. It may be added that during the past year there have been several implementations of optical sensors on chemical production lines for real-time process control, and the results have shown considerable enhancement in the yield. It should be noted that traditional chemical analytical laboratory instruments often use optical or spectroscopic techniques to measure or detect the presence of trace compounds. The market for atomic spectroscopic instruments, for example, is shown in Figure 3.8. However, it is difficult to compare market sizes because data are not widely available.

The use of optical chemical sensors is growing in many major manufacturing sectors, including pulp and paper, semiconductors, petrochemicals, pharmaceuticals, steel, and glass. Many of these industries have their own professional or industrial societies that coordinate, promote, or fund sensor-related technology in their area. For example, the American Chemical Society (1996) has made several recommendations for the future competitiveness of the chemical industry, including the development of high-performance, real-time spectrometric instrumentation and the promotion of centers of excellence focused on chemical

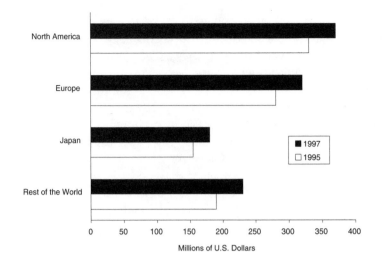

FIGURE 3.8 Recent growth in demand for atomic spectroscopic optical instrumentation. (Data from Strategic Directions International.)

measurements and process control. There is little coordination or dissemination of information between industrial sectors, however, and there is much duplication in the development of sensors by various institutes and industrial research laboratories. The Department of Energy (DOE) and the National Institute of Standards and Technology (NIST) have programs to assist in such cross-sector transfer, but they are small to moderate in size and often emphasize a particular segment of manufacturing, such as industries that require large amounts of energy (steel, aluminum) or ultraclean gases (semiconductors, pharmaceuticals). There is thus a current need to better coordinate and disseminate information and results. The private sector will probably accomplish this eventually through trade publications and industrial trade shows, but it could be accomplished faster if such coordination was fostered by a government or industrial program or institute.

Finally, new laser and optical sensors have been developed for metrology (remote measurement of position and flow). For example, laser sensors have recently been used to measure the depth and flow of molten aluminum in a large aluminum production plant in Norway. These sensors have a depth accuracy of 0.25 mm and have increased process yield considerably. In addition, a photoacoustic instrument utilizing a wavelength-tunable CO_2 laser and acoustic sensor is operating in an air conditioning-refrigeration factory to monitor refrigerant leaks down to 0.1 ounce per year. Such metrological sensors are covered in more depth and breadth in Chapter 5.

Digital, Video, and Thermal Imaging Cameras

Video or television cameras have been in use for more than 50 years, but they have become household items only within the past 15 years as a result of the development of new, compact silicon semiconductor video detector array chips (as opposed to older vacuum tube video detectors). Mass production of these video CCD array chips has brought their cost down to about $10 each. They are now used in video phones, camcorders, and surveillance systems, with an overall annual U.S. market for video cameras on the order of $1 billion. This market is fairly mature and stable at present, although advances are being made in the technology (e.g., image stabilization, electronic zoom, and lower light level detectors). The number of resolution elements (pixels) in these video cameras is usually about 500×500 at present.

A new, rapidly growing segment of imaging detectors is the development of digital cameras that use CCD detectors instead of photographic film. These cameras will be used as a replacement for traditional photographic film in digital cameras, photocopiers, and the printing industry. At present, several companies (Kodak, Casio, Chinon, Fuji) produce moderate-resolution 300,000-pixel digital cameras that allow for digital storage of pictures and their transfer to a computer or floppy disk. They cost only about $300 to $1,000 each at present, with cost reductions expected in the future. Manufacturing rates approach 10,000 per month at some larger camera companies. A new line of high-resolution CCD digital cameras is now being produced by Kodak that uses a 6,000,000-pixel CCD chip to detect the video image. These are equal in resolution to ISO 400 photographic print film and produce a picture equal that of the best single lens reflex (SLR) cameras. Such digital cameras cost on the order of $15,000 but provide software that allows for off-line image manipulation. These digital cameras are expected to be used in conjunction with laser color copiers in the printing industry, with projected camera sales on the order of $1 billion per year. This marriage of high-resolution optical CCD cameras with widespread home computer image manipulation is expected to completely change the photographic and printing industry from analog photographs toward digital computer images.

There is a large (several $100 million per year) worldwide commercial and industrial market in thermal imaging cameras and radiometer (temperature) instruments. These can be used in a wide range of applications, including the imaging of hot machinery due to friction or misalignment (see Figure 3.9), the location of overloaded circuit breakers in an electrical distribution network that may be on the verge of failure, and monitoring the uniformity of cooling of the output from a paper pulp mill. The thermal imaging cameras used in these applications can be cooled or uncooled, with cooled detectors used to image smaller temperature differences (ΔT). Many of the cameras use PbZr detector focal plane arrays and cost from about $20,000 to $50,000 for hand-held systems used in industry to $50,000 to $200,000 for airborne systems used to detect humans on the ground or their warm temperature trails. Many of the most sensitive cameras are cooled using a closed-cycle Sterling cooler so that no external cryogenics (liquid nitrogen) is required. Significant improvements in these commercial thermal imagers have been made recently, including smaller ΔT values (on the order of 0.1°C) that can be detected, higher spatial resolution that can be displayed so that a separate video camera does not have to be used, and lower production costs. The market for thermal imaging cameras is growing rapidly because of these factors and an increasing number of applications in industry.

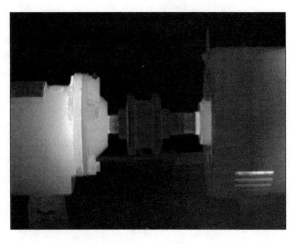

FIGURE 3.9 *Infrared thermal image of a motor and drive assembly, showing the increased heat generated in the bearing as a result of misalignment between the driveshaft and the bearing. This real-time imaging technique is used for testing motors and machinery in industrial plants while in operation and under load. (Courtesy of FLIR Systems, Inc.)*

Significant progress has been made recently in the development of infrared focal plane array detectors for low-intensity and small-ΔT applications, led by significant DOD investment in HgCdTe arrays for very low-intensity applications and in high-resolution PtSi and Si:As infrared arrays approaching 1000 × 1000 pixels. These have been developed with DOD and NASA funding for military night vision cameras and IR signature studies. The Si:As detectors are sensitive to long-wavelength (30 μm) IR, but they have to be cooled to 10 K. The Pt:Si detectors operate at cryogenic temperatures, are sensitive to 3- to 5-μm radiation, and can be used in surveillance to detect the warm temperature of a person hiding in foliage or to image relatively warm aircraft against the colder background of the sky. The cost of these last two detectors is still high, between $20,000 and $100,000 each, but this is expected to fall as increased production occurs. Also of importance is the recent development of uncooled silicon microstructure bolometric array detectors that have wide use as thermal and IR imagers. Their performance approaches that of staring cryogenic IR imagers, and they cost far less.

Law Enforcement and Security

Optical sensors play an important role in law enforcement and security. They are used in night vision scopes for border patrol and surveillance, in motion sensors for home security alarms, and to detect physical specimens at crime scenes. Examination of physical evidence is usually done with the ubiquitous microscopes and standard laboratory analytical chemical techniques. As in the case of industrial chemical sensors, a significant amount of new research is being conducted in the potential use of optical sensors in law enforcement, with a large fraction of new applications using newly developed sensor technology from DOD. However, analysis kits are needed that are portable and can be used easily at a crime scene.

Surveillance

Most common home and business security surveillance systems involve the use of video cameras placed at remote locations. Recently, new cameras have become more readily available that operate at very low light levels (starlight) or in the infrared for nighttime surveillance. For instance, Figure 3.10 shows a camera or scope used by recreational fishermen to navigate at night; these are also being used for border surveillance. Current optical monitoring of the U.S. border by the Border Patrol relies primarily on the use of thermal FLIR (forward-looking infrared) imaging systems and night vision goggles. As the new room-temperature IR video cameras (described earlier) become more readily available and affordable, they will enhance border patrol and civilian police surveillance efforts. Additionally, the next generation of CCD and innovative microchannel plate technologies will also place greater capabilities for visual surveillance in night operations in the hands of law enforcement officials. Further, as multispectral imaging cameras (hyperspectral) being developed for military applications become more affordable they will similarly enhance law enforcement surveillance capabilities. They can be used on low-cost, small, remotely controlled or autonomous airplanes to provide three-dimensional data at a moderate distance (2 to 5 km) in heavy rain, under most fog conditions, and under camouflage and smoke obscurants. This may be especially useful in providing an "eagle-eye" view of urban sites.

Several DOD agencies have recently started laser and optics application groups directed toward the application of new optical sensor technology for law enforcement and security. The programs include, for instance, the use of an invisible infrared searchlight and camera

FIGURE 3.10 Night vision cameras are used by recreational boaters to navigate for night fishing. (Image of Night Mariner 210 courtesy of ITT Night Vision.)

to locate people at night and the use of the "cats-eye" returns from people's retinas to locate potential snipers. Other potential applications being studied include high-power (1-W) visible lasers that can illuminate a criminal at night to indicate his impending apprehension and helicopter-borne 10-W IR laser "spotlights" coupled to a night vision scope and telescope, which have been shown to double search capability.

The common household motion sensor used in home security alarms employs infrared detectors that sense the movement of a warm body against the colder background of the room. These motion sensors are relatively cheap ($30) and have sales of several million units per year. They have no spatial resolution (i.e., they cannot locate the detected body) but serve well as a general warning of a security breach. They would have to be augmented with the new room-temperature IR camera to determine the actual position of the intruder within a room. Such a system could be augmented with a time-domain radar using micropulses to determine the exact location and distance of the intruder.

Drug and Explosives Detection

It has long been thought that optical and laser sensors could be used to detect the presence of drugs or explosives, since in the laboratory, laser-induced fluorescence and frequency-modulated spectroscopic detection techniques have been used for the detection of very small amounts of gases and molecular substances. However, in practice, the sensitivity of these techniques for the detection of minute or trace amounts of drugs or explosives has been shown to be much reduced in the presence of other chemical or background compounds often found in everyday situations. As such, the optical detection of drugs and explosives in their final bulk form has not proven a useful means of detection compared with analytical chemical detection techniques such as x-ray computed tomography, thermal neutron activation, mass spectrometry, neutron thermalization, ion-trap time-of-flight mass spectrometry, gas chromatography/mass spectrometry—or even trained dogs (SPIE, 1996a,b). However, it may be possible to use specific sprayed reagents that form colored compounds when they come in contact with explosive and/or drug vapors to detect the presence and concentration of each; another technique being analyzed is to place a tracer element in explosive agents that can be more easily detected and identified. Also, there is the potential use of IR radiometry techniques that would detect explosives by the difference in their heat capacity relative to their surroundings (e.g., in the case of explosive mines). This technique requires long-term observations and either thermal or microwave illumination before observation and is not yet deemed viable for general use.

Often it is important to locate a drug manufacturing site. In theory, this can be done by various techniques using the optical and laser remote sensing technology described earlier in the discussion of environmental and atmospheric monitoring. One can detect a drug factory's effluent of chemical solvents and reactants and then identify it either by differential absorption light detection and ranging or by hyperspectral infrared imaging. Further, the physical location of a facility hidden by foliage or camouflage can be detected by a compact airborne multispectral imager or by a laser detection and ranging (lidar) seeker. The sensitivity of such a device has not yet been demonstrated at the detection levels required for practical remote surveillance, although progress is being made.

Forensics and Evidence Examination

The use of optical techniques in forensics and law enforcement for the examination of physical evidence is growing but is still not an accepted practice in general, except for optical microscopy, photography or video, and more recently, DNA analysis. In general, optical sensors are not used in forensics, because they cost too much, they are not portable enough to be taken easily to a crime scene, or other more established techniques are already being used. Most physical evidence is currently examined by standard chemical analysis, with little use of optical instrumentation except optical microscopes. Confocal optical scanning microscopes with high spatial resolution are now being used in Canada but not the United States. Two exceptions are the enhancement of fingerprints by laser fluorescence and the detection of some bodily fluids (blood, semen) using filtered lamps to excite fluorescence of the specimens (Menzel, 1989). The latter technique is employed at crime scenes where portability is important, in contrast to the laser-enhanced fingerprint system, which is laboratory based because of the large laser size. This situation is expected to change as new, efficient, compact diode-pumped UV lasers and time-resolved imaging systems are developed.

The analysis of DNA evidence uses laser tagging of the DNA sequencing segments; it is just starting to be employed by law enforcement agencies. However, as always, care must be taken to ensure that the evidence is not contaminated with additional DNA specimens. This requirement shows the need to develop self-contained portable laboratories for the analysis of evidence.

It is felt that low budgets and lack of training are hindering growth in this area in the United States. Research support for new optical sensing techniques for forensics is only about $1 million to $10 million per year in the United States but is higher in Europe.

Finally, optical systems will play a greater role in the detection of counterfeit money since most anticounterfeit sensors will use optical or holographic techniques to detect hidden patterns and embedded markers in the bills. This will be a growth area as U.S. bill designs become more resistant to counterfeiting and detection devices become cheaper and more prevalent. It should be noted that an NRC report addressed counterfeit detection techniques (National Research Council, 1993).

Common Everyday Optical Sensors

A variety of optical sensors and instruments are used every day and manufactured in large quantities; in some cases, these represent significant financial markets. Although some of these devices are discussed in Chapter 6, they are listed here because they rely on optical sensors as the basis of their function. Examples include optical bar-code readers, garage and elevator door safety sensors, television remote controls, "ear" infrared thermometers, night-activated photoelectric light switches, infrared coupling of personal computers to keyboards and printers, and infrared switches that automatically turn on and off the water faucets in public restrooms. Bar-code readers use either an invisible infrared optical beam or a red, low-power helium-neon laser (supermarket scanner) to scan and detect the reflected light from the pattern of the bar code. Bar-code scanners are now used on most manufactured goods for inventory control, including rental cars, railroad cars, packaged food, parcel delivery, and most items sold at retail; Figure 3.11 shows one common type. Garage door and elevator door safety stops often use an invisible infrared beam directed across the bottom of the door, so that someone in its path can be detected and closing of the door stopped; a similar photoelectric eye is used to record and time track, ski, and race events. Television remote control devices ("clickers") work by emitting a very low power, pulse-coded infrared (LED)

FIGURE 3.11 A handheld optical bar-code scanner used for inventory control and point of sale. (Courtesy of Metrologic Instruments.)

beam that is detected and interpreted by the television set as an instruction to change channels and so forth. Infrared thermometers have recently become available for less than $30 that measure IR thermal emission from the eardrum and deduce the blackbody temperature of the patient. These devices have proven accurate and are very important for use on children. Photoelectric switches have been used for more than 50 years to turn city streetlights on and off according to how dark it is; they have also

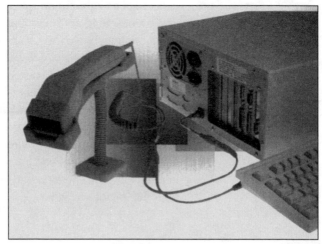

been used in homes to conserve battery life in accent lights powered by solar cells. Automated infrared proximity sensors are used in many public restrooms to turn on faucets and lavatories.

The market for these sensors is estimated at several hundred million dollars per year, and for bar-code scanners about $1 billion per year.

A variety of optical sensors are used in common devices related to communication and information processing. Fax machines use a long linear array of optical sensors to measure and image the scanned input paper for subsequent transmittal over phone lines. Optical copiers use either a traditional light source (a xenon lamp) for black and white prints or a laser scanner to image and form a color print. The laser computer printer has been widely available for the past 10 years and uses a semiconductor diode laser to write extremely small dots onto a toner cartridge and paper.

DOE, NIST, and industry, in cooperation with the technical and professional societies, should pursue a program to enhance the coordination and transfer of optical sensor technology among industry, academia, and government agencies.

Lighting

One of the most important uses of optics by society is in the form of lighting. Lighting systems represent a major economic market, with annual U.S. sales of lamps and fixtures worth about $10 billion and related electricity costs of about $40 billion each year; the latter figure represents about 19% of total U.S. electricity consumption. Because these numbers are so large, a small increase in lighting efficiency represents a large savings of energy and cost. There have been steady efficiency improvements as a result of better fluorescent lamps, better reflective and radiative coatings in incandescent bulbs, and new concepts such as light-emitting diodes. This section presents some of the new and exciting advances that are being made in lighting sources and distribution technologies. It should be noted that several U.S. government agencies, university research institutes, and industrial associations have major programs in energy and lighting efficiency improvements. For example, the Department of Energy has a consortium of several major research programs, funded at about $210 million in 1996, with about $110 million for weatherization assistance and $9 million for research in lighting. The Environmental Protection Agency has the highly successful applied Green Lights program to assist companies in reducing their energy costs, including lighting, with some companies reporting savings near 50%.

Lighting History, Future Directions, and Standards

Artificial lighting is a rather recent technological advance in terms of human history, having become a useful technology only within the past 50 to 100 years. Historically, artificial lights existed for thousands of years in the form of fire (torches), but this changed radically with the invention of the incandescent lamp in the 1840s by DeMoleyns and Starr, followed by the first successful commercial (mass-produced) lamp in 1880 by Edison. Similarly, low-voltage fluorescent lamps were invented by Meyer in 1926, followed by their commercial production in 1938 by the General Electric Company. Since that time, several other light sources have been invented and used, such as the mercury vapor, metal halide, high-pressure sodium, and halogen lamps, and most recently the sulfur-dimer (microwave discharge) and LED light sources.

One measure of the utility of these different light sources is the output light generated (lumens) per input electrical energy (watts), including any ballast electronic losses. A lumen is a radiometric (physical optics) power unit related to the total light intensity (UV, visible, and IR or heat) given off by a blackbody radiator of white-hot platinum at 2,000 K (3,100°F) and human-eye spectral sensitivity data collected under bright lighting in the 1920s. Figure 3.12 shows the efficiency of various light sources and indicates the theoretical maximum value of 220 lumens per watt. Although the figure is interesting and shows the improvements expected in several new light sources, it is only a part of the story and does not indicate several very important, photometrically related visual parameters such as (1) color index efficiency (ability to view different colors); (2) changes in human visual color response at

FIGURE 3.12 Efficiency of several lighting sources in terms of output lumens per watt of input power. Note that the color rendition of the human eye, night-time visual sensitivity, and the ability to distribute or focus light are not taken into account in this plot.

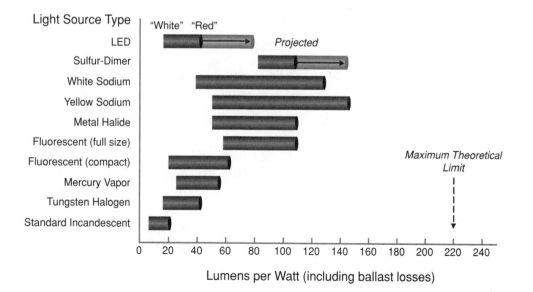

low nighttime light levels; (3) ease of distribution and ability to focus light; (4) role of the photic system (human response-biological clock); and (5) degradation of the light source. All of these factors influence the efficacy (effectiveness) of the light source, including its efficiency, the directional nature of the ability to focus light (lumens per steradian of solid angle), the efficiency of distributing light to a distant place, and the human visual response (color and contrast) at different light levels. For instance, although sodium lights have a high efficiency in terms of lumens per watt, they are poorer for viewing colors at night than metal halide sources. Also, although incandescent or halogen lamps have a lower lumen efficiency than fluorescent lamps, they are more effective in delivering light to a specific location (e.g., as in an automobile head-lamp) because they can be focused better into a beam. On the other hand, the sulfur-dimer lamp has very high efficiency, but it is more diffi-cult to deliver the light efficiently to distant locations where it can be used by people to see.

For these reasons, it is important that light sources be judged not only on their energy efficiency but also in terms of human visual response and ease of distribution. A set of standards encompassing all of these effects is being worked on by universities, industry, and govern-ment agencies, but it has yet to be universally adopted. Such standards would be extremely helpful in light source selection and would help to direct future research in this important area. It is therefore extremely important that there be coordinated activity in this area from the major research institutes and associations, including university research insti-tutes, DOE, EPA, the Electric Power Research Institute (EPRI), and the National Electrical Manufacturers Association (NEMA).

New Lighting Sources and Distribution Systems

New lighting sources and distribution systems are being actively studied, with considerable work being done on making existing light sources more efficient, developing new kinds of light sources, and developing new distribution or high-efficiency transmission systems. This section discusses some recent exciting developments.

Conventional Lighting (Incandescent, Fluorescent, and High-Intensity Discharge) and Fixtures

The majority of light sources used today are incandescent, fluores-cent, or high-intensity discharge (HID) lamps. (HID lamps include high-pressure "white" sodium, metal halide, and mercury vapor.) Although many of these kinds of light sources have been around for more than 50 years, lamp manufacturers continue to make improve-ments in terms of efficiency, lower cost, and better maintenance. Half of the lamp products on the market today did not exist 4 years ago.

These new lamps include parabolic aluminized reflector (PAR) incandescent lamps, IR coated lamps with a ceramic cylinder to reflect heat back to the tungsten filament, and metal halide lamps for residential table lamps and automobile headlights. Metal halide lamps are the fastest-growing category, since they produce a white light at night; they are also rapidly becoming the main choice for ceiling lighting in large discount stores. Fluorescent lamps were the focus of development just 5 or 10 years ago, but metal halide lights are expected to see the most activity and advancement in the next 5 years. Considerable evolutionary progress has been made in this area and will continue to be made.

Such progress has not been seen, however, in the related area of lighting fixtures (luminaires), on which the lighting industry spends little for research. Most innovations in light fixtures took place in the first half of this century; relatively little has been done since. Because lighting is a system, it is important to have not only an efficient light source but also an efficient way to distribute or place the light where it is needed. Light produced by the most efficient lamp in the world is wasted if the lamp is placed in a black box. There should be more research and development on lighting systems overall, in the context of how a person will use the system to see.

Sulfur-Dimer (Microwave Discharge) Lamps

One of the more promising new light sources at present is the sulfur-dimer (S_2) lamp that mimics the color spectrum of sunlight and uses an electrodeless inductively coupled (microwave) discharge with an overall efficiency of 50%. A 100-W sulfur-dimer lamp has the same light output as a 1,500-W incandescent lamp and currently costs about $500, although prices are expected to come down as the technology becomes more widely used. The S_2 lamp may offer significant enhancements in building lighting, using a single high-intensity lamp and light pipes to distribute the light to other rooms. Lifetime tests are now being conducted with these lamps, and the results are quite encouraging. Distribution of their light is difficult, however, since they require that light pipes be built into new buildings. Such high-efficiency distributed lighting may profoundly change the architectural design of future commercial buildings, since lighting and the associated electricity costs have a major impact on the overall cost of such buildings. Their utility for residential buildings remains to be determined.

Routing of Light by Fiber-Optic Cables or Hollow Pipes

Another new form of lighting uses plastic fiber cable bundles to transmit or route light from an efficient metal halide lamp to distant points. These systems do not transmit heat or UV and can be changed in color. However, a drawback of fiber-routed sources as a main light source is that the efficiency of transmission of visible light and the ability to focus it effectively at the distant end are not very high. Such lighting is therefore rarely used for general illumination but is rather employed for accent lighting, remote walkways, and architectural accents. For example, the 42-foot-high Coca Cola sign in New York's Times Square uses 30 miles of fiber-optic cables to distribute the light used to form the sign's image. Such signs are rapidly starting to gain a share of the worldwide $40 billion annual market for neon signs, since the plastic-routed light uses 50% of the electricity of a standard neon sign. At present the worldwide market for these types of lights is about $60 million per year, but it has been growing by a factor of two each year for the past 3 years and is expected to reach several billion dollars per year in the next decade.

Controlled Light-Reflecting Windows

Another area related to lighting is controlled-reflectivity windows for homes and businesses. Low-transmission or low-thermal-loss glass windows have been sold for years, but a related technology involves electrically controlled windows in which the visible or IR reflectivity of the glass can be changed by a controlling voltage. Materials and systems for such electronically controlled light-reflecting windows (electrochromics) are being studied worldwide by several groups. Electrochromics are being used in small-scale applications, such as rearview mirrors for cars and trucks. Coupled with a light-level sensor, they can be used for rapid control of the mirror's reflectivity at night. For use in large windows, however, large-scale, cost-effective, durable coatings have not yet been reliably produced. Their successful development could have far-reaching consequences in terms of reducing overall energy needs (lighting and heating or cooling) for residential and business use.

LED Lighting

Light-emitting diodes are a relatively new (past few decades) light source produced by semiconductor manufacturing techniques; they are used for displays in a wide range of consumer electronics. Until now, they have been used mainly as indicator lights in clock radios and TV remote controls, not being bright enough for use in lighting. Two recent breakthroughs, however, have significantly enhanced the prospects for LED use in a much wider range of applications: Blue and green were added to the spectrum of efficient LEDs through the discovery of improved semiconducting properties in gallium nitride, and better designs have led to much improved efficiency. At certain wavelengths, better than 20% efficiency is commercially available. Internal efficiencies are close to 100%, so the only question is how much of the light can be extracted before it is parasitically absorbed in the device structure. It can be projected that design improvements will allow commercial LED efficiencies to exceed 50% in due course. Because of these high efficiencies, LEDs can potentially be used as lighting sources. In addition, since LEDs have a lifetime of 30 years in intermittent operation, their use eliminates the sometimes expensive labor cost of changing light bulbs or handling the waste (burned-out light bulbs). These factors are extremely important in many applications.

These recent developments would seem to portend considerable growth into new markets. Displays, architectural lighting, and many other traditional markets for LEDs are expected to expand. In addition to LED market growth, there are benefits in energy conservation. For example, room-temperature blue and green LEDs have been commercially produced with high efficiency so they can be used with existing red LEDs to produce "white" light. Another white-light LED lighting concept uses a blue LED coated with a broad-wavelength emission phosphor; these glow with a white light at an intensity strong enough that looking at them is painful. These white-light LEDs are now equal in efficiency to incandescent lamps and should compete in efficiency with fluorescent lamps in 5 years. However, at present, their cost is prohibitive—about $300 for a white-light searchlight—but it is expected to come down as manufacturing costs are reduced. For example, several large companies (such as Hewlett Packard and Phillips) have formed a joint venture for the commercial development of white-light LEDs as competitors to light bulbs.

As an example of the large market potential, several U.S. companies have started to sell replacement LED red traffic lights for as little as $200 each (see Figure 3.13). Current sales are about 60,000 units per year, with a much larger potential market estimated, given that there are about 250,000 traffic intersections in the United States with an average of 10 red traffic lights at each. The electrical cost of a red light bulb traffic light is about $6 per month, versus $0.85 for the LED alternative. At present, the total annual electrical cost for conventional red traffic lights is about $200 million, which would be reduced to about $25 million by using LEDs, for a savings of $175 million each year in the United States alone. The city of Philadelphia has started to replace all red traffic lights over the past 4 years and expects to save several million dollars per year in electrical costs as well as significantly reduced replacement costs. In addition, China and many other developing countries are starting to use LEDs in traffic lights and related signs because their use will reduce the need to build new electrical generating plants in the future.

FIGURE 3.13 New LED traffic bulbs are being used to replace conventional red traffic lights in many cities because of their much lower operating costs and longer (30-year) lifetimes. (Courtesy of Electro Tech, Inc.)

There is now ample reason to believe that manufacturing improvements will allow the cost of producing LEDs to come down, much as it has for other semiconductor products. New types of reactors are now available, as mentioned later in the discussion of solar cells, that will make metallo-organic chemical vapor deposition (MOCVD) epitaxial growth a much cheaper process. Moore's law, the exponential decline in semiconductor product costs, can be expected to apply to high-efficiency LEDs as well. In addition, similar advances are expected in the GaN family of semiconductors and in the development of longer-lived organic LEDs. Thus, although the current LED market worldwide is about $2 billion per year, it can be expected to grow considerably, perhaps by an order of magnitude within the next decade.

Lasers and Lighting in Live Entertainment

The entertainment field uses optics in a variety of ways, from movie projectors, to television and video game displays, to optical illusions and laser light shows. This section emphasizes the use of optics and lasers in live performance entertainment and laser light shows, since most of the other areas are covered elsewhere in the report (e.g., the discussion of displays in Chapter 1).

Live entertainment such as rock concerts and theme park stage shows uses a considerable amount of lighting and laser light shows. The use of specialized lighting and laser light shows represents an enhancement of the primary entertainment of these performances. The laser or light show itself is usually considered secondary to the main performance, however, although in some cases, such as in Las Vegas and at Disney World, the laser light show is a major nightly production. The overall market for the use of lasers in this field is about $20 million to $50 million per year, with an expenditure of $500,000 to $2 million per tour not uncommon for a major rock band. The use of optics and lasers in live entertainment is stable, with significant growth not expected at present. It should be added that eye safety is always a concern for laser light shows and is often handled by making sure that the laser beams are directed above and away from people in the audience. In an outdoor laser show where the beam is shot upward into the sky, some production companies are now using a radar system to detect aircraft flying close to the laser beam so that the laser can be blocked as appropriate.

DOE, EPA, EPRI, and NEMA should coordinate their efforts and create a single program to enhance the efficiency and efficacy of new lighting sources and delivery systems, with the goal of reducing U.S. consumption of electricity for lighting by a factor of two over the next decade, thus saving about $10 billion to $20 billion per year in energy costs.

Optical Sensors and Lighting in Transportation

Optical sensors, lighting, and optical instruments are used to a large extent in automobiles and to a smaller degree in aircraft. This section covers the use of optics and the potential advances in their use in the area of transportation (aircraft and automobiles) because these represent some of the largest industrial sectors in our economy.

Aircraft Applications

Considerable research has been conducted on the use of optics in aircraft, ranging from the fly-by-light concept, in which aircraft surfaces are controlled by fiber-optic control systems, to optical sensors for precise measurements of airspeed and the pressure in front of the aircraft. In general, however, conventional mechanical and electrical techniques still have cost and reliability advantages, especially because of the wide range of operating temperatures in aircraft (–50 to +120°C). Some uses are advantageous for optics, including the use of optical fiber to distribute audio and video entertainment in passenger aircraft and the use of laser gyroscopes for stand-alone navigation. Laser gyroscopes use a small laser beam inside a stable ring cavity to measure the acceleration and position of the aircraft with an accuracy of several feet, thus permitting the aircraft to navigate over continental distances. The current market for laser gyroscope systems is about $400 million per year, of which $100 million is the cost of the ring lasers. Figure 3.14 shows a picture of a ring laser gyroscope (RLG) like those used in many commercial aircraft today. It is anticipated that satellite-based Global Positioning System (GPS) instruments may eventually supplant laser gyroscopes, but not for the next few years.

Other potential uses for optical instruments include Doppler laser radar to detect wind shear in front of landing aircraft or wake vortices at airports. At present, most commercial aircraft are planning to use Doppler microwave radar for wind shear detection, based on recent flight tests by NASA and the Federal Aviation Administration (FAA), but these agencies are also testing new solid-state Doppler laser radars for airport wake vortex detection, which offer advantages in dry air turbulence conditions.

FIGURE 3.14 A ring laser gyroscope used in commercial aircraft for accurate position location, navigation, and autopilot flight control. (Courtesy of Honeywell.)

Finally, for high-speed or stealth aircraft where mechanical pitot tube air speed sensors are inaccurate or create an undesirable radar return, optical sensors using Doppler laser radar are being developed to measure air speed and air density at distances 5 to 10 feet in front of the aircraft, outside the aircraft's shock front. Such a sensor is also being studied by NASA for use in the high-altitude Super Sonic Transport, which will use ramjets for propulsion. Ramjets require very precise information on the density and velocity of the incoming air to avoid flameouts and optimize combustion at high altitudes.

Automobile Applications

Optics has been used in automobiles for the past hundred years, ever since someone needed to drive at night, and even before that in horse-drawn carriages. Today optics is used extensively in automobiles, and its use is growing as cheaper, more compact light sources and optical devices become available. Traditional lighting applications still dominate, such as headlamps, taillamps, and instrument displays. A number of new, non-lighting applications are also being found, however, such as optical fiber strain gauges embedded in roadways and bridges (Box 3.2) and optical sensors for the control functions required for precise computer control of modern engines (e.g., sparkplug timing and sensing of engine rotation and position). The cost of these optical items is important, because traditionally the individual unit component cost in automobiles is on the order of a few dollars to a few tens of dollars. The total cost of optics in each vehicle sold today is estimated to be about $100 to $200. Annual U.S. production is about 15 million vehicles, so the total U.S. market for vehicular optics is on the order of $1.5 billion to $3 billion per year.

New applications of optical materials and systems for sensors and lighting are being pursued vigorously by the auto companies. For example, high-intensity (blue-white) arc discharge lights will be used as headlamps on many vehicles since they last longer and are more efficient than the conventional incandescent sources they will replace. The efficiency gain is particularly important for electric vehicles. HIDs are already available in some luxury car lines. LEDs are starting to replace conventional lamps in rear and brake lights because of their longer lifetime, smaller package size, and greater optical efficiency (15% for red light compared with 1% for a filtered incandescent bulb). Simple reflective heads-up displays, which display information on the windshield, are already available in some vehicles and more sophisticated versions are being planned. The next generation of instrument-panel displays is expected to use full-color, fully reconfigurable, flat-panel devices with sufficient brightness for daytime use in the passenger compartment. Vacuum fluorescent technology is the leading contender for this application, but there are many competitors.

BOX 3.2 OPTICAL FIBER STRAIN GAUGES IN HIGHWAYS

An emerging application of optical sensors in transportation is the use of embedded optical fiber strain gauges in bridges and roadways. Optical strain gauges are more durable than conventional foil and piezoelectric gauges and are immune to the electromagnetic interference that plagues electronic gauges. Because optical fibers are small, they can be embedded in concrete or asphalt without disrupting the structural properties of the surrounding material.

The Federal Highway Administration and other federal and state agencies are exploring various uses of these sensors. Bridge beams can be monitored for stresses during their fabrication and transport to the construction site. Bridge design ratings can be tested. Arrays of sensors placed below a pavement surface can improve the prediction of design lifetimes by collecting data on the loads imposed by passing vehicles. Sensors that monitor the deterioration over time of structures and pavements can help to prioritize maintenance tasks and increase safety by giving early warning of impending structural failure. Sensors embedded in highways can count and weigh passing vehicles while they are in motion to help in traffic planning and to enforce laws that restrict truck weights.

A system based on Bragg gratings has been developed that permits as many as 64 strain gauges to be put on a single optical fiber. This multiplicity greatly simplifies installation and reduces cabling requirements and the cost per sensor, and because the Bragg grating system senses wavelength shifts, it is more reliable and repeatable than approaches based on amplitude or shifts in interference fringes. Bragg grating sensors were first used in the laboratory to study the performance of beams and bridge decks. This research culminated in the installation of 48 sensors in a full-scale section of a reinforced-concrete bridge deck. The sensors monitored loads during the casting of the concrete, strains due to the subsequent curing and shrinkage, and strains under static and dynamic loading. Bragg grating sensors are now being applied in real field situations. An array of 32 sensors has been installed to monitor an interstate highway bridge in New Mexico that has developed significant fatigue cracks, and there are plans to install 128 sensors on the heavily traveled Woodrow Wilson Bridge outside Washington, D.C.

Bragg grating strain sensors illustrate the value of materials science in optical engineering. Because germanium-doped glass is sensitive to certain wavelengths of light, it is possible to produce the Bragg gratings in the fiber by photoengraving. Furthermore, because the Bragg gratings can be manufactured in a continuous-draw tower process, sensors can in principle be put on an arbitrarily long fiber. As a result, the unit cost of Bragg grating sensors is competitive with that of conventional strain gauges.

It is now realistic to consider installations of 1,000 or more optical fiber sensors. That possibility will create opportunities for advanced mathematical analysis of structural response and challenges for data management and visualization. It may lead to a radical rethinking of how and where strain gauges are placed in civil engineering structures.

Optoelectronic sensors using laser, LED, or microwave radar have the potential to be a major growth area for use in collision avoidance, laser ranging, and blind-spot warning of nearby vehicles or obstacles. New optics-based chemical sensors are being used to monitor catalyst efficiency by measuring the concentrations of such exhaust gas species as unburned hydrocarbons, carbon monoxide, and oxides of nitrogen using miniature spectrometric real-time sensors. An important concern for these optoelectronic systems is that they must operate over a wide temperature range, from -40 to $+85°C$; they must be vibration tolerant; and they must last the life of the vehicle (typically 10 years or 10,000 hours of operation).

Energy

Advances in the efficient generation of electricity can have a significant impact on the energy consumption of our society. It should be noted that the energy (including electricity, gas, oil, and coal) used in the United States each year amounts to about $600 billion, of which $200 billion is used in business and residential buildings.

This section covers several optics technologies that may have significant impact in this area, including the use of lasers to produce inertial confinement fusion as a future source of energy and for new basic science; the use of lasers to enrich uranium for reactor power plants; and recent developments in solar cell technology.

Inertial Confinement Fusion Using Lasers

Laser-induced inertial confinement fusion (ICF) is an approach to producing controlled nuclear fusion on earth (Lindl et al., 1992). As an integral part of the pursuit of a zero-yield nuclear test ban treaty, the United States has committed to the design, development, and construction of the glass-laser-based National Ignition Facility (NIF). The NIF is intended to use inertial confinement fusion for weapons studies, but it will also provide insight for future energy applications.

Figure 3.15 shows a drawing of the NIF facility, which is based on flashlamp-pumped neodymium-doped glass technology and frequency conversion. It is scheduled to be completed in 2002 and to produce 1.8 MJ at 350 nm, with a total project cost of approximately $1.1 billion. From 1995 to 1998, the United States will also invest about $170 million in this project for the development of laser technology, large-scale precision optical components, and low-cost advanced optics manufacturing methods. NIF will be the largest optical system in the world and will develop and employ state-of-the-art adaptive optics systems on each of its 192 laser beam lines. To further control the beams'

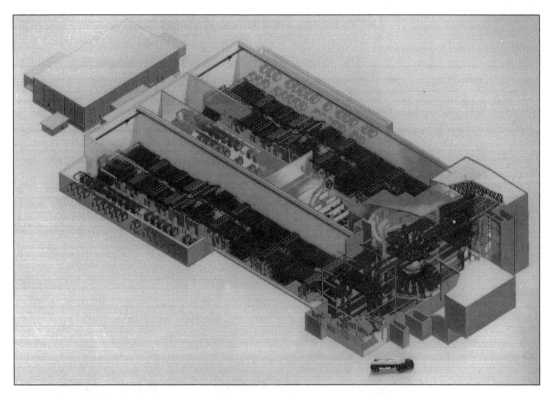

temporal and spatial coherence, NIF will use diffractive optics and phase modulation technology to produce optical bandwidths up to 0.5 THz.

Recent scientific and technological advances in efficient, powerful semiconductor laser diode arrays (as laser pump sources), in specialized crystalline laser gain crystals, and in sophisticated gas flow cooling techniques now permit the conceptualization of an efficient, multimegawatt, all solid-state laser suitable for driving a central electric power plant based on inertial fusion energy (IFE). The development and demonstration of a highly modularized 1-kJ unit beam line for such a laser system over 5 to 10 years represents a grand challenge that would drive the envelope of diode-pumped solid-state laser technology and lay the foundation for the timely pursuit of IFE after laboratory ignition is achieved at NIF early in the next century. A major advance required for IFE is to reduce the cost of laser diode arrays to less than 10 cents per watt.

Overall, ICF now amounts to about a $400 million per year project worldwide ($240 million in the United States in 1996). All funding for ICF currently comes from governments.

Although ICF has not yet been demonstrated, NIF is the next critical step. The spin-off value in its optics and laser advances is great, and the potential payoff as a new energy source is enormous, although still uncertain and many years in the future. The consequences are also wide-ranging for astrophysics and other sciences.

FIGURE 3.15 Drawing of the National Ignition Facility (NIF), to be used for laser-controlled fusion. The system will have 192 laser beams and will be the world's largest laser and optical system. (Courtesy of Lawrence Livermore National Laboratory.)

Laser Isotope Separation of Uranium for Nuclear Energy

Atomic vapor laser isotope separation (AVLIS) is an economical, environmentally improved method to enrich natural uranium for light-water reactor fuel. AVLIS is based on technology demonstrated over the past 20 years and uses a precisely tuned laser to selectively excite and photoionize uranium-235 (^{235}U). The selectively ionized ^{235}U is then collected to generate a product enriched in this isotope. It should be noted that the United States controls about 40% of the world's uranium enrichment market (currently several billion dollars annually) using technology developed 50 years ago. AVLIS will be able to produce enriched uranium at a much lower cost and will enable the United States to capture a significantly larger fraction of the world market.

The U.S. Enrichment Corporation, a government corporation formed in 1992 with plans to privatize, is refining the technology and designing a large AVLIS system for this purpose. Key plant systems consist of separators to vaporize the uranium and collect the selectively photoionized ^{235}U. There will be several identical separator lines. Dye lasers generating 50 kW of process light are optically energized by 160-kW copper vapor lasers (or possibly solid-state lasers) distributed via a fiber-optic network.

The overall goal is to construct an AVLIS enrichment facility and bring it to full production early in the next century. The AVLIS project may become the largest technology transfer effort from DOE to the commercial sector.

Space Solar Cells

Solar cells, which convert light to electricity, have been used as the primary power in communication, defense, and weather satellites for the past 35 years and can be considered part of the satellite and spacecraft manufacturing industry. The worldwide space solar cell business is approximately $150 million per year, with about two-thirds of the total dominated by two American companies. The United States thus has a strong international position in this field, with a total annual market of about $100 million.

In the past 5 years, however, there has been a technological and materials revolution in this field. For 30 years before 1994, the technology was stable, using standard crystalline silicon solar cell technology. Then in 1994, gallium arsenide (GaAs) cells grown on germanium substrates became mature for space use and the market switched largely to GaAs cells. The inherent leverage in reduced area and weight associated with the higher-efficiency GaAs cells influenced the launch booster design and the overall system. Therefore, the added cost of the new technology was more than warranted, to the degree that it is now difficult to sell any but GaAs-based solar cells for space use. In fact, by early 1996 the space market had moved to begin production of dual-junction and triple-junction InGaP-GaAs cells.

Because of the great demand for new space-based communications technologies, this market is booming. Telephone systems requiring 800-satellite constellations have been proposed, and several communications systems using up to 66 satellites have already been started. A major concern for the industry is the capital investment required to meet a demand surge that may not last long enough to amortize the equipment.

This industry is playing a leadership role in optoelectronics and optical materials, being the first to apply many technologies that will become more widespread later. For example, there is an opportunity to apply new high-speed MOCVD equipment, which is likely to allow low-cost, high-speed epitaxial growth. This equipment will eventually become standard in many other areas of the semiconductor manufacturing industry, eventually making high-quality epitaxial films much cheaper and creating new opportunities for their use in many fields.

Terrestrial Solar Cells

The terrestrial uses of solar cells can be classified into two general categories: on the power grid and off the power grid. Off-grid applications include supplying power for hand calculators, remote instruments, stand-alone communication gear, remote mechanical pumps, and refrigerators. Other important off-grid uses are for industrial and residential general-purpose power, particularly in remote locations. On-grid solar cell systems make up only a tiny fraction of the available electric power capacity of the United States. The fraction is expected to become much larger as the cost of solar cells declines, the conversion efficiency of sunlight to electrical energy increases, nonrenewable fuel becomes scarcer and more expensive, and greenhouse gases begin to severely impact the environment. To quote the senior managing director of the Royal Dutch/Shell Group concerning renewable energy usage in general (Herkströter, 1997):

> . . . our various scenarios suggest that renewables could provide some 5% of the world's energy by 2020. They also suggest this could rise to over 50% by mid-century—a shift as fundamental as that from coal to oil in this century.

The United States currently spends approximately $600 billion per year on energy—about 8% of the $7 trillion annual gross domestic product (GDP). By the middle of the next century, renewable energy is expected to meet a significant fraction, perhaps 50%, of U.S. energy needs. Solar energy, particularly photovoltaic solar cells, could play a central role in renewable energy, especially for electric power generation. It is reasonable to expect that the U.S. solar cell industry could reach $100 billion per year by the middle of the next century.

Although a number of competing sources of renewable energy exist, there are obvious advantages to photovoltaics if costs can be reduced. Solar cells can produce electrical energy at a rate of about 200 kWh/m^2 per year (Zweibel, 1990). The current electric energy usage of the United States is 3 billion kWh per year, requiring solar cells covering a land area 120 km × 120 km. With a cost of $150 per square meter of installed solar energy panels and associated infrastructure, including power conditioning and storage, the capital investment in solar energy required to meet current needs is $4.5 trillion. By spreading conversion over 30 years, which is also the approximate lifetime of a panel, the cost would be $150 billion per year. The United States currently spends approximately $200 billion a year on electrical energy; this is about 3% of the $7 trillion GDP per year. For comparison, shipments of photovoltaic modules in 1996 were at the worldwide level of $500 million per year (Strategies Unlimited, 1996). The industry grew 10% in 1996 and was expected to grow 20% in 1997.

There have been steady, incremental reductions in the cost of photovoltaic power over the past 15 years. In quantity, terrestrial solar panels can now be purchased for $4.50 per watt (this figure is for conventional crystalline silicon-based panels.). This represents a decline by a factor of 33 from the 1970 cost of $150 per watt. There have been suggestions that certain manufacturers have an internal cost of production of $2.75 per watt, which is a credit to their streamlining of the manufacturing process. The $4.50 per watt figure translates into a price of about $500 per square meter for solar panels. To provide solar electric power at a cost comparable to the present cost of electricity, silicon solar panels must drop in price by a factor of 5 to 10, if the remaining systems cost is equal to the panel cost (Zweibel, 1990). This is a reasonable expectation in the next 10 to 15 years, given the steep decline noted above and the even more remarkable decline in the cost of silicon electronic devices.

The power output of photovoltaic modules shipped in 1996 was 82.5 MW, divided according to technology as indicated in Table 3.2.

About 4 MW of the thin-film figure went to consumer products such as calculators. There have been a number of attempts over the years to replace single-crystal silicon with lower-cost forms such as polycrystalline or amorphous silicon, but the cost advantages are somewhat offset by the reduced efficiency. Through the 1980s, the single-crystal market share dropped from 70 to 40%; it has since recovered to the level shown in Table 3.2.

Silicon photovoltaics are still in a fairly early stage of development as a practical power technology. A number of complex issues surround the various technologies (Partian, 1995; Zweibel, 1990). These include the cost of materials, growth and fabrication of wafers or films,

TABLE 3.2 Market Share of Photovoltaic Technologies for Terrestrial Use, 1996 Shipments

	Power (MW)	Percentage
Single crystal	45	54.5
Polycrystalline including ribbon	26	32.5
Thin film	11.5	14

Source: Strategies Unlimited.

manufacturing technology, and process control. The industry is quite contentious and competitive, which are both healthy signs. The use of silicon by the solar cell industry is approaching 10% of its use by the electronics industry. There have recently been shortages of silicon for solar cells. The sale of electronic-grade silicon is much more profitable for silicon manufacturers than sale of the lower-grade silicon needed for solar cells. There is concern that the portions of boules discarded by chip makers may no longer provide an adequate supply for solar cell manufacturers.

The cost of single-crystal compound semiconductor films, such as GaAs epilayers, is expected to come down in the future. Advanced MOCVD reactors for growing high-quality epitaxial material are now available with growth platters 40 cm in diameter and cycle times of 2 minutes. Several techniques are available for reusing the compound semiconductor growth substrates from these reactors. There are great opportunities for research on solar cells that combine the best of both worlds, high efficiency *and* low cost.

Research is being carried out on other solar cell materials such as $CuInSe_2$, CdTe, GaAs, and GaSb. Depending on the material, questions remain concerning issues such as stability and the feasibility of low-cost, large-scale manufacturing. The development of new materials has been and will continue to be a major driving force in solar cell advancements.

Solar Thermal Energy

Of the other forms of solar energy (e.g., wind, hydropower, and so forth) only solar thermal energy uses optics. Solar thermal energy is widely used to provide hot water for domestic and commercial use, process heat for industry and agriculture, and space heating and cooling. With sun-tracking parabolic concentrator mirrors, sufficiently high temperatures can be reached to drive a turbine generator. Using this technology, more than 350 MW generating capacity was installed in the Mojave Desert in the 1980s.

World leadership in optical science and engineering is essential for the United States to maintain its dominance in energy-related technologies such as laser-enhanced fusion, laser uranium enrichment, and solar cells. The Department of Energy should continue its programs in this area.

Summary and Recommendations

Optical Sensors and Imaging Systems

Optical gas sensors are beginning to make a major impact in the field of air quality and pollution emission monitoring and offer real-time quantititative advantages (remote, in situ) over standard chemical analytical techniques. Open-path air monitoring is often used to measure environmental emission levels for compliance with environmental regulations. There are several lidar research programs for global mapping of greenhouse gases and environmental emissions. Lidar is starting to be practical as a commercial instrument, but commercial uses at present are mostly in traffic laser radars, wind shear sensors, and precision range finders and mappers.

Industrial optical chemical sensors are just starting to be used for process control and have shown significant potential in several cases. Optical methods are used in only a minority of chemical sensors, but the fraction is growing as sensors and lasers become smaller and cheaper. Optical biosensors are important trace detectors in the pharmaceutical and medical laboratory industries and are the basis for a wide range of sensitive medical diagnostic tests and DNA sequencing instruments. New photooptical materials, sensitive to specific trace chemicals or biological species, will enable the development of new families of optical sensors.

Satellite-based optical spectroscopic instruments have been used to detect the ozone hole and gases involved in global climate change. Optical and infrared camera sensors in Landsat and weather satellites are used to provide important agricultural and weather data on a daily basis. Future weather and Earth-viewing satellites will be cheaper, smaller, more numerous, and more often commercially financed. Ground-based telescopes using atmospheric compensation and optical interferometric techniques will revolutionize ground-based astronomy at visible and IR wavelengths. Planetary and space probes use optical and microwave sounders to detect and image heretofore unknown chemical species on different planets and have discovered water on the moon.

New high-resolution (high pixel count) optical imaging arrays and CCD video detectors are increasingly used in commercial digital cameras. The most advanced digital cameras have a resolution comparable to photographic films and are expected to revolutionize and computerize the photographic film and printing industries. New infrared detector arrays are providing improved high-resolution detection of thermal images and will be increasingly used in industry for real-time monitoring of manufacturing lines.

Sophisticated optical and laser sensors are not used to a large degree in forensics and law enforcement because of their cost and lack of portability, but video surveillance and infrared motion security sensors are used extensively. Infrared LED and laser spotlights coupled to IR video cameras are being developed for night vision surveillance. Law enforcement will greatly benefit from advances in materials and systems for IR lasers and room-temperature IR cameras, while decreased size and cost of lasers will increase the use of optical and laser spectroscopic sensors.

Optical sensors are used in many everyday devices, including bar-code readers, proximity switches for water faucets, safety shutoff beams for elevator and garage doors, and new ear-type thermometers for children. Laser printers and fax machines use optical sensors and imaging systems.

Optical sensors are employed in a wide range of important industries and fields and will be an important factor in future industrial growth and competitiveness. However, at present, there is little coordination or information exchange of optical sensor technology between industrial sectors.

DOE, NIST, and industry, in cooperation with the technical and professional societies, should pursue a program to enhance the coordination and transfer of optical sensor technology among industry, academia, and government agencies.

Lighting

New light sources and delivery systems will offer a large improvement in lighting efficiency. "White-light LEDs" made by coating a blue LED with a phosphor are now as efficient as incandescent lamps and are expected to become competitive with fluorescent lamps within 5 years. High-efficiency, room-temperature red LEDs are being used in red traffic lights and are expected to save $175 million per year in electrical costs. New light delivery systems, such as plastic fiber cable light bundles and light pipes, are being used in advertisements and other neon-sign-type applications. New materials for light sources, light delivery, and controlled reflectivity have led to increases in lighting efficiency, lifetime, and utility. There is a need for light output measurement standards that account for the response of the human eye and the delivery efficiency of the lighting system.

Laser light shows and optical staging are used in many live performances, although they are often considered secondary in importance to the performance itself. The annual market for lasers and optics for laser light shows and live performances is steady.

New high-efficiency lighting sources and light distribution systems will have a significant impact on the country's electricity use. There is only moderate coordination of research and standards setting on new, efficient, and effective lighting sources among government, academia, and industry.

DOE, EPA, EPRI, and NEMA should coordinate their efforts and create a single program to enhance the efficiency and efficacy of new lighting sources and delivery systems, with the goal of reducing U.S. consumption of electricity for lighting by a factor of two over the next decade, thus saving about $10 billion to $20 billion per year in energy costs.

Optical Sensors and Lighting in Transportation

With the exception of fiber-optic gyroscopes, optical sensors are not used extensively in aircraft, where they do not yet offer advantages over conventional mechanical or electrical sensors.

Optical sensors and lighting are used to a great extent in automobiles, where they play an integral and important role. New high-efficiency headlamps, LED taillights, and optical collision avoidance systems are being introduced into automobile lines.

Energy

The world's largest laser and optical systems are being developed for nuclear energy-related programs. The National Ignition Facility will be the largest sophisticated optical system in the world and a major new research tool for the United States. The AVLIS program will provide economical separation of uranium reactor fuel.

Solar cell efficiency has increased and cost may have decreased to $2.75 per watt. A decrease of a factor of 5 to 10 in solar cell cost would make the price of solar photovoltaic electrical energy comparable to that for nonrenewable sources. If the cost of solar photovoltaic cells continues to decline, solar cells could begin to impact the electric power industry by 2020 and could provide as much as half the world's electric power by 2050. As a source of renewable energy with low environmental impact, efficient, low-cost solar energy could have a great impact on world energy consumption. If successful, such programs could have a significant effect on future energy programs and the cost of energy in the United States.

World leadership in optical science and engineering is essential for the United States to maintain its dominance in energy-related technologies such as laser-enhanced fusion, laser uranium enrichment, and solar cells. The Department of Energy should continue its programs in this area.

References

American Chemical Society. 1996. *Technology Vision 2020: The U.S. Chemical Industry.* Washington, D.C.: American Chemical Society.

Bruton, T.M., et al. 1997. Multi-megawatt upscaling of silicon and thin film solar cell and module manufacturing. Paper presented at the European Community Photovoltaic Conference, Barcelona, June. (Author's address: BP Solar, P.O. Box 191, Sudbury-on-Thames, Middlesex TW16 7XA, United Kingdom.)

Herkströter, C.J.A. 1997. Speech at Erasmus University, Rotterdam, March 17. Available from Shell Oil Company, 712 Fifth Avenue, New York, NY 10019.

Janata, J., M. Josowicz, and D.M. DeVaney. 1994. Chemical sensors. *Anal. Chem.* 66:207.

Lindl, J., R. McCrory, and E.M. Campbell. 1992. Progress toward ignition and beam propagation in inertial confinement fusion. *Phys. Today* (September):32.

Menzel, R. 1989. Detection of latent fingerprints by laser excited luminescence. *Anal. Chem.* 61:557a.

National Research Council. 1993. *Counterfeit Detection Features for the Next-Generation Currency Design.* Washington, D.C.: National Academy Press.

Office of Technology Assessment. 1990. *Technology for a Sustainable Future.* Washington, D.C.: U.S. Government Printing Office.

Partian, L.D., ed. 1995. *Solar Cells and Their Applications.* New York: Wiley.

Rea, M., ed. 1993. *The Illumination Engineering Society Lighting Handbook.* New York: Illumination Engineering Society of North America.

Rogers, K.R., and C.L. Gerlach. 1996. Environmental biosensors: A status report. *Environ. Sci. Technol.* 30:486A.

SPIE. 1996a. *Physics Based Technologies for the Detection of Contraband.* Conference No. 2936. Boston: SPIE Conferences.

SPIE. 1996b. *Chemistry and Biology Based Technologies for the Detection of Contraband.* Conference No. 1937. Boston: SPIE Conferences.

Strategies Unlimited. 1996. *Five-Year Photovoltaic Market Forecast: 1995-2000.* Report PM-43. Strategies Unlimited, 201 San Antonio Circle, Suite 205, Mountain View, CA 94040.

Vo Dinh, T., K. Houck, and D.L. Stokes. 1994. Surface enhanced Raman gene probes. *Anal. Chem.* 33:3379.

Warner, I.M., S.A. Soper, and L.B. McGown. 1996. Molecular fluorescence, phosphorescence, and chemiluminescent spectrometry. *Anal. Chem.* 68:73.

Zweibel, K. 1990. *Harnessing Solar Power: The Photovoltaics Challenge.* New York: Plenum.

4
···

Optics in National Defense

We are not the only nation with competence in defense science and technology. To sustain the lead which brought us victory during Desert Storm . . . recognizing that over time other nations will develop comparable capabilities, we must . . . invest in the next generation of defense technologies.

William J. Perry (1996)

Combat is repugnant to Western civilization, but it is a reality that we must face squarely because alternatives to victory are not an option. The stark realism of Desert Storm emerged from the television coverage with a technological perspective never before seen. Did anybody miss the striking footage of a precision laser-guided bomb zeroing in and then obliterating a military headquarters building in Baghdad? Could anyone have possibly not seen an entire field of Iraqi tanks, personnel carriers, and artillery units being devastated by the allied forces in just a few minutes, with hardly a response from the enemy? Also, there were very few allied casualties, unlike previous conflicts such as Vietnam and Korea. (See Box 4.1.) From this it is clear that a higher technological level has been achieved in modern warfare, serving as a powerful deterrent to potential aggressors. Technology is the centerpiece in modern warfare, enabling us to deploy and obliterate enemy forces while sustaining minimal casualties.

Throughout history, new technology has had a profound effect on how wars are conducted. Usually, the victors were those best able to apply the new technology. Over the course of the past 50 years, nuclear weapons, microwave radar, guided missiles, and other developments have led to major realignments of defense strategy. Today, the traditional modern strategy of massing large numbers of military personnel and materiel to engage enemy forces is giving way to high-tech methods of conducting warfare that minimize casualties. The U.S. military mission now requires a versatile fighting force capable of both conventional field and urban warfare in a global venue. To improve the effectiveness

BOX 4.1 OPERATION DESERT STORM

"Operation Desert Storm was primarily a sustained 43-day air campaign by the United States and its allies against Iraq between January 17, 1991, and February 28, 1991. It was the first large employment of U.S. air power since the Vietnam war, and by some measures (particularly the low number of U.S. casualties and the short duration of the campaign), it was perhaps the most successful war fought by the United States in the 20th century. The main ground campaign occupied only the final hours of the war."

U.S. General Accounting Office (1966)

of the combatant while reducing casualty rates, the military has a number of efforts under way that include reliance on speed and stealth to overcome opposing forces; a better equipped land warrior; rapid detection and control of nuclear, chemical, and biological threats; and dissemination of real-time intelligence on enemy targets. Optics plays a key enabling role in these plans. For the future, optical systems are sure to be the basis for entirely new classes of defense applications that will change yet again the way wars are conducted.

Since World War II, the U.S. technological approach has featured defense sponsorship of leading-edge research and development at levels necessary to maintain strong defense leadership in the world. Defense support of research has included activities that range from basic research at universities to system-level developments in industry. When the Cold War was at its peak and the Soviet Union presented a severe threat, the Department of Defense (DOD) actively pursued R&D in virtually every technical discipline. The overarching goal was the anticipation of potential breakthroughs that could upset the balance of power. To be sure, this strategy was expensive, but it has been effective and has served society well. U.S. defense capability today is preeminent in the world. Much of the resulting technology has eventually found its way into everyday life in various forms. It is arguable that national defense has been the mainspring carrying basic research discovery into applications and driving our society into the technological era in which we now reside.

Since the end of the Cold War in the early 1990s, U.S. defense strategy has undergone a seminal change. The need for the U.S. military to stay at the forefront of technology has diminished considerably. The strategies for Defense Department R&D and weapons system acquisition are consequently being realigned as a result of comparatively weak threats to U.S. security, the national desire to direct more resources toward improving U.S. competitiveness in the global marketplace, and the increasing complexity and cost of military systems.

The current major defense trends affecting optical technology include increasing reliance on commercial components to reduce costs, use of lower-risk technology to cut system development time in half, and pursuit of only those technologies judged to have the greatest potential impact on national defense (*Defense News*, 1996). The cost advantages of using commercial components for niche military applications arise from their large market base and designs for minimal product cost. For example, the use of commercial computer displays in military field equipment could result in considerable savings, provided the displays are fully functional in the military environment. A grand challenge for developers of military systems is the incorporation of commercial designs, components, and test capabilities into military systems that must not only work in extreme conditions, but also interoperate reliably with other military equipment. Economic considerations also favor small suppliers with lower overhead costs and in some cases the use of technology developed with government support. These are profound changes for system acquisition.

Even in this age of reduced threat levels, there is still an overriding requirement for DOD to invest in technology that provides unique military advantages. Some argue that the return on R&D investment, in this time of declining defense budgets, is larger than for any other investment and essential for DOD to preserve its edge. This argument would tilt investment to new technology to preserve the U.S. edge, since the use of commercial components in current systems levels the playing field for our competition. Systems and system components must meet military field requirements; these are severe environments in which commercial equipment often experience high failure rates because commercial design requirements do not encompass severe stress conditions. There is little overlap between optical systems for defense and commercial markets. Commercial industry has little incentive to include DOD requirements in its designs since DOD makes up a small part of its market, particularly when the changes would increase product cost. Compounding the problem are reductions in R&D funding discretion by the services, with greater reliance on work at small commercially oriented firms funded through the Small Business Innovative Research (SBIR) program.

Optics has matured during the past 30 years to the extent that it is now on a par with electronics and microwaves in defense systems. Optics is the nucleus of entirely new systems and system concepts that are essential to U.S. national defense. Measuring the stature and effectiveness of a technology is always problematical. Since fieldable (operational) military capability is the "bottom line" for the military, implementation of a technology is one measure of its effectiveness. Figure 4.1 is a broad depiction of the impact of optical technology on DOD systems. The optical items cited in Figure 4.1 are those that provide the warfighter with significant leverage. The trend indicates dramatic improvement of field effectiveness in both tactical and strategic areas. Examples are laser designators that have enabled a generation of "smart" weapons to surgically strike a specific target and high-resolution surveillance of military capability unobtrusively from space. Desert Storm pilots stated that they "owned the night" with their night vision capability; the strategy of shunning daylight flights in Iraq and Kuwait during the war contributed greatly to their low casualty rate.

The level of activity in a technology can be gauged by the amount of DOD resources devoted to potential future system applications, as well as to exploratory and advanced development. Annual U.S. defense expenditures for optics are large, exceeding $5 billion per year, consistent with the capability depicted in Figure 4.1. Microwaves, optics, and electronics R&D for fiscal years (FY) 1996-1998 had a budget of $600 million per year;[1] approximately 27% of this is devoted to optics. Within this "electronics technology area," which includes funding for the Defense Advanced Research Projects Agency (DARPA), the amount of resources devoted to optics technology is on a par with its sister technologies, but erosion of these resources is anticipated. For instance, DOD reprogrammed resources from this budget category in FY 1996 and 1997 to cover shortfalls in the Bosnia peacekeeping operations, and DARPA, the principal organization conducting R&D, is downsizing by 20% (*Military and Aerospace Electronics,* 1996).

The role of optics in defense is pervasive and ubiquitous. Its roles in military systems are so broad that simply to categorize and describe them is difficult; they span the range from low-cost to expensive and from enabling to dominance of system designs. Laser propagation in free space can be used like radio waves and microwaves for radar and communications. Data can be transmitted over communication cables and imaged on displays. Unlike radio waves and microwaves, light is ideal for beam weapons, targeting (designating), and passive surveillance applications. Also, totally new applications have emerged that have no direct electronic analogue, such as laser gyros for navigation and gigabit cable communications.

The remainder of this chapter discusses optical advances that are important for national defense in surveillance, night vision, laser systems, fiber optics, displays, and special technologies. Examples are presented in terms of the way technologies are applied to various defense missions. However, the demarcation between defense-unique and civil

---

[1] From the Office of the Secretary of Defense, FY 1997 science and technology budget. This estimate understates the total DOD investment in these technologies. Not all R&D funds for optics, microwaves, and electronics can be included here for security reasons.

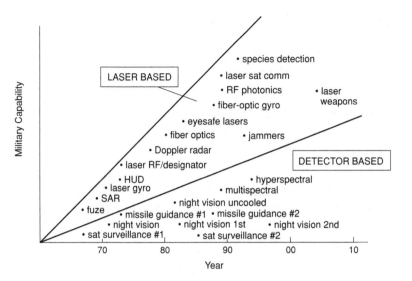

applications is becoming blurred. For example, the National Ignition Facility will advance our understanding of fusion energy, as discussed in Chapter 3, and will make use of the large laser facilities developed for the Department of Energy's (DOE's) nuclear weapons mission. The optics field is still relatively young, and these examples are only a small sample of the possibilities that the next 30 years will bring.

Surveillance

Surveillance has played a central, critical role in detecting and assessing hostile threats to the U.S. homeland and to its forces stationed around the world. For instance, as discussed in Box 4.2 and Figure 4.2, high-resolution imaging satellites have been deployed for three decades to provide data to U.S. defense experts in a broad spectral band record-ed over land and sea that could be obtained by no other means.

For the purposes of this chapter, surveillance is defined as the process of collecting optical data, usually in the form of images. This can be done with an inexpensive throw-away camera or a multimillion-dollar sophisticated reconnaissance satellite system. Warfighters and government decision makers rely on these sources of on-line informa-tion to provide input to their assessments and to assist in identifying strategy options. They not only demand the best-quality data, but also often need it immediately. Thus, older film-based surveillance systems have generally been replaced by real-time electronic image formation systems installed aboard airplanes, ground vehicles, and satellites.

The heart of current surveillance systems is a very sensitive focal plane array (FPA) of detectors that capture very weak optical and

BOX 4.2 MILITARY USE OF OPTICS THROUGHOUT HISTORY

Historical uses place in perspective the accelerating rate of change that
is occurring in the military uses of optics:

DIRECTED-ENERGY WEAPONS

Archimedes, advisor to Hiero the tyrant of Syracuse, reportedly used polished metal sunlight reflectors to set fire to Roman ships. This concept
was given new birth with the development of high-power lasers.

SURVEILLANCE

Telescopes have been used by the military for centuries. Today high-resolution imagery can be obtained on call from airborne and spaceborne surveillance systems.

COMMUNICATIONS

Simple mirror sunlight reflectors and semaphores, used in the nineteenth
century for communication, presaged the advent of laser communications
in development today.

· ·

infrared (IR) emissions arising from a multitude of sources. Also, common to most systems is an optical collection telescope structure operating in the visible or infrared spectrum. Advanced signal processing methods use the timing, location, and spectral signature of events to identify
those of potential military interest. These satellite systems are generally
expensive (more than $100 million) but provide worldwide surveillance.

Major advances have recently been made in manufacturing FPAs to
achieve pixel densities of approximately 10^6 per chip that parallel the
advancements made in achieving ever higher densities on silicon integrated circuits. This has enabled greater sensor sensitivity over a wide
field of view. Optical collection and detection technology operates
close to fundamental theoretical limits to the extent that image data
processing and transmission are now the bottleneck for achieving the
next major improvement in system capabilities. System designers face
the awesome challenge of selecting affordable technologies and platform
configurations to best take advantage of these advanced capabilities.

Consistent with current acquisition policy for all new systems, the
cost-performance-requirements envelope must be optimized for new
surveillance programs. As a result, DOD is now designing new surveillance systems with today's available optics technology that emphasize
system demonstrations to mitigate development risk. Opportunities to
insert new technology exist at each of the primary data collection steps:
wide-area survey, target acquisition, target characterization, target tracking, and system parameter updating.

To implement and optimize these observational sequences, the following system-level challenges for airborne and spaceborne sensor optics must be addressed:

- Innovation of solutions to technology insertion to obtain the most "bang-for-the-buck";
- Fast dissemination of data to the digital battlefield;
- Use of many spectral bands, multispectral or hyperspectral segments extending across the visible and IR, to penetrate enemy camouflage and locate targets (Marmo, 1996); and
- On-board fusion of other sensor data for more accurate target identification and tracking.

Keep in mind that optics-based surveillance is usually passive (unlike microwave radar systems, which require a pulse and return echo). Consequently, the technology insertions are mainly to improve

FIGURE 4.2 An optical surveillance umbrella helps to protect the United States.

DOD eyes in the sky have been watching the world for decades. The best known and arguably the most important satellite surveillance system is the recently declassified Defense Support Program (DSP) system, which detects infrared emission of radiation from ballistic missile launches and other IR events from a geosynchronous orbital post 22,300 miles above Earth. The sequel Satellite Based Infrared Surveillance (SBIRS) system, a $22 billion mix of other optical surveillance satellites to be launched into various orbits over the next 20 years, constitutes the next generation of global IR eyes for DOD.

Highly classified photographic satellite surveillance details have also recently emerged (Wheelon, 1997). High-resolution visible-light photography, anticipated by writers of fiction, has been used for some time to monitor events on Earth with great precision. The first successful CORONA satellite flight on August 10, 1960, initiated a series of missions that constituted the backbone of U.S. intelligence for 12 precarious years during the Cold War. Shown is the Pentagon viewed from more than 100 miles in space. Newer versions team up with their IR cousins to furnish strategic and on-line tactical battlefield support data.

(Courtesy of the National Reconnaissance Office.)

very sensitive surveillance systems. These technologies include materials and systems for IR FPAs, charge-coupled devices (CCDs), lightweight optics, compact coolers, staring arrays, and efficient electronic readouts and processors. These technologies are operated on board satellites, uninhabited airborne vehicles (UAVs), and aircraft. UAVs can relay real-time images of battlefield troop deployments to field commanders as will future satellite systems.

Night Vision

During World War II, the use of radar by Allied forces to see through clouds and inclement weather was invaluable. This capability was literally the difference between losing and winning battles. Now, some 50 years later, optical devices are available that can see at night with such important advantages over radar as high spatial resolution (as good as ordinary eyesight) and lack of detectable radiation emanation. U.S. soldiers and fighting machines equipped with this night vision capability have had a unique tactical advantage in the post-World War II era. U.S. military forces have essentially "owned the night" and hence have been able to fight under most favorable conditions.

Early night vision units amplified reflected starlight and demonstrated considerable tactical advantages. Battlefield night vision devices now use passive detection of IR, which senses the heat radiated from objects in the scene. The challenge is to discriminate objects such as tanks, which may be only a few degrees hotter than the background. Older devices produced images that resembled bad, noisy television signals; today, the devices have been improved to the level of quality television pictures. These devices, produced in large volume, have little in common with one-of-a-kind complex surveillance systems, although the underlying physical principles are the same. It should be noted that this is not all-weather capability since heat radiation is absorbed by rain and fog, imposing well-understood operational military limitations.

Night vision units, often termed FLIRs (a historical acronym meaning forward-looking infrareds) use FPAs in a wide variety of formats for tactical battlefield applications (Lerner, 1996). Night vision designs are mass produced (in quantities of more than 10,000) at low cost; for example, more than 100,000 first-generation units, known as Common Modules, have been built. This is in contrast to the surveillance units discussed in the previous section, which have ultrahigh performance but are produced in limited numbers (1 to 100) at high cost.

The first night vision devices used cooled detectors to gain the sensitivity required to detect weak thermal radiation. The availability of detector materials has largely driven system design. Early materials such as lead sulfide (PbS) were hand-made and suffered from a number of technical problems. Most modern FLIRs use mercury cadmium telluride (HgCdTe, or MCT) because the composition can be varied to afford detection over different regions of the IR spectrum and the elements can be mass produced with high purity. Cooling to the vicinity of 100 K requires a mechanical device and dewar (thermos bottle) with a window and optical elements to admit thermal radiation. Either the scene is scanned over a linear array of detectors with about 10^5 elements or a large array of detectors with about 10^5 elements stares at the scene to be imaged. Cooled detectors feature excitation of electrons as photons are absorbed (photodetectors), with signal processing in chips followed by conventional display of the image.

Figure 4.3 shows a cooled thermal imaging roadmap that details the progress of this class of night vision devices for Army applications. Basic materials research drives the progression as poor-quality bulk material gave way to liquid-phase epitaxy (LPE) material amenable to mass production at low cost. Future generations of detectors may employ molecular beam epitaxy (MBE) or smart FPAs with specially designed readout integrated circuits (ROICs) to provide multifunction elements. The older rotary coolers and bulk elements that made up the Common Module class of devices for the first-generation FLIRs used in Army vehicles were characterized by "noisy TV picture" quality and poor reliability. LPE elements in Standard Advanced Dewar Assembly (SADA)-class FLIRs use a linear drive cooler to produce a quality picture with 10 times greater reliability. These second-generation FLIRs are being introduced into the vehicles indicated through a program called Horizontal Technology Insertion (HTI). Third-generation devices are now in the R&D phase.

This discussion cannot cover every type of night vision system insertion; a brief synopsis of the types of fielded units is provided here.

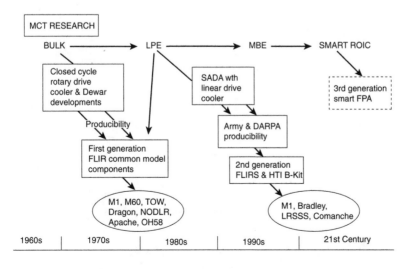

FIGURE 4.3 Progress in cooled thermal imaging for Army applications. (Courtesy of J. Ratches, CECOM Night Vision and Electronic Sensors Directorate, U.S. Army.)

Early uses focused on tank target detection and missile guidance. As the technology progressed and the sensitivity increased, aircraft and helicopters began to rely on FLIRs for targeting and navigation. Today, even higher-sensitivity FLIRs are in development for long-range threat detection, termed Infrared Search Track (IRST). The progression to lower-sensitivity, lower-cost units has permitted vehicle driving at night and use by individual soldiers. The most compact imaging IR devices are found in missile guidance units for Javelin and Stinger, for example. Obviously, for expendable munition guidance applications, very low cost is paramount, but the performance is usually limited to short-range targeting. An affordable cost for FLIRs ranges from $10,000 to $200,000, depending on sensitivity and other parameters.

The most recent FLIR product breakthrough is in the area of uncooled detectors. Unlike the cooled photodetector class, uncooled detectors rely on the use of a silicon microstructure upon which a thin film of material is deposited. The temperature increase brought about by the absorption of IR changes some property of this film. The three detection methods used are (1) resistive bolometric, (2) pyroelectric, and (3) thermoelectric. The low cost of silicon microstructure devices is key to the widespread use anticipated for this detector class.

In Figure 4.4 a roadmap for this uncooled class of detectors is presented showing the progression from ferroelectric element research to bolometer element research into FPAs for short-range thermal sights. Consistent with this class, a sensitivity sufficient to image up to about 1 km is possible; material advances will improve this range. As Figure 4.4 shows, the increase in size and producibility correlates with wider use in soldiers' thermal weapon sights and in driving Army vehicles, with the ultimate use in missile seekers as detectors improve. Since the units do not require a cooler, they are much lower in cost with projections of less than $10,000 for high-volume manufacturing.

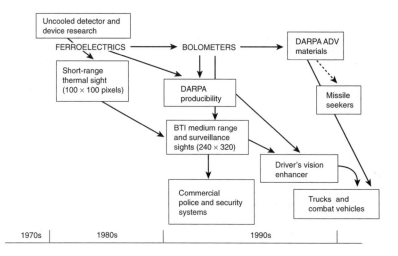

FIGURE 4.4 Progress in uncooled thermal imaging for Army applications. (Courtesy of J. Ratches, CECOM Night Vision and Electronic Sensors Directorate, U.S. Army.)

Figure 4.5 shows a small uncooled hand-held FLIR device with a 320 × 240 pixel array and a range of ~1 km. With excellent pixel-to-pixel uniformity yielding imagery at an affordable cost, this type of unit is suitable for deployment to individual soldiers and is also expected to see widespread commercial use.

FIGURE 4.5 A handheld uncooled infrared imaging system. (Courtesy of Raytheon Systems Company.)

Essential to the success of this technological thrust have been the cooperative efforts of the military services and DARPA to take the "black magic" out of detector producibility and to create a quality manufacturing infrastructure. Future DOD thrusts are to develop even larger staring arrays, multicolor FPAs for fusion of detector information, wide deployment of more sensitive uncooled imagers, and development of effective automatic target recognition systems.

Laser Systems Operating in the Atmosphere and in Space

Lasers have become such a key part of our life, at the grocery checkout counter and in compact disk (CD) players, that military uses are easily anticipated. Laser light projects long distances in very narrow beams because of its short wavelength, unlike radio waves that spread out. So optical power is more efficiently delivered to a target, and this simple idea is the essence of most military uses of lasers. Early laser research workers dreamed of destroying targets at the speed of light (a million times the speed of sound) and a host of other applications. DOD supported the early work on masers that led to the discovery of the first (ruby) laser at Hughes Research Laboratory, but it took many years of concerted Defense Department R&D to build lasers with the required efficiency of tens of percent (early versions were 0.0001% efficient) needed in fielded systems. Even after lasers were improved, it was quite a chore to make laboratory units work reliably in the field.

This section includes the analogues of microwave systems that operate in the atmosphere and space, namely, laser radar, jammers, target designators, communications, and laser weapons. Solar cells, environment sensing, law enforcement, transportation, and so forth, also operating in the atmosphere or in space for civilian use, are treated in Chapter 3. In military applications we are usually concerned about preserving the power in narrow near-diffraction-limited beams over long ranges (many kilometers), which are accurately pointed and controlled to eliminate both platform jitter and atmospheric beam distortions. The technologies common to this class of laser systems include acquisition, pointing and tracking, fieldable optics with domes or aerodynamic windows, bore-sighted detectors, and laser operation somewhere in the visible to 10-μm IR spectral ranges where the atmosphere is transparent to laser radiation.

Laser Range Finders, Designators, Jammers, and Communicators

This category of lasers was developed first, then laser power levels were gradually increased to the weapons class range discussed in the next section. The first laser range finder using ruby lasers was demonstrated less than a year after the laser's discovery and marked the introduction of widespread practical use. As improvements in the technology, especially new laser materials, and new classes of lasers came along, both range and performance were greatly improved.

Typically, a tank laser range finder is used to illuminate (with great haste) an enemy tank; the range is calculated from the received laser return pulse to determine the ballistic trajectory of a tank shell. The tank's gun is elevated and fired while the vehicle is moving. The objective is to be faster than the enemy and to kill the enemy tank by delivering highly accurate munitions. With better laser materials, especially Nd:YAG (neodymium-doped yttrium aluminum garnet, a crystalline solid with outstanding performance), field units have greater reliability and performance. In simplest terms the key to successful systems of this class is the capability to design, manufacture, and deploy lasers that are affordable and reliable. Early laser systems suffered from internal degradation of optics, which blocked their widespread deployment and required a major effort to resolve. Even more years of development were needed to field other laser subsystems after the technical community sorted through thousands of options to find the right combination of power output, efficiency, reliability, and so on. Today, DOD generally utilizes the following:

- Light-emitting diodes (LEDs) or laser diodes (LDs) for very short-range illumination;

- D array pumped solid-state pulsed lasers for laser range finders, target designators, jammers, and so on; and
- Nd:YAG and other eye-safe solid-state lasers.

The many thousands of battlefield range finders in tanks and designators for ground and aircraft were deployed using older flash lamp excitation (or pumping) of YAG in first-generation laser technology. New technology utilizes LDs to convert electrical energy into light energy for pumping solid-state crystals such as Nd:YAG with greater efficiency and reliability. In addition, many more applications have emerged for this type of laser that depend on the cost-performance trade-off. The cost of LDs has been driven down by technology advances and volume manufacturing and is expected to drop even more. Simple LD illuminators for personal weapons are the basis for the important Multiple Integrated Laser Engagement Systems (MILES) combat training system with commercial spinoff to war games for entertainment. The sportsman will recognize this illuminator as the basis for the rifle spotting beam. Retrofit and upgrade of the entire class of pulsed range finder and target designator units so important to our success in delivering precision munitions in Desert Storm are proceeding. Modern battlefield doctrine is, in fact, profoundly shaped by laser-guided bombs and missiles, enabled by our ability to cost-effectively make the approximately 100,000 reliable laser designator sources that have been fielded.

The above low-power class of laser systems operates in the <1 W average power range. Countermeasure lasers for jamming and sensor blinding require 1 to 100 W, a range difficult to achieve until LD array pumped solid-state lasers became available. In combination with various wavelength shifting schemes [optical parametric oscillators (OPO), Raman effect] to avoid sensor selective rejection filters, jammers and blinders constitute a new escalation in the "optical" battlefield. It should be noted that Secretary of Defense Perry in September 1995 announced a prohibition against the "use of lasers specifically designed to cause permanent blindness of unenhanced vision and supports negotiations prohibiting the use of such weapons"—a position probably established, in part, because an eye safety device effective against all types of lasers generally found on the battlefield has not yet been developed and deployed. However, the use of lasers for target destruction is still permissible. R&D is continuing, and future thrusts are mostly in support of laser sources and subsystems with emphasis on wavelength diversity and agility. The U.S. Air Force study *New World Vistas*[2] anticipates that this class of systems will be fielded in the next decade.

---

[2]Air Force Scientific Advisory Board, New World Vistas, study commissioned by Secretary of the Air Force Dr. Sheila E. Widnall and Chief of Staff Gen. Ronald R. Fogleman (dated Nov. 29, 1994; study reports were completed early in 1996).

Chapter 4

It is interesting that early forecasts of the significant use of laser communications in free space have not materialized. Development of ground, ship, aircraft, and satellite terminals has been extensive (about $1 billion) over the past two to three decades, with active work still in progress, especially for satellite-to-satellite relay links that would permit elimination of a ground relay station. This status is due largely to technical problems with lasers combined with advances in the capability of competing microwave links. (The better cost-risk trade-off of microwave links is described in Chapter 1.) The maturation of LDs as a product seems likely to result eventually in fielded commercial links, especially in the 0.01- to 1-gigabit-per-second regime for satellites and other airborne vehicles.

Laser Weapons

The concept of near instantaneous destruction of airborne and space-based targets is quite appealing. To this end, high-power lasers have been under development since the 1970s. Weapons-class laser development began with the discovery of the high-power carbon dioxide (CO_2) molecular laser operating at 10-μm wavelength. There have since been many technology advances in high-energy lasers, and both ground and airborne demonstrations have validated the basic weapons concept (Figure 4.6).

Laser weapons designs and matching mission roles, such as the destruction of sensors in imagers, missile guidance, and surveillance systems, are understood well enough that advanced system development could proceed if national security required it (Knowles, 1996). The *New World Vistas*[2] study advocates extensive use of laser weapons against missiles, satellites, and other ground assets in the next decade.

Today a number of gases have been discovered with the right properties to generate high power in the 1-MW regime and offer the flexibility of excitation by electrical or chemical means. Entirely new areas of physics and optics are encountered in this high optical power regime. Technical challenges include the following:

- Maintaining stable optical resonators under high thermal loading;
- Extracting diffraction-limited beams with high efficiency from the laser cavity;
- Suppressing nonlinear optical effects along the propagation path;
- Delivering beam power on target via optical control to correct for beam distortions during propagation; and
- Solving operational issues such as environmental factors and lethality for different target classes.

At first glance, high-power lasers would seem to serve only military needs. However, advances in this technology have provided many other uses for scientific and commercial applications. For example, adaptive

optics technology for atmospheric compensation of laser weapons is revolutionizing the design of astronomical observatories (discussed in Chapter 3). The distortion of optical beams along an atmospheric propagation path is highly complex. It is compounded by high-power nonlinear effects, which can also break up the beam.

During the 1960s and 1970s, DOD mounted a major technical campaign to understand and resolve these problems. In addition to elucidating the physics of beam propagation, adaptive optics and phase conjugation were developed to solve the problem. These techniques have been reduced to practice and are now employed as essential elements in system designs. Adaptive optics control of laser beams constitutes a major recent breakthrough in laser weapons system developments. Ground-based lasers can now potentially negate satellite threats. However, not all elements of the technology are complete. Continued developments are required in terms of power, size, weight, reliability, and beam quality. Future activities should address reducing the manufacturing cost of high-power laser systems to make them more affordable. Adaptive optics is effective in this area as well, providing compensation for lower-cost, relaxed optical tolerance, laser resonator designs.

Laser weapons can revolutionize battlefield strategies, but they raise a new class of system engineering and battle operational issues. Knowing how and when to use this new capability takes careful planning. For example, the laser weapon must work in concert with existing defenses. A shipboard laser weapon with short time response may be the last line of defense against an incoming enemy missile that has passed through layers of conventional weaponry; hence, the tolerable range for destruction of missile guidance or munitions is important to survival of the ship. Many such issues, involving the agility of the laser weapon and targets in comparison with the allowable time for weapons use, are still unresolved.

The 1972 Anti-Ballistic Missile Treaty constrained work on space-based platforms with lasers designed for use against space targets. An airborne system is in active development that would intercept tactical (theater) ballistic missiles such as Scuds during their boost phase and blow them up via laser heating. This is an important deterrent since the munitions would be destroyed over enemy territory. A compact chemical oxygen iodine laser (COIL) has been selected for this mission. The Airborne Laser (ABL) program, employing a COIL weapon against theater-range ballistic missile threats launched from a mobile platform, was approved in 1996 to proceed to the Program Definition and Risk Reduction phase. This action ushers in a totally new weapons system concept, capping three decades of intensive R&D. This entirely new dimension in threat deterrence cannot easily be duplicated by other nations and underscores the breakthrough potential of optics to respond to national defense needs.

Laser weapon technology and military needs seem to be converging in this decade. Ongoing programs like the DARPA/Tri-Service Mid Infrared Laser development for the 2- to 5-µm range and the Space Based Laser (SBL) project sponsored by Ballistic Missile Defense Organization enjoy broad support for multiple missions. As a result of the Air Force *New World Vistas* study, systems and operations analysts are very active in matching laser concepts to today's problems. For example, the notion of a "frugal kill" optimizes laser fluence on target to just the right amount for destruction without overdesigning the system. Future R&D directions feature efficient laser designs such as high-power DL-pumped solid-state lasers and better beam quality for frugal kill.

Fiber-Optic Systems

Fiber-optic (FO) systems represent an area in which commercial investment has led the way for DOD applications. For basic data transmission similar to commercial telephone service, DOD has adapted and improved the hardware for battlefield environments, with commercial organizations leading the R&D effort. For DOD, desirable attributes include freedom from electromagnetic interference, low power consumption, small size and weight, enhanced physical security, and high available data transmission rates.

The use of fiber optics on aircraft, satellites, ships, and submarines has proceeded at a somewhat slower pace as the design and maintenance issues associated with the military field environment have been solved. Digital FO communications are now used both to connect military communications terminals and within military facilities. The center of gravity of DOD development and use has been in the Navy, although the Army push to digitize the battlefield will bring more FO systems into play. Air Force work has featured conventional systems for large ground installations and special FO links for aircraft and satellites. Much of this work has proceeded in parallel with the maturation of commercial FO networks. DARPA has undertaken a major development program, Broadband Information Technology (BIT), to gain greater flexibility and performance for DOD use of commercial networks through test-bed field trials and component development, especially wavelength-division multiplexing (WDM). Special FO and photonic techniques have been developed by DOD. The extension of digital techniques into the terabit region, development of special sensors by the Navy, parallel and serial local area networks for avionics, and both backplane and chip-to-chip communication projects are under way.

Other nondigital uses of fiber optics include the FO gyro (discussed below) and the propagation or control of radio-frequency (RF) signals via fiber, which is important because of the widespread military use of the RF spectrum. Compared with atmospheric RF propagation, fiber offers the inherent advantages of ultrawide bandwidth and much lower propagation loss, but the laser must be modulated and the RF signal faithfully extracted at the receiver terminal. FO usually offers a lower-loss alternative to coaxial cable. At 30 to 100 GHz it will probably prove to be superior to conventional waveguide or microstrip for which the propagation loss is prohibitive at long distances (i.e., >1 m). Today the technology is practical at 20 GHz for direct diode laser modulation and at 50 GHz for external modulation (at higher cost), with further improvement to 30 and 100 GHz expected within 10 years. This technical area has spurred a thriving commercial activity (valued at $100 million) for cable television FO cables, L-band Earth terminal links, and wireless personal communications.

DOD has ground-based communication needs that range from RF transmission to and from remote antennas and within communication terminals, to shared aperture transmit/receive antennas, to very complex phased array antennas for radar. An oft-cited example of the effectiveness of this technology is the avoidance of antiradiation homing missiles launched by the enemy to disable the antenna by guiding on the outgoing radar signals. Remote placement of the antenna, enabled by low-loss FO transmission of RF signals, secures the safety of personnel and the field shelter housing the control equipment.

FO permits a "true time delay" phased array radar by switching lengths of fiber (see Figure 4.7). This capability is not possible at

FIGURE 4.7 Photonic phased-array conformal antenna.

A grand challenge for military aircraft manufacturers is the development of conformal antennas for radar, communication, and electronic warfare. Integrating antennas into the airplane skin saves space and weight and improves the aerodynamic profile. To make this happen, the antenna beam patterns must be steered electronically.

Photonics offers a new approach to the vexing problem of building a compact, flyable, stationary antenna with an electronic beam steering mechanism. Fiber-optic elements of different length, which by themselves take up very little space, are the key to the solution. By switching in different lengths of optical fibers, the microwave signal phase can be adjusted to each phased array element and beam direction can be changed at will. This concept has been demonstrated, and efforts are proceeding to make a workable version for aircraft systems.

(Courtesy of B. Hendrickson, Air Force Research Laboratory.)

microwave frequencies; emulating time delays in RF causes a form of radar degradation termed "squint." The first full FO radar operating at 850-1400 MHz was demonstrated in September 1995. It had a wide spur-free dynamic range (no extraneous signal channel noise) nearly adequate to meet today's desired radar performance requirements. Major Navy projects are well under way to use this technology to make order-of-magnitude improvements in surface ship antenna structures, with tests of readiness planned for 1998. There are no obvious technical impediments to further improvement of FO-RF technology. Widespread future use and high payoff to DOD are likely.

Displays

In warfare, timely acquisition and distribution of information is essential. Military planners, very conscious of this basic tenet, are "digitizing the battlefield." This will allow combatants to take advantage of the veritable explosion of information gathering and distribution capability to give a tactical advantage. The Air Force *New World Vistas* study calls for better cockpit displays to relieve overworked pilots; hence, display technology must keep pace with the means to rapidly process and analyze data. DOD has identified flat-panel displays (FPDs) as a critical technology. Older cathode-ray tube (CRT) displays are being replaced with superior FPDs. Commercial off-the-shelf (COTS) displays are used wherever possible in DOD. However, many military requirements cannot be satisfied by COTS displays. Since the size of the DOD market is small, accounting for about 2% of the worldwide display market, DOD has sponsored display development and manufacturing necessary to meet its specific needs.

Since 1992, DARPA has led DOD display development programs for high-definition systems (HDs) and head-mounted displays (HMDs) with triservice support. Structured as dual-use programs for commercial and military technology, these DARPA programs have spurred a U.S. presence in a technology otherwise dominated by offshore suppliers. Today, a small number of U.S. suppliers (about 10) are active in serving the special needs of the U.S. military listed in Table 4.1. All displays require tolerance to $-54°$ to $71°C$ storage temperatures and worst-case shock. Pixel counts vary approximately from 500×500 to 2500×2500, and most applications require full color.

The DARPA HMD program is now over with reports of mixed success. Active-matrix liquid crystal (AMLCD) and active-matrix electroluminescent (AMEL) systems were both developed. Both technologies had a goal of providing 1280×1024 monochrome pixels in approximately 1 square inch. Placing color filters on these elements provides a 640×480 color display. Production and device yield issues remain to be solved before fieldable hardware results.

The DOD FPD technologies include plasma, electroluminescence, and AMLCD. Displays utilizing these technologies are operable under high ambient light levels and can be made rugged for field use. Newer FPD technology includes field emission, MEMS (microelectromechanical systems), and three-dimensional displays requiring no viewing aids (e.g., glasses). The U.S. manufacturers still rely on overseas sources for key materials such as polarizer-retardation films, color filter material, and phosphors.

A consortium of manufacturers works on the HDS program in partnership with the government. The following results have been reported:

AMLCD	6.3×10^6 pixels for a 13-inch display
Plasma	1024×768 full-color video tactical monitor
AMEL	4-inch × 5-inch panel
MEMS	2-million-element digital color micromirror

This work has been done under the aegis of the National Flat Panel Display Initiative. Efforts are being made to determine these products' producibility and develop manufacturing processes for them. It should be reiterated that COTS displays cannot satisfy all military display needs (Mentley, 1996).

Future DOD R&D activities include high-definition displays with acceleration of field emission display technology and organic luminescent materials for the mobile user.

TABLE 4.1 Military Display Requirements

Category	Use	Size
Large-area multiviewers	Command and control workstations	19-60 in.
Vehicle and cockpit	Sensor, tactical, instrumentation	5-16 in.
Miniature and low power	Head mounted	0.5-4 in.
	Weapon mounted	
	Body worn	

Special Techniques

This section provides an overview of techniques that address special military applications and/or mission requirements. The major technology initiatives are described in detail below. Many other niche applications use optics to advantage, attesting to the many dimensions that this technology can bring to bear on solving specific problems. Examples include the following:

- Use of LEDs to mark friendlies during Desert Storm;
- Laser scanners that convert synthetic aperture radar (SAR) signals into images;
- Moderate-power, handheld medical lasers for cauterizing battlefield wounds; and
- Optical radar for tracking satellites.

Chemical and Biological Species Detection

The 1995 release of sarin in the Tokyo subway once again brought us face-to-face with the destruction and loss of human life that a small terrorist group can cause. Some believe that the open U.S. society is particularly vulnerable to this type of attack. The Oklahoma City and World Trade Center bombings are reminders of our need for greater vigilance and better technology for early detection of this type of threat.

Weapons of mass destruction involving nuclear, biological, and chemical (NBC) species are a significant new emerging global threat and have a high priority within DOD. Because chemical and biological weapons can be produced with relatively low technology and are easily acquired, their number is increasing. More than 30 countries are now suspected of having chemical weapons capability and more than a dozen of having biological weapon competence. There are also about a dozen confirmed nuclear-capable countries.

Electro-optic technologies are central to meeting mission area requirements arising from these threats. The technologies of interest include the following:

- Active detection devices, such as backscatter lidars at eye-safe wavelengths, differential absorption (DIAL) lidars over the band between 2 and 11 mm, resonance Raman lidars, and laser-stimulated biological fluorescence;
- Passive detection devices, including imaging and nonimaging spectrometers and FLIR and Fourier-transform infrared (FTIR) systems; and
- Lasers to decontaminate and kill biological species.

Long-range chemical detection using lidar or other stand-off techniques would give the greatest tactical advantage. An effective detection scheme must meet many practical mission requirements. The objective of achieving an ultralow species concentration detection capability is important, but what is also required is an extremely low rate of false alarms for detection for a wide range of chemical species. This places multidimensional requirements on the sensor suite, making electro-optics an essential part of strategies for addressing this threat.

In FY 1996, DOD made major investments in electro-optics for non-proliferation ($20 million), strategic and tactical intelligence, battlefield surveillance ($9 million), counterforce ($18 million), active defense ($2.5 million), and passive defense ($20 million). Furthermore, DARPA announced a new initiative funded at the $100 million level in a multi-faceted approach to inject new technology into this critical problem area, and DOE efforts continue at a significant (> $20 million) level. A well-planned, well-coordinated, cohesive effort is required to greatly advance the probability of success. This would include a multiyear strategy overseen by a strong single manager to coordinate activities within DOD, DOE, and so forth.

Laser Gyros for Navigation

Laser gyros are important as inertial navigation sensors. These devices run laser light around in a closed path in both directions; if the platform rotates, the tiny differential time delay can be detected and the rotation rate deduced. Early gyros used helium-neon ring resonators, which oscillated at different frequencies; platform rotation is proportional to the differential frequency. Many years of development have yielded ring laser gyros (RLGs) with a low bias drift (< 0.005 degree per hour) and random walk (< 0.0015 degree per hour), low cost ($10,000), and long life ($10^5$ hours). With these parameters, a 4-pound, 80-cubic-inch RLG is suitable for aircraft navigation. For other uses such as missile guidance, smaller, lower-accuracy units are necessary.

Development of the fiber-optic gyro (FOG) has created another alternative, especially for lower-accuracy applications. A FOG with a size and weight of 1.3 cubic inches and 0.3 pound, having a hundredfold lower accuracy, is still adequate for missile guidance and costs only $4,000. These units are now in production as replacements for conventional mechanical gyros that have to be spun up for each mission. Inertial guidance units are still necessary, even in this age of global positioning systems, since these systems can be jammed during wartime.

Optical Signal Processing

Many signal processing functions can take advantage of the unique properties of optics. Mathematical calculations can be performed in real time using analog optical techniques. Some computations are extremely tedious when performed on a digital computer but relatively easy on an analog optical computer. For example, Fourier transforms can be readily accomplished by placing a suitable lens in an optical telescope. Since Fourier transforms are often used to obtain the frequency spread of short-pulse radar signals, an optical subsystem could provide an important capability.

Image and electronic signal processing have long been fostered by research organizations within DOD, such as the Office of Naval Research. Acousto-optic modulator-based signal correlators have been designed and incorporated into military products, although widespread field deployment has not yet occurred. Prototype vector-matrix, matrix-matrix, and neural network optical processors have all been applied to DOD problems with varying levels of success. Much of the work describing actual field tests remains classified. (Further discussion of these techniques for commercial applications can be found in Chapter 1.)

Summary and Recommendations

Acquisition reform and the use of commercial items are important strategies for reducing DOD system cost, but many commercially available optical products require special adaptation or improvement to meet unique DOD needs. There are special DOD field requirements for military systems that are not required for commercial applications. Examples are (1) displays that must work in high ambient light levels, (2) devices that will withstand large temperature excursions, and (3) IR imaging devices with long-range detection and surveillance requirements. Commercial industry is reluctant to invest development effort in these low-volume military products. However, some products such as diode lasers, developed by the military to exacting specifications, have a large commercial market potential.

Special DOD operational requirements and low-volume production will necessitate continuing DOD support for core optical competencies from basic research to manufacturing technology. For example, DOD is financially supporting the development of a manufacturing plant that produces active-matrix LCD display products for aircraft and other military platforms.

Past producibility efforts directed at key, high-cost system components have successfully lowered costs and improved performance. The optics assembly process is currently the most expensive system manufacturing step. DOD system suppliers are trying to attack this problem with concurrent engineering and assembly-friendly designs. Examples are the triservice-DARPA FPA producibility program for night vision, extensive DOD support of diode laser arrays, and more recent work to enable cost-effective production of vertical cavity surface emission lasers (VCSELs) for photonic applications.

Funding of new technology developments for DOD systems is declining. There is considerable interest in using existing or commercial optical technology. This is a result of DOD's acquisition reform initiated in 1994-95, which stresses an affordability-system performance trade-off. DARPA senior management and industry sources confirm the trend.

DOD R&D funding practices and budgets have resulted in greater optical industry reliance on SBIR grants and cooperative R&D agreements (CRADAs) with federal laboratories. Numerous small firms rely totally on SBIR grants (which are increasing) and CRADAs (which are decreasing) to develop a technology into viable products. The DOD objective of keeping the military technologically sharp so that our nation will not be blindsided by a foreign power adds a longer-range dimension to R&D that firms find indispensable to innovation.

Small companies are becoming increasingly important as a source of advanced optical technology for rapid insertion into new DOD systems. Defense prime contractors have noted that they are increasingly dependent on small companies for innovative solutions as a better economic alternative to in-house work.

The downsizing of DOD programs is eroding the defense manufacturing base. DOD programs must address the producibility needs of high-leverage optical components and systems that provide strategic advantage. COTS items cannot satisfy all military requirements. There is a tendency to associate optical technologies with the use of commercial electronic computers and components. Optical systems do use many commercial parts, and 25% of DOD's optics are imported. However, most military optical systems could not be assembled solely from commercial parts. Companies have been extremely reluctant to change commercial specifications to satisfy peculiar DOD requirements. Also, it is not necessarily in the national interest to design advanced optical systems that provide U.S. fighting forces with a strategic advantage and are also readily available to potential adversaries. For this reason and to ensure a supply of key components during periods of crisis, DOD must sustain selected manufacturing infrastructures.

Even with post-Cold War macrochanges in military doctrine, DOD will continue to favor optics for surveillance, surgical strike with precision guided munitions, and so forth. High-leverage technologies that offer crucial advantages to the nation must be maintained. It is a daunting task to bring to completion the high-leverage military systems needed in limited conflicts, antiterrorist activities, and rogue nation situations as the DOD budget shrinks. The greatest return on investment has historically come from R&D. This imperative becomes even stronger as the budget shrinks.

Maintaining a proper R&D balance is difficult as DOD downsizes. Very close coordination among all involved parties in optics R&D (e.g., DARPA, the military services, DOE) is essential. The committee's assessment of R&D balance and long-range planning leads to concern. With downsizing, DOD plans become more vulnerable to unexpected uses of R&D funds, such as peacekeeping in Bosnia. The result has been unplanned, short-notice reductions in the R&D budget, which leave key projects incomplete. Since DARPA accounts for roughly 70% of DOD's science and technology budget, very close cooperation with the military services and other agencies is needed to avoid the disruptions that have occurred in the past. With science and technology projects undertaken by many university, government, and industry laboratories, planning must be closely coordinated and continuity maintained to extract maximum benefit from these activities. The critical R&D support that DARPA provides to warfighters with ultimate technology transfer to the services dictates early, close coordination of all parties.

The confluence of a number of DOD and congressional acquisition policy changes has resulted in greater COTS use, lower-cost system design compromises, and greater dependence on small companies for new technology. This chapter has discussed the acquisition policy of cost-performance-specification optimization for new systems. Economic factors have also forced reliance on commercial and other small business enterprises for R&D and manufactured products. This trend is particularly strong in the optics field, with hundreds of small companies supplying government's needs.

DOD should ensure the existence of domestic manufacturing infrastructures capable of supplying low-cost, high-quality optical components that meet its needs via support for DARPA and the Manufacturing Technology Program.

It is fundamental that a well-founded manufacturing process saves money on the ultimate product, so this recommendation could seem highly generic. However, with increasing reliance on innovative small suppliers, affordability and quality as new program trade parameters, lower-volume system production, unique mission requirements, and new mission objectives for DOD, the manufacturing paradigm has changed. New manufacturing techniques are being developed to improve commercial productivity, and DOD should take full advantage of them.

A central, coordinated DOD-DOE time-phased plan should be developed and conducted to enable worldwide optical detection and verification of chemical species that threaten civilians and military personnel through hostile attacks.

Much has been said about this problem, and its implications are clearly very serious in both civil and military scenarios. After many years of work, partial solutions have emerged. The committee believes that concerted R&D activity with a focus on optics has the best chance for a technical solution to airborne detection in view of the rich molecular spectrum accessible by optical methods. A single federal authority should be placed in charge of these crucial programs.

A coordinated multiyear DOD plan should be conducted to develop RF photonic phased antenna-array technology for radar and communications.

An L-band version of this photonic system and many of the components for higher-frequency systems have been demonstrated. With improved modulator and switch designs, Bragg grating fibers, and higher-power diode lasers, new and better approaches are likely and it is time for a major push.

Key technologies such as high-power laser activities and new optics should continue to be pursued by DOD.

Erosion of the optical technology base through benign neglect or lack of federal agency coordination must not be allowed; the return on investment is very high. After decades of work, fieldable laser devices can now provide the power levels, spectral diversity, and adaptive optics configurations necessary as countermeasures to extreme threats, especially from missile attack. Recent program awards—the Airborne Laser program ($1.1 billion from 1996 to 2002), the Space Based Laser ($100 million per year), and two conformal optics programs ($24.6 million)—carry forward some of this essential work. New laser sources and optical technology innovations offer solutions and totally new system concepts that can provide our nation with true strategic and tactical defense advantages.

References

Balcerak, R., P.W. Pellegrini, and D.A. Scribner, eds. 1992. *Infrared Focal Plane Array Producibility and Related Technology.* SPIE Vol. 1683. Bellingham, Wash.: SPIE Press.

Council on Competitiveness. 1996. Pp. 94-103 in *Endless Frontier, Limited Resources.* Washington, D.C.

Defense News. 1996. Pentagon reaches out to private sector for new ideas. November 4-10.

Kitfield, James. 1994. The Pentagon's plan. *Government Executive* (November).

Knowles, J. 1996. Early morning DEW: Directed energy weapons come of AGE. *J. Electron. Def.* (October):48-49.

Lerner, E.J. 1996. Infrared array detectors create thermal images. *Laser Focus World* (September).

Marmo, Jay. 1996. Hyperspectral imager will view many colors of Earth. *Laser Focus World* (August):85-92.

Mentley, D. 1996. U.S. must display its support for flat panels. *Electron. Eng. Times* (October 14):58.

Military and Aerospace Electronics. 1996. Downsizing at DARPA. November 6.

National Research Council. 1996. *Conflict and Cooperation in National Competition for High-Technology Industry.* Washington, D.C.: National Academy Press.

Perry, W.J. 1996. Preface to Science and Technology Review. Washington, D.C.: Office of the Secretary of Defense, May.

Scribner, D.A., M.R. Kruer, and J.M. Killiany. 1991. Infrared focal plane array technology. *Proc. IEEE* 79(1):66-85.

U.S. Department of Defense. 1996. Defense Science and Technology Strategy. Washington, D.C.: Deputy Director, Defense Research and Engineering, May.

U.S. General Accounting Office. 1996. *Operation Desert Storm: Evaluation of the Air War.* Washington, D.C.: U.S. Government Printing Office, July.

Wheelon, A.D. 1997. Corona: The first reconnaissance satellite. *Phys. Today* (February):24-30.

5

Optics in Industrial
Manufacturing

Modern manufacturing is being revolutionized by the use of optics, which can both improve current manufacturing capabilities and enable new ones. Light can be used to process or probe materials remotely, even through windows isolating harsh or vacuum environments. With no surface contact, there is no contamination of the process by the probe beam and no wear of tool edges. Scanning provides action over large areas. Light can be used to induce photochemistry, for example, in photolithography to produce submicron features in thin films of photoresist or in rapid prototyping where liquid polymers are solidified by lasers to form a three-dimensional piece from a computer-aided design database. Light can cast images, making it possible to inspect a part or use the image to guide the working tool to the correct area of the workpiece. Images of the surface topology can be compared to the topology of the "perfect" image captured in a database or the topology of an identical piece to ensure consistent component fabrication. For these many reasons, optics has reached into every aspect of manufacturing and promises to increase in use with improvements in speed, control, precision, and accuracy.

Numerous optical techniques are used throughout industry and are critical to the manufacture of such diverse and basic products as semiconductor chips, roads and tunnels, and chemicals. Optical techniques, grouped by function, fall into two broad classes:

1. *Performing manufacturing:* Light interacts directly with the finished or intermediate product to change its physical properties, as in the case of photolithography or materials processing.

2. *Controlling manufacturing:* Optics is used to provide information about a manufacturing process, as in the chemical industry's use of optical sensors for in-line process control, or to inspect a manufactured

product, as in the semiconductor industry's use of optical inspection tools to characterize particulate contamination.

Some applications may be relatively familiar, such as the use of high-power lasers for cutting, drilling, or welding steel. Others are less familiar, such as the use of optical sensors to monitor chemical processes in real time or the use of lasers for alignment and control in the construction industry. Some of the challenges that these applications face are unique to a particular industry, but others, such as the need for trained optics technicians or the importance of making equipment robust and reliable, are universal.

Table 5.1 shows the most important uses of these optical manufacturing techniques for five major U.S. industries—automotive, semiconductor, chemical, aerospace, and construction. These industries in aggregate account for approximately $1 trillion, or 17% of the 1992 U.S. gross domestic product (GDP). Each has a critical dependence on one or more optical manufacturing techniques.

Because of the diversity of U.S. industry, this chapter cannot address the use of optics in every single branch of manufacturing. It endeavors instead to cover a representative sample, including those applications

TABLE 5.1 Major Uses of Optics in Industrial Manufacturing

	Automobiles	Semi-conductors	Chemicals	Aircraft and Aerospace	Construction
Value of shipments (billions of 1992 dollars)[a]	152.9	32.2	305.4	131.9	391.2
Photolithography	—	Critical	—	—	—
Laser materials processing	Critical	Major	—	Major	Significant
Rapid three-dimensional prototyping	Emerging	—	—	Emerging	—
Metrology (location, position, dimension, and alignment)	Major	Critical	—	Critical	Critical
Machine vision (features, orientation, and defects)	Emerging	Significant	—	—	—
Optical sensors (composition, temperature, pH, etc.)	—	Significant	Critical	—	Major

NOTE: Critical means that a technique is used pervasively and cannot be replaced by alternative nonoptical techniques without major negative economic impact to the entire industry. Major means that a technique is used pervasively and adds significant economic value to the entire industry. Significant means that a technique is used for specialized niche applications within an industry and adds significant economic value to those niche sectors. Emerging means that a technique is being put to increasing use in an industry and has the potential to be of at least significant importance.

[a]All shipment values are from the 116th Edition of the *Statistical Abstract of the United States*, 1996.

that represent large markets for optical systems and devices. An illustrative selection of other applications with significant potential for growth is also given.

This chapter is organized in five sections. Two explore the use of light to perform manufacturing and the use of optics to control manufacturing, respectively. Industry-by-industry examples follow to highlight the interplay between the various applications of optics to perform manufacturing in each industry. Prospects for increasing the use of optics in manufacturing are discussed in the next section. Findings, conclusions, and recommendations are gathered in the last section.

Use of Light to Perform Manufacturing

Because of the many unique properties of light and the manner in which light interacts with matter, optics offers a rich variety of application options for manufacturing processes. The imaging properties of light and its ability to induce photochemical reactions allow highly complex mask patterns to be transferred to photoresist in the optical lithography process. Tightly focused laser beams can deliver thermal energy to the workpiece for cutting, welding, or drilling with a precision and accuracy unmatched by any other technique; they can also induce localized photochemical reactions to generate solid three-dimensional prototype parts. Additional advantages are the ability to deliver this energy at a distance in a noncontact manner through windows and in various atmospheres. Some of light's diverse range of utility is illustrated in the following applications.

Photolithography

Photolithography plays an essential enabling role in integrated circuit processing. Photolithography requires both an optical system—the step-and-repeat camera (stepper) that is the workhorse of the integrated circuit (IC) industry—and an optical material—the light-sensitive photoresist used to transfer the desired pattern to the silicon substrate or thin film of interest (Figure 5.1). As the demand for faster processing speeds continues, increasing pressure will be put on photolithographic processes to produce smaller feature dimensions, requiring new photolithographic tools, new materials, shorter wavelength light sources, and other more advanced optical system designs.

At present, photolithography requires the use of three elements:

1. The mask, which defines which areas of the film to be patterned will be exposed to light;

2. The exposure tool, which images the pattern from a mask onto the substrate; and

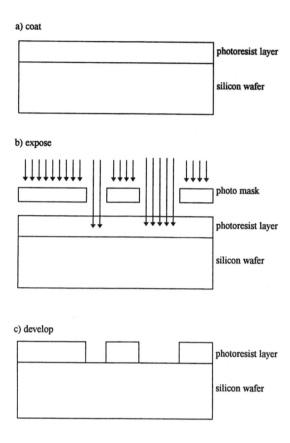

a) coat

photoresist layer

silicon wafer

b) expose

photo mask

photoresist layer

silicon wafer

c) develop

photoresist layer

silicon wafer

FIGURE 5.1 The photolithography process: (a) coat, (b) expose, (c) develop. Advances in the resolution and depth of focus of photolithography systems drive increases in the speed and performance of computers and computer-controlled systems.

3. The photoresist, which changes solubility when exposed to light and transfers the pattern on the mask to the film or layer below the photoresist.

Effective combination of these three elements, with appropriate integrated circuit design, has resulted in tremendous decreases in the minimum size of features and increases in the number of elements on a chip, allowing for increased speed and number of computational operations.

In the early 1980s, state-of-the-art IC devices contained as many as 8,000 transistor elements and had minimum feature dimensions of 5-6 µm. Today, devices with several million transistor cells are commercially available and are fabricated with minimum features of 0.5 µm or smaller. Indeed, the decrease follows an almost perfectly exponential trend known as Moore's law. The steady decrease in integrated circuit linewidths or feature size has largely been fueled by improvements in the resolution of optical lithography. This improved resolution, in turn, has been enabled by the use of shorter and shorter wavelengths for the exposure tools. Deep ultraviolet (UV) lithography using 248-nm wavelength light is just coming into production use for chips with minimum dimensions as small as 0.25 µm. A lithography roadmap prepared by SEMATECH (1997) projects the minimum feature sizes desired in the future and the technologies that must be developed to achieve them.

Exposure Tools

The workhorse of photolithography is the step-and-repeat camera. The optical imaging system of this device is the most demanding application of commercial lens design and fabrication today and can cost in excess of $1 million. Four competing demands on lens performance are (1) increasing resolution, (2) increasing depth of focus, (3) increasing field size, and (4) decreasing aberrations.

Maximum resolution and depth of field are determined primarily by the wavelength of the imaging light and the numerical aperture of the projection lens, with changes that increase resolution and result in decreasing depth of field. The trade-off of resolution with depth of field has driven many process changes, for example, the development of planar metallization.

Industry anticipates a transition to an exposure wavelength of 193 nm by a change in the excimer laser light source from KrF to ArF. Generations of exposure tools have relied on high-quality fused-silica refractive lenses. Due to issues of compaction and color center formation with fused silica, which are not adequately understood, the 193-nm exposure tools will likely use, for the first time, some reflective elements as well as CaF refractive elements. The lack of materials that are adequately transparent at 157 nm or 126 nm is a barrier to further reduction in wavelength, necessitating all-reflective exposure tools for use at these wavelengths. All reflective optical systems with high numerical aperture (0.6) are prohibitively difficult to fabricate because of the large number of aspherical reflectors required and the stringent specifications for these reflectors. This situation speaks to the issue of the manufacture of optics covered in Chapter 6.

Step-and-scan systems offer another alternative to the step-and-repeat equipment common today. Because of the difficulty of making bigger lenses, an alternate approach is to combine modest-sized lenses with scanning systems to increase the field size. By synchronously scanning the mask and the wafer through an illuminated area corresponding to the corrected field of the lens, it is possible to achieve patterning over large areas. The synchronization between the mask and the wafer stages must be kept well under 100 nm, which is not easy. However, for 256-megabit DRAM chips and beyond, step-and-scan technology will likely prove more cost-effective than step-and-repeat because of the smaller optical system employed.

Photo Masks

In the past 10 years the transition from 1x to 4x and 5x optical systems has provided a technology respite to the mask-making industry, but the recent emphasis on optical proximity correction combined with the relentless trend toward smaller geometries and more complex structures has accelerated mask-making requirements. The mask-making industry generates insufficient revenue to cover the cost of developing new generations of mask-making tools. Given the current direction, mask making will almost certainly be a major impediment in only a few years, although there are some initiatives under way aimed at alleviating this.

Mask alignment is also a critical issue. Subsequent masks must be precisely aligned with patterns on the silicon wafer with a precision far beyond that of the minimum feature dimensions. New metrology will be required for next-generation systems.

One interesting possibility is the conversion to maskless systems that have a large micromirror array or similar device in the lens focal plane. In this case the mask pattern is simply a data file stored on an array of hard disks or other high-speed data storage device, which feeds pattern data to the mirror array. The flexibility of an electronic mask would be unprecedented and could correct for small imperfections in the imaging system.

Photoresist

The pattern on the photomask is transferred to the silicon wafer by means of a light-sensitive polymer that is spun uniformly onto the wafer surface. Exposure to UV light changes the solubility of the polymer such that the exposed (positive photoresist) or unexposed (negative photoresist) regions can be removed in a solvent after exposure. Optimum materials exhibit high photosensitivity and uniform absorption of the UV light for uniform solubility and contrast.

The key to developing an effective photoresist is to develop a material with excellent etching resistance *combined* with good imaging characteristics. This combination presents a significant challenge and is the focus of several research efforts today.

Present conventional photoresists are not appropriate for use with the nonconventional lithographic technologies that will be necessary for sub-0.5 µm lithography. The most notable deficiencies of the conventional novalac-quinonediazide resist are the exposure sensitivity and absorption properties of the materials. New photolithographic tools in general have low-brightness sources, and high-sensitivity resists are highly desirable. Additionally, the absorption of conventional photoresists is too high to allow uniform imaging through practical resist film thicknesses, usually on the order of 1 µm. For 248-nm lithography, these challenges were accommodated by application of chemically amplified resist technology, which greatly enhances photosensitivity. However, hydroxystyrene polymers, which form the basis for this technology, are effectively opaque at 193 nm. Thus, new polymer materials are required for 193-nm single-layer resists that possess high optical transparency at the exposure wavelength, combined with good etching resistance and functionality that will effect a change in solubility of the exposed regions.

No matter what technology becomes dominant when today's photolithography capabilities have reached their limits, new optical materials and processes will be required, necessitating enormous investments in research and process development. The introduction of new resist materials and processes will also require a considerable lead time to bring them to the performance level currently realized by conventional materials, as has been the case with new photolithography techniques. For example, the printing of 0.5-µm features, common in manufacturing since 1993, was the result of more than 12 years of development of new photoresists, designed to respond to 248-nm light, and new step-and-repeat cameras producing that illumination. The next big decrease in resolution, which is expected to be in production in 2001, is the production of 0.18-μm features; this technology will use photoresists that have been under development since 1992, as well as new steppers operating at 193 nm.

Future of Photolithography

What are the alternatives for future advances in photolithography? There are currently several possibilities:

- *Wavefront Engineering.* Because integrated circuit design uses a limited set of objects with limited dimensions, the limitations of classical imaging can be overcome by appropriate design of a mask feature, use of phase-shift masks, or modifying the illumination to change the amplitude and phase of the optical wavefront.
- *Extreme Ultraviolet (EUV).* At wavelengths as short as 14 nm, small numerical aperture reflective systems can provide high resolution and depth of focus. Hurdles to overcome include EUV-robust and reliable x-ray sources, defect-free EUV masks, aspheric reflective optics, and surface imaging photoresists.
- *Electron Beam.* Electron beam projection lithography offers promise for resolution as fine as 30 nm with a depth of focus as high as 75 μm for 0.25-μm features. This approach would, however, require a significant departure from current industry processing; for example, electron beam lithography requires processing under vacuum. High cost and low throughput continue to limit the use of this technology.

At this time, considerable progress can still be made with optical lithography—previous predictions of its demise have proven completely wrong. It is important, however, to recognize that the risk is too great for a new technology to be introduced in a single generation of devices. Whatever the technology of choice, it must be developed and put in limited production with operational experience well before full implementation. Appropriate metrology tools for process control and evaluation must be developed in parallel with improved lithography equipment.

Laser Materials Processing

Laser materials processing offers many powerful advantages for manufacturing applications. Unlike competitor technologies such as resistance welding, plasma arc cutting, and flame hardening, lasers deliver energy to the workpiece without physical contact, provide high localized energy densities, and are remarkably versatile in their energy delivery. Although capital equipment acquisition costs can be high, once installed the ease of application, high-speed processing, reproducibility, reliability, greatly reduced distortion especially in thin sheets, ability to interact with complicated shapes (joints in restricted areas can be welded provided a line of sight to the weld is available), and environmental advantages (especially compared with chemical processing) make laser materials processing increasingly attractive for commercial applications. Such applications range from macroscopic processing (e.g., metal welding, cutting, drilling, slitting), where the thickness of material processed can be several millimeters, to new advances in micromachining where dimensions range from 1 mm down to 1 µm, and finally to the submicron processing of semiconductors.

Although laser types abound, there are two that dominate the laser materials processing field (Bell and Croxford, 1995): CO_2 and Nd:YAG. At present only these two types provide sufficient power and a usable beam as a package that can be integrated economically into a production line. Of the two, CO_2 lasers have tended to dominate the higher-power market, whereas Nd:YAG is favored for high-precision, low-heat-input applications. Another type of laser, the excimer laser, is beginning to make an impact in industrial processing (Weiss, 1995). These lasers operate in the ultraviolet part of the spectrum (as contrasted with the infrared), and favored types include KrF (krypton fluoride) and XeCl (xenon chloride). Besides the potential for ultrasmall feature sizes due to the short wavelengths, an advantage of the excimer is the way it interacts with materials. Materials processing using an infrared laser is a thermal process, whereas the laser-materials interaction with high-power pulses of UV radiation is a "cold process" that uses energy to break chemical bonds rather than heat the material. Thus, excimer lasers are particularly useful for processing polymer-based materials and ceramics to avoid problems of ablation, charring, or gasification that often accompany the heating of these materials to high temperatures. Box 5.1 notes the use of excimer lasers to clean ancient metal art objects.

There are two areas in which the general field of laser materials processing could benefit from advances in optical technology. The first is

· ·

BOX 5.1 LASERS FOR ART RESTORATION

Cleaning ancient and antique art objects can be a tricky task. Chemical cleaning can harm surfaces and be hazardous to the environment. In one alternative method, an excimer laser is used to clean items such as ancient Roman coins by exciting particles in microcontaminants and breaking their bonds with the surface. A flowing inert gas then blows away the contaminants, leaving the surface undisturbed. This process is expected to reveal details by removing oxidation products.

improvements in optomechanics to solve beam delivery problems for applications such as pipe welding. A second area is adaptive optics to make possible laser surface treatments of arbitrarily shaped surfaces through obscuring, turbulent, or aberrating intervening regions between the laser and the workpiece.

Welding Applications

In the automotive industry, lasers have been used to join stamped steel panels to form underbodies (American Society for Metals, 1983). The process is computerized, and welding is performed at a rate of 1,000 to 1,150 cm/s. Laser welds are continuous, which results in high structural integrity and eliminates the need for a sealing operation. The ability to program a laser welding system has the advantage that underbodies for any car model can be welded by calling up the correct program from the computer memory.

A growing application for laser materials processing is the welding of zinc-coated (galvanized) steel for car bodies. The low melting point of zinc (419°C versus 1535°C for steel) greatly changes the characteristics of the process (Bell and Croxford, 1995). If allowed to stay in the molten weld pool, zinc can alloy with steel during the welding process to produce unacceptable welds. The zinc also forms vapor pockets that, when trapped, expel molten material out of the weld, thus resulting in weld porosity. The solution to this problem includes the use of a pulsed laser that rapidly vaporizes the zinc out of the weld zone before joining the steel.

In the electronics industry, laser welding is used to seal electronic devices that are either high-value, low-quantity production devices or welds that must meet stringent reliability and other special end use requirements (American Society for Metals, 1983). Examples of the latter type include hermetically sealed devices for commercial and military aircraft applications. These devices must maintain highly reliable operating performance under extremely severe environments. For instance, the laser welding of relay containers processed according to military specifications has proven an effective way of sealing each package. Laser welding has been quite useful in such applications because of its ability to produce welds near heat-sensitive, glass-to-metal seals.

The production of heart pacemakers is another application requiring high-quality welded construction (American Society for Metals, 1983). Today, laser welding has become a widely accepted technique for producing hermetic welds in titanium and stainless steel pacemaker cases. A principal power source for pacemakers is the lithium battery, which because of its highly reactive nature must be hermetically sealed. Lithium cells have also entered the consumer and industrial markets as long-lived power sources for such applications as watches, calculators, and backup power for computer memories. The small size, coupled with requirements for a fusion welded seal in close proximity to the reactive contents of the cell, again make the laser ideal for this job.

Although most industrial laser applications are autogenous, lasers are increasingly used in the production and refurbishment of components by adding material during processing (Azer, 1995). For example, prealloyed or mechanically mixed powders can be added to a weld pool. By scanning a laser beam across the workpiece, a weldment is created that is metallurgically bonded to the base metal. The automotive, aerospace, oil, and nuclear industries are benefiting from such welding and cladding techniques.

Cutting and Drilling Applications

In laser beam machining and drilling, material is removed by melting. Such melting does not involve mass material removal since only a very thin layer is actually melted. The technique has the advantage of rapid material removal with an easily controlled, noncontact, nonwearing tool. A major application of lasers is in metal cutting, primarily two-axis profiling of sheet goods that otherwise would be blanked out by punch presses or fabricated by hand after laborious layout of the pattern (American Society for Metals, 1989). Laser cutting and drilling is ideal for batch processes, just-in-time, or low- to medium-volume production. Sheet thicknesses up to 13 mm can be processed.

A recent publication describes one of the first uses of an Nd:YAG laser (instead of the CO_2 laser) for cutting sheet metal (*Industrial Laser Review,* 1997). The application is for assembling burner systems and high-pressure cleaning machines. Other applications include the production of continuous-flow oil heaters and exhaust mufflers. Burst disks for hydraulic systems represent another product. If a fault develops, these disks are designed to burst at a given pressure to vent the system. They are made of high-grade steel, 1 mm thick, and are laser-cut at a precise location to break under a specific pressure.

Laser beams are used also to drill small-diameter holes in stainless steels. Advantages compared to other techniques include lower aspect ratios, less deformation of hole walls, higher accuracy, less taper, and most important of all, high production rates (one hole per second).

Micromachining with excimer lasers is becoming increasingly popular (Weiss, 1995). Applications include using an excimer laser as an alternative to ion milling to pattern thin films onto disk-drive heads. Areas other than semiconductor and electronic applications are growing as well and include flow orifices, such as nozzles for inkjet printers and automobile fuel injection; optical fiber positioning ferrules and waveguides; devices for DNA and other biomedical or biotechnology research; and medical devices. Lasers may be the only technology that can process such materials as chromium- and titanium-based metals for orthopedic implants without unacceptable levels of corrosion.

Laser cutting and welding techniques are used for nonmetallic materials as well. They are highly effective in cutting hard workpieces with low electrical conductivity such as cubic boron nitride, an ultrahard tool material. Alumina likewise can be cut or drilled with lasers rather than diamond saw blades and drills that rapidly dull or wear out.

Surface Hardening Applications

The flexibility of laser delivery systems has made lasers very effective in selective hardening of steel surfaces, especially those subject to wear or fatigue. Although in such applications the heat generated by the laser at the surface is controlled to avoid melting, a steep temperature gradient is set up between the surface and the interior. Selective austenization (change from body-centered iron to face-centered iron) occurs at the surface, which transforms to martensite (a hard form of iron resulting from a diffusionless phase transformation) as a result of rapid quenching (self-quenching) through the conduction of heat into the workpiece. Because the process is all solid state, no change in chemistry is produced at the surface by this laser transformation hardening.

Laser transformation hardening is often used to harden the surfaces of automobile components such as camshafts and crankshafts. High hardness and good wear resistance with less distortion result from the process. Also, the laser method differs from induction and flame hardening in that the laser can be located some distance from the workpiece (American Society for Metals, 1991).

Molian has tabulated the characteristics of 50 applications of laser transformation hardening. The materials hardened include plain carbon steels, alloy steels, tool steels, and cast irons. Because the absorption of laser radiation in cold metals is low, laser surface hardening often requires energy-absorbing coatings on surfaces (Molian, 1986).

Industrial Lasers Market Perspective

A special class of lasers, known as "industrial lasers," has evolved to serve the needs of laser materials processing for manufacturing and exists as an industry in its own right. The United States once dominated this industry, but in recent years has dropped to a minority share (Box 5.2).

The three main applications of industrial lasers today are (1) sheet metal cutting, (2) automotive welding, and (3) component marking and product coding. By mid-1995 more than 62,000 industrial lasers had been installed worldwide. The annual worldwide market for industrial lasers has grown to more than $400 million per year (Table 5.2), with the worldwide market for systems that use industrial lasers at approximately $1.5 billion, supplied by 500 separate companies that employ

BOX 5.2 RESTORATION OF U.S. COMPETITIVENESS

The United States has slipped from its leadership position in industrial laser technology. A comprehensive plan is needed to restore U.S. competitiveness, including promotion of this technology, education of laser-aware engineers, and making information more available to potential users.

• •

more than 18,000 people (Belforte, 1995). By the year 2000, however, at least 25% of the new industrial lasers installed are expected to be used for applications that are not yet in routine use, such as precision hole drilling in the aerospace industry. Emerging technologies being developed for these applications in Europe and Japan include high-power Nd:YAG lasers coupled with robotic beam delivery (mostly in Britain and Germany), combination systems for metal fabrication (mostly in Japan), and compact high-power CO_2 and Nd:YAG lasers (mostly in Britain and Germany). No similar scale development effort is under way in the United States.

Until the 1980s, industrial laser technology and the industrial laser market both developed slowly and without direct government funding. U.S. suppliers dominated the world market for many years, although a few European companies began to produce CO_2 and solid-state lasers as early as 1970. Between 1970 and 1985, however, there was a significant net outflow of laser technology from the United States, and in 1985, industrial laser activity in Europe and Japan began a period of rapid development in both technology scope of application and market

TABLE 5.2 Annual Worldwide Market for Industrial Lasers

Year	Market ($ millions)	Increase (%)
1970	8	—
1975	10	25
1980	40	300
1985	100	150
1990	250	150
1995	410	64
2000*	820	100

* Projected

Source: Belforte, 1997.

TABLE 5.3 World Share of Industrial Laser Production

Source	1980 Share (%)	1996 Share (%)
United States	85	37
Europe	10	29
Japan	5	31
Other	0	3

Source: Belforte, 1997.

share (Belforte, 1995). Fifteen years ago the United States supplied more than 90% of all industrial lasers worldwide, but by the early 1990s the U.S. market share of the industrial laser business had slipped to only 37% (Belforte, 1995; see Table 5.3). U.S.-owned companies now enjoy a leadership position only in low-power sealed CO_2 lasers; intermediate-power fast axial flow CO_2 lasers; low-power, diode-pumped Nd:YAG lasers; and laser marking and coding systems (Belforte, 1995).

Industrial lasers must compete with many other materials processing technologies. To be accepted by industry as a new replacement option, they must show excellent cost-benefit performance. Acceptance also requires the education of potential users. European and Japanese universities have graduated numerous trained applied laser engineers who now occupy decision-making positions in industry in their countries. No similar scale effort has occurred in the United States.

The Precision Laser Machining Consortium

The Precision Laser Machining (PLM) Consortium is a government-industry-academia alliance designed to spur new technology development and insertion to reestablish U.S. leadership in industrial lasers and capture a larger share of the total market. The PLM alliance was formed under the DARPA Technology Reinvestment Project (TRP). Boeing, Caterpillar, Chrysler, Cummins, Ford, GE Aircraft Engines, General Motors, Newport News Shipbuilding, TRW, and United Technologies are among the 20 organizations to join forces under this $38 million project. The above industries, which represent 8% of GDP, drive development teams to achieve the PLM alliance goals.

The alliance plans to build on the Department of Defense (DOD) high-brightness diode pumped solid-state laser (DPSSL) program, exploiting this U.S. technical advantage as a dual-use technology. High average-power diode-pumped Nd:YAG can be scaled to power levels beyond the current generation of lamp pumped rod geometry lasers. Two lasers are in development: a 6-kW quasi-continuous wave (cw) or cw laser and a 2.5-kW Q-switched laser at 25 MW peak power, both with beam divergence less than three times the diffraction limit. PLM also focuses on flexible fiber-optic laser-system power delivery with a net intensity increase of 30 to 100 times that of current industrial laser systems and on extensive machining process studies.

Figure 5.2 shows an example of the preliminary results that have interested the alliance users. Most of the benefits accrue from shorter wavelengths, higher beam intensity, and sharper focus. These subscale tests will be expanded to two beta-test sites and eventually to six to eight Laser Application Centers. The sites will feature machining stations in a factory setting with computer numerically controlled robotics, positioning, and diagnostics. Information on laser machining will be made available through outreach programs to educate users and designers in the application of these laser tools.

Key challenges for PLM and beyond include the following:

- Providing highly reliable lasers at an affordable cost;
- Achieving lower diode laser array costs and increasing their life to 20,000 hours or so;
- Achieving near-diffraction-limited performance from multimode fibers; and
- Developing a "smart machine tool" with real-time process control.

Rapid Prototyping and Manufacturing Using Optics

The pressure to bring new products to market on an ever-decreasing design-to-manufacture cycle time has driven the development of techniques to produce prototype parts in the early stages of development,

FIGURE 5.2 Technology developed by the Precision Laser Machining Consortium can cut composite materials for airframes with "polished edge" quality (left). Conventional lasers (right) leave charring and delamination. (Courtesy of L.J. Marabella, TRW.)

BOX 5.3 NEW APPROACHES TO MANUFACTURING

Techniques for creation of solid materials from CAD tapes can now produce tooling directly, revolutionizing the manufacturing process and creating a business worth approximately $500 million per year.

••

where adjustments can be made to part size and shape easily and inexpensively.

Previously, firms spent several months generating drawings, using extensive hands-on tool or pattern makers. In the new paradigm being employed by major automobile, aerospace, digital equipment, plastic product, and other manufacturers, designers and modelers interact, exchanging part descriptions digitally. Many smaller firms use service centers that own, operate, and maintain the solid manufacturing capability for the "manufacturing middle," the time between completion of the design and manufacture of the first part based on that design. More than 200 service centers now exist worldwide. Worldwide sales are expected to exceed $1 billion by 1998, at an annual growth rate of 50%. This reflects the desire of major firms to save money and shorten the product time-to-market.

Some advantages of rapid prototyping are as follows:

- Optimizing design before commitment to hard-tooling;
- Reducing cost and lead time in the product realization cycle;
- Producing prototype models more efficiently and quickly; and
- Quickly providing high-quality patterns for investment casting.

Computer-aided design (CAD) and manufacturing (CAM) started the manufacturing industry on the path to a totally digital process. Computer numerically controlled machining equipment translates the digital information directly to create the physical product. All of this technology makes the manufacturing process more efficient, has tremendously improved the quality of final products, and allows for cost-effective, flexible, agile, and portable manufacturing approaches (Box 5.3).

Laser modeling (Figure 5.3) is a new addition to this suite of digital, automated manufacturing process steps that addresses the manufacturing middle. Solid models can be created in several different media by using laser irradiation to build a solid design directly from the information stored on a CAD tape, layer by layer. Relay and scanning optics are used to position and focus moderate-power visible lasers to photopolymerize a liquid or by high-power CO_2 lasers to thermally activate powder or paper. The first use of rapid prototyping in commercial, industrial equipment was based on photopolymers; systems based on this technique

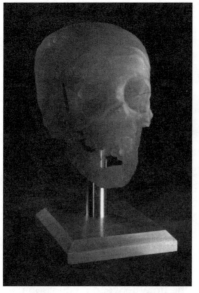

now constitute about half of the market for rapid prototyping equipment. With newer epoxy resins, surface roughness accuracy better than 50 micron inches [root mean square (rms)] are achieved for models and finished forms for casting. In a competing rapid-prototyping technology, powders of thermoplastic-coated particles are sintered into a solid, or adhesive-coated paper is patterned by CO_2 laser radiation with powers of approximately 50 W. Other variants of these basic photopolymer and powder techniques are under development or are manufactured in limited quantities, especially in Germany and Japan. The United States dominates the industry with 60% of the systems in use worldwide.

To maintain a competitive edge, the industry is moving away from visible gas lasers toward doubled or tripled diode-pumped YAG lasers. A major goal is a solid-state 1-W laser at 350 nm. In all wavebands, the cost of beam delivery and modulator optics is too high and has to be reduced. Beam shaping and positioning also require improvement to achieve RMS accuracy of less than 2.5 micro-inches. Although this technology is creating a manufacturing revolution, individual equipment costs exceeding $100,000 are an impediment to even higher growth. Cost reductions starting with the incorporation of laser diodes will lead to market expansion.

Use of Optics to Control Manufacturing

Optics has immense advantages for providing real-time information that can be used to control manufacturing processes. Optical probes for in-line process control offer the ability to perform noncontact remote sensing in hostile environments with high-speed response and high spatial resolution; optics also provides other important benefits such as spectroscopic analysis of composition. The imaging capabilities of optics can be exploited by machine vision for automated (robotic) manufacturing processes. Optical metrology techniques allow control of critical dimensions and layout or positioning; infrared imaging allows rapid determination of temperature profiles of semiconductor chips and preventive maintenance in many manufacturing scenarios. The text below discusses selected examples of such techniques.

Metrology

Metrology, by definition, is the science of weights and measures, but the term is more broadly applied to the entire field of measurement and inspection. Optically based metrology and inspection systems are found throughout industry for the manufacture of a broad range of products. Industry's use of optical metrology is widespread and growing, but limited at present by the need to customize a solution for each specific application. Some of the optical metrology techniques that are in use today are based on conventional imaging including video and microscopy systems, displacement measurement and ranging, interferometry, scanned imaging systems such as confocal microscopes, structured lighting for large structure shape profiling, optical microsensors using fiber optics and integrated optics, scattering, spectroscopy, and polarization, to name a few.

As shown in Figure 5.4, the optoelectronic sensing equipment market for industrial metrology applications is expected to double every 5 years.

The usual optical metrology system consists of an illumination source, optics to focus the illumination and direct it where needed, a detector to collect information about the part under inspection, and electronic processing to extract the desired property from optical data. Uses of metrology systems include the following:

- Inspecting products and components to identify defects;
- Measuring products for dimensions and conformance to specifications;
- Monitoring manufacturing process conditions; and
- Positioning and aligning pieces for subsequent processing.

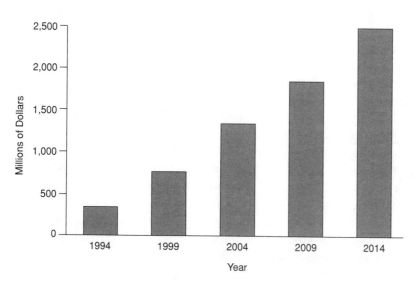

FIGURE 5.4 *Projected worldwide market for optoelectronic sensing equipment. (Source: Optoelectronics Industry Development Association.)*

A major economic benefit of optical metrology systems is in the use of measurements to provide real-time manufacturing process feedback control. The advantages of optical metrology are that it is fast, it involves no contact with the part, and it is nondestructive. Optical inspection systems can be sensitive to a variety of parameters simultaneously, such as chemical composition, electronic state, mechanical stress, temperature, size, flaws, or texture. Many optical metrology systems are built using visible light, but specific applications may use wavelengths ranging from the vacuum ultraviolet (10 nm) to the far infrared (300 μm). This broad wavelength range is of great utility in metrological applications because of the ability to measure relative dimensions with equal accuracy over a wide range of scales.

Measurement problems facing industry include measuring absolute dimensions over long distances to an accuracy of less than 100 μm, high-accuracy measurement of aspheric surfaces, three-dimensional data visualization, inspection of particle contamination as small as 10 nm on semiconductor wafers, and waveguide sensors for biological and chemical sensing. In one such waveguide sensor, embossed gratings control the light path, and the sample to be measured is analyzed spectroscopically as it flows over the waveguide.

Another general trend in optical metrology system development is the use of optics to encode additional information about the part in the image captured by the machine vision system. An example is to use structured lighting to allow height or profile information to be determined from an image. Other examples are dark-field illumination for defect enhancement or multispectral characterization.

Measurement standards and practices are another area in which optical metrology systems play a significant industrial role. A current problem is dimensional metrology for lithography systems. As integrated circuit feature sizes decrease, the linewidths on the wafer must be measured to an accuracy of 1 nm. It is not surprising that no suitable linewidth standards exist for these narrow lines, since an industry rule of thumb is that a measurement system must have resolution approximately one order of magnitude finer than the accuracy required (Brueck, 1995). Standards for film thickness measurement are also being pushed to their limits. Specifications for oxide layers ranging from 40 to 6000 Å (angstroms) call for film thickness accuracy better than 2% and uniformity over a wafer of better than ±5%. As with measurements of lateral dimensions, standards and measurement systems for vertical thickness must be significantly more accurate than the manufacturing dimensions they are designed to address.

Other areas that require development of standards are high-definition imaging systems and colorimetry. A goal of the latter is that the appearance of a color on a cathode-ray tube (CRT) display of a computer-aided design or layout should match the produced color on a textile or printed hard copy under the chosen illumination.

Future advances in optical metrology systems that are likely to have a large economic impact on manufacturing include the development of integrated optical sensors, more optics-based sensors for process monitoring and feedback control, smarter sensors with advanced data processing techniques, holographic three-dimensional imaging, and optical figure measurements to 1-nm accuracy.

One often-observed barrier to the introduction of optical metrology systems is the apparent need for custom solutions for each specific application. This customization has an impact on the selection of a particular optical configuration, the illumination system, and the processing algorithms. This situation implies that after new metrology technologies are developed, the technology must often be customized before implementation, which increases cost and the time for system delivery. More broadly applicable "plug-and-play" systems are needed, especially modular units and mounting apparatus.

Machine Vision

Machine vision in the manufacturing environment means the extraction of useful information from the imagery of products or product-related scenes, using image processing techniques and pattern recognition algorithms. Machine vision is a collection of image-based applications that use sensitive, high-resolution, visible and infrared cameras for detection purposes. Rapid data processing then allows accurate establishment of location and dimensions for fixtureless machining, high-sensitivity flow detection for quality control, metrology, and other applications. The goal of machine vision is to allow automatic adjustment and optimization of the manufacturing process, quality control, and inspection ultimately in real time. Machine vision is currently limited to use in expensive, multistep processes and industries that require high throughput but simple inspection. This will remain true unless machine vision algorithms improve dramatically.

Most machine vision applications are being developed to replace human inspectors. One aim is cost reduction. Another is the need for a more objective, quantifiable assessment that provides information in a form suitable to control or modify a process or to perform acceptance testing and results in enhanced reproducibility and precision.

Replacing human visual inspection can have several advantages. Human performance varies from inspector to inspector and degrades after inspecting continuously for an extended period. Human performance also degrades as the complexity of the inspection task increases. For example, one particularly successful application of machine vision has been the replacement of human inspectors of printed circuit boards and silicon wafers, a highly complex task given the industry's stringent manufacturing requirements. Machine vision with appropriate signal processing and process control can replace a human on a high-speed assembly line, allowing even higher processing speeds and performing inspections when an item is available for only a fraction of second. However, the image analysis capabilities of human inspectors are unsurpassed for a wide variety of inspection tasks such as cosmetic appearance. Human inspectors are generally insensitive to small variations in lighting or orientation. The complete range of remarkably robust and flexible capabilities of human vision cannot presently be duplicated.

Beside its obvious and frequent use in robotics, machine vision can provide a wealth of information, including the size, position, and orientation of products; spectral characteristics; texture; and defect detection and classification. Some of the tasks for which machine vision systems are used in a manufacturing setting include the following:

1. Product inspection
   - To catch defects at intermediate stages of a process,
   - To detect hazardous defects in a final product, or
   - To verify the cosmetic appearance of the packaging;
2. Part identification and location on an assembly line;
3. Gauging (process monitoring);
4. Process control feedback; and
5. Sorting and grading.

A complete machine vision system typically requires carefully designed lighting, video hardware, user-computer interfaces, powerful computer hardware to carry out computationally intensive processing rapidly, and a set of algorithms to analyze the imagery. As defined here, machine vision systems represent a subset of the more general category of industrial metrology and inspection systems. Machine vision systems analyze and extract information from video imagery and do not include systems, such as proximity position sensors, that require little elaborate image analysis. About 7,000 machine vision units were sold in North America in 1990, according to a survey by the Automated Imaging Association. Various manufacturers have developed a similar number of units in-house.

What has limited the further application of machine vision technology? By far the most important limiting factor is the poor performance of the image processing and pattern recognition algorithms. Although many sophisticated algorithms have been developed, each is designed for a specific task and can perform effectively only in highly constrained or defined situations. In a defect detection application, for example, changes in lighting, changes in the reflectance of the product, or changes in position or orientation—to which human inspectors are relatively insensitive—can be interpreted incorrectly by the computer system as a flaw in the product. There are also some hardware limitations, such as the limited resolution of charge-coupled devices for gauging systems. However, machine vision technology can be designed, on a case-by-case basis and under very tightly controlled conditions, to handle many useful tasks. There is no widely applicable generic approach, and the necessary sophistication of the design and equipment can make machine vision quite expensive.

Two areas might lead to improved algorithms and hence to broader use of machine vision technology in manufacturing: (1) Interdisciplinary vision science research between psychologists researching human vision and engineers and scientists researching machine vision would help to understand and mimic human visual perception; and (2) techniques employed in satellite image analysis, including the use of extended wavelength or multispectral information and recognition of high-dimensionality patterns, would provide advantages that machine image analysis might have over human vision.

Sensors
The topic of sensors is covered in depth in Chapter 3.

Specific Industrial Applications

This section provides examples of the use of optics in several large manufacturing industries. In these industries, optics is used to perform and control manufacturing. The specific examples that follow are the automobile, semiconductor IC, chemical, aircraft, construction, and printing industries. There are many other important uses of optics in manufacturing, but they are so diverse and widespread that this report can address only a few representative examples.

Automobile Manufacturing
The utility and insertion or adoption of optical methods in automobile manufacturing is ultimately determined by basic economics. Because a cost differential of even $100 per car is significant in this industry, greater manufacturing cost-effectiveness is required. The potential for optics to lower automotive manufacturing costs is largely untapped in the United States, and fuller exploitation requires immediate changes in the optical engineering infrastructure.

The use of high-power lasers for materials processing functions such as cutting and welding has now become routine in numerous automobile manufacturing applications. Despite a decade-long head start in the United States, this area is now dominated by Japan, which currently has

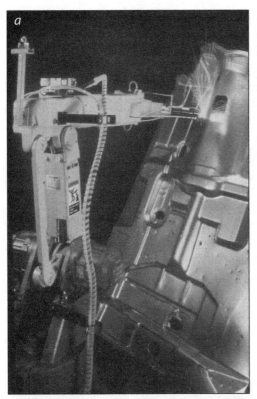

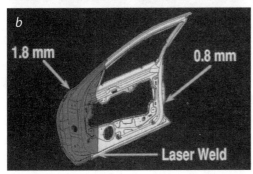

1.8 mm   0.8 mm

Laser Weld

FIGURE 5.5 (a) Fiber-optic-guided Nd:YAG laser beam cutting holes in a truck floor. (b) Design for a tailored blank for a car door panel. (Courtesy of D. Roessler, General Motors R&D Center.)

about 60% of the world's automotive lasers. Some current U.S. applications are a belated recognition of the successes enjoyed overseas.

Figure 5.5(a) shows the use of a fiber-optic guided Nd:YAG laser beam to cut various option holes in a truck floor pan. The flexibility of the laser allows a drastic reduction in the number of different floor pans that must otherwise be stamped out. On average, about half of the sheet metal used to make automobile bodies ends up as scrap. Figure 5.5(b) shows how a door panel can be made from a "tailored blank"—a sheet of metals made by welding together smaller pieces of varying thicknesses, alloys, and coatings to give properties appropriate to different parts of the panel. The use of tailored blanks greatly reduces scrap and also ensures that expensive or heavy materials are used only where necessary in the final component.

It is important to note that in the area of automotive manufacturing, there is an important technical difference between U.S. auto manufacturing and foreign auto manufacturing. U.S. automotive production lines have a significantly higher throughput than those of the leading foreign adopters of industrial lasers for materials processing, such as Volvo and Daimler-Benz. Installation of industrial lasers on these high-volume automotive assembly lines will be on a step-by-step basis, achieved only after cost benefits are proven for each step (Dinda, 1996).

Machine vision systems are being used in a wide range of applications in automotive manufacturing and help improve quality by generating information used to adjust and optimize manufacturing processes (Box 5.4). The principal needs of machine vision system users in the automotive industry are more robust systems configured from low-cost, standardized components. There is a need for an accurate three-dimensional machine vision system that would serve as a foundation on which to build a variety of fully automatic manufacturing systems that can automatically adapt to their changing environment, leading to the fixtureless manufacturing line.

Although the use of tailored blanks and other such advanced laser materials processing technologies is not fully used in U.S. manufacturing plants, several other modern optical techniques are being pursued

BOX 5.4 IN-LINE VISION GAUGING FOR AUTOMOTIVE BODY ASSEMBLY

The dimensional quality of an automotive body greatly influences the downstream assembly operations (panel fitting and general assembly), quality and functionality (wind noise, water leakage, and door closing effort), and customer perception (gaps and flushness) of the vehicle. Accurate and rapid measurement of these assemblies has been difficult to accomplish. However, in-line vision gauging systems have now been developed as the keystone checking device to measure dimensional variation. These systems are known as in-line optical coordinate measuring machines (OCMMs). These optical gauges are essential in the effort to improve the fit-up of automotive body panels to a gap tolerance within 2 mm (see Figure 5.6).

An OCMM can measure the dimensions of each automotive body produced. It can also rapidly measure as many as 100 points on the body, providing high-dimensional multivariate data. By investigating the relationship among the various measurement points obtained by the OCMM, the sources of dimensional variation can be readily identified, and appropriate corrective actions can be taken.

The basic principle is cross-correlation among measurements. An OCMM measures critical points on the subassembly or component employing several sensors. During assembly, these points move together in some pattern. Analysis of the degree of correlation and relative variation in the pattern of these points (e.g., translation or rotation) leads to systematic identification of workstations that are introducing dimensional variation (Hu and Wu, 1992). The reasons for the dimensional variation can then be addressed to improve the assembly tolerance.

This methodology has been successfully applied in automobile assembly plants to reduce variation and shorten the launch times for new body assemblies. Figure 5.7 shows the reduction of variation in one automotive assembly plant.

· ·

actively in the automobile industry's development and systems integrating laboratories, frequently at auto industry component suppliers' facilities. Some, such as laser ultrasound for nondestructive testing of solid materials, offer great potential, but their success will be determined only in the relatively distant future. Thermal imaging systems offer unique capabilities in the area of predictive and preventive maintenance, where hidden problems in machinery or defects in products can often be located and detected prior to failure due to their inducing some form of localized heating. Adapting semiconductor processing techniques for scribing, labeling, and trimming to the automobile industry will focus particularly on cost effectiveness. Other technologies are likely to become important in the nearer term, such as those illustrated by the following three examples:

1. Machine vision for drilling, welding, and aligning;
2. Microelectronics to control automobile functions; and
3. Rapid prototyping to reduce the time from design to production, now typically on the order of 4 years.

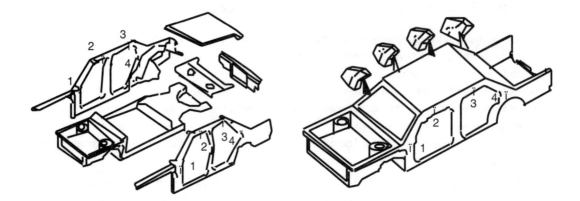

FIGURE 5.6 Schematic of an optical coordinate measuring machine (OCMM) measuring body dimensions. (Courtesy of D. Roessler, General Motors R&D Center.)

The committee believes that the use of modern optics in automobile manufacturing will continue to increase but that the pace of introduction of these methods in the United States is slow in comparison with many foreign competitors. Improved training in modern optical manufacturing methods is important for efficient insertion and technical development of these new applications. This training should involve all levels of the industry, from the assembler to the plant engineer to plant management to the user and field installation or repair sectors.

The Semiconductor Integrated Circuit Industry

The immense semiconductor integrated circuit manufacturing industry is powered by optics, optical systems, and optical materials. A modern electronic integrated circuit is a complex three-dimensional structure of alternating patterned layers of conductors, dielectrics, and semiconductor films. This structure is fabricated on an ultrahigh-purity wafer substrate of a semiconducting material such as silicon or gallium arsenide. The speed and density of the devices is, to a large degree, governed by the size of the individual circuit elements. As a general rule, the smaller the elements are, the faster is the device and the more functions or operations it can perform per chip. The device structure is produced by a series of steps used to precisely pattern each layer. The patterns are formed by the photolithographic processes discussed earlier in this chapter. For each processing step performed, there are a variety of measurement and inspection techniques (many of them optical) to ensure that the step was performed correctly.

The worldwide market for semiconductor components was estimated to be $100 billion in 1993 (Brueck, 1995) and is expected to grow by 20% or more per year for the rest of the century. To meet the demand, the capacity for semiconductor chip manufacture will continue to increase. Semiconductor manufacturing tools were a $12 billion market

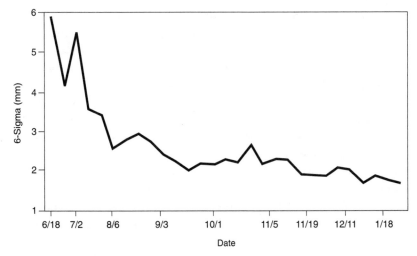

FIGURE 5.7 Reduction in body variation. (Courtesy of D. Roessler, General Motors R&D Center.)

in 1993; with an expected compound annual growth rate of 17%, the 1997 market was projected to exceed $20 billion (Brueck, 1995).

Photolithography is a key enabling technology for the industry. The total costs for photolithography are approximately 35% to 40% of each chip fabricated, because each wafer must pass through the photolithography process many times to build a complete integrated circuit. For example, a 64-megabit chip requires about 21 photolithographic steps.

In addition to photolithography, there are numerous applications of optics in inspection and process control. The very first step in IC manufacturing, an initial inspection of the silicon wafers as they are removed from their packages, is performed by shining light at an oblique angle onto the wafer while viewing it from above. Surface debris or flaws will reflect light into the viewing lens and will appear in contrast as light or dark spots against the gray background of the clean silicon wafer. Previously, this inspection was performed by a human operator, who determined if the number of defects on the wafer was fewer than an acceptable number. This image analysis is now performed by an automated system, which will produce a defect map of each wafer inspected and will sort wafers without operator intervention into "acceptable" and "unacceptable" bins according to the user's criteria. This process is a very simple application of machine vision to integrated circuit manufacture. Similar inspections are made after almost all of the processing steps, from oxide growth to thin-film deposition to etching. Some of these inspections are automated, but many still require the image analysis powers of the human operator to accomplish because of the increasing complexity of the images as the wafers proceed through the processing line.

Another area in which optics and optical systems contribute greatly to manufacturing process control is in noncontact film characterization—for example, thickness measurements. Integrated circuit manufacturing requires controlled growth or deposition of a wide variety of films, including oxides ranging in thickness from 100 Å to a few microns, polysilicon films for gate structures, photoresist for patterning, and silicon nitride films for vertical isolation of conductive lines. Controlling the thickness of many of these films involves empirically characterizing the deposition system by growing a set of films under different processing conditions (e.g., varying time and temperature) and measuring ex situ the film thicknesses to determine the processing conditions necessary. For transparent films, thickness and other characteristics are routinely measured using ellipsometry. For reflective films, a mechanical step-measurement system is more common. With each wafer lot processed, the film thickness is measured and processing conditions are adjusted (controlled) to maintain the correct film thickness for the subsequent lot.

Diagnostic uses for optical sensors in semiconductor manufacturing can be thought of as addressing three areas of manufacturing: contamination-free manufacturing, adaptive process control, or environmental safety and health. Contamination-free manufacturing includes the detection and analysis of trace impurities, particles, and submonolayer surface contamination. Adaptive process control includes the use of sensors and feedback systems for photolithography, dielectric film deposition, etching, and metallization processes. Measurement and analysis of trace impurities and "smokestack" monitoring contribute to improved environmental safety.

Common optical measurements used in semiconductor manufacturing include scatterometry for measuring critical dimensions, thicknesses, and sidewall profiles, and interferometry for measuring the alignment of upper layers with those below and for measuring temperature in rapid thermal processing applications. Particles and contamination are commonly measured using infrared spectroscopy.

These many requirements for measurement represent a tremendous impetus for the development of in-line process sensors for immediate feedback and process adjustment. To improve process yield and reduce defects requires sensors that are fast enough to measure and analyze pertinent process parameters during processing and that are as reliable as the processing equipment itself. Along with the development of sensors that can work in the often hostile atmospheres required for IC processing, nonoptical requirements include increased computational power at the individual tools and throughout the manufacturing facility and improved modeling of manufacturing processes to identify the most effective measurements for process control (Brueck, 1995).

Optical systems that can replace the human operator for inspections or can provide in situ film thickness measurements offer great potential for integrated circuit manufacturing improvement.

Display Manufacturing

Optical techniques are used extensively in making active-matrix liquid crystal displays (AMLCDs). The first step is to use optical lithography to produce the thin-film transistor arrays and color filter arrays. Optical inspection is then used to monitor the quality of bare substrates, patterned substrates, color filter arrays, and the final display product itself. Diagnostic procedures during processing make use of optics for particulate control. Ultraviolet light is often used to cure the seals of the liquid crystal cell. Finally, lasers are often used to locate and repair manufacturing defects.

Because of the extensive use of such techniques, incremental improvements in the tools for optical lithography, optical inspection, and optical diagnostic procedures during processing have considerable leverage in manufacturing large-volume throughput and hence in the cost and performance of devices such as future flat-panel display products. Some challenges for the improvements of lithography tools include achieving higher resolution (1 μm versus today's 3 μm), better overlay (1-μm precision over a 12-inch lateral distance), better linearity, and faster exposure times. Optical inspection tools should be more intelligent, easier to use, cheaper, faster, and better able to operate in concert with other tools. Processing diagnostics must be able to count lower levels of particulate contamination and allow optical testing of starting materials both before and during manufacturing.

In addition to these incremental changes, there are several ways in which new inventions in optical technology can have a profound effect on the display industry. For example, if optical inspection tools for display components could be made sufficiently intelligent, the need for costly and more error-prone human involvement in the inspection process could be further reduced. Optical techniques for liquid crystal alignment could be developed that would diminish the yield-limiting rubbing step in the manufacturing process, or new optical components could be developed, such as compensation films or structured backlights, that could significantly enhance display efficiency.

The Chemical Industry

Optical diagnostic instrumentation, especially instrumentation for absorption spectroscopy and spectrometry, has long been a mainstay of the chemical industry (Box 5.5). Off-line analytical laboratories were the primary users for applications such as quality control and trouble shooting. Today, optical techniques are used to provide information for process control in-line (within the process tool) and at-line (within the processing area), which allows up-to-the-second evaluation of chemical streams during the manufacturing process, in turn allowing real-time process control. This ability to maintain optimum operating conditions

BOX 5.5 OPTICAL PROBES FOR MONITORING AND CONTROL

Optical instrumentation for monitoring and control is critical to efficiency and quality in the chemical industry. A key goal is the development of in-line optical probes that can withstand adverse environments without degradation.

• •

in real time to meet specific customer requirements results in a measurable reduction in waste, which in turn leads to a desirable reduction in environmental degradation and a reduced need for waste management. The improved speed and sensitivity that have led to this expanded range of applications within the manufacturing line have been made possible principally by the advent of optical fibers, lasers, or other improved light sources, and new optical detectors, along with advances in computer technology.

In the chemical industry, optical techniques are critical parts of a multifaceted approach to chemical process issues to advance quality, quantity, customization, and reductions in waste while enhancing yield. This has resulted in part from excellent collaboration between industry and the academic sector, where optics have been integrated into process developments and design. A concept known as Plant 2000, being developed at the University of Washington, the University of Tennessee, and other centers, will use model-based control and on-line simulation to establish expert system supervision of the chemical stream.

The use of optical probing techniques has expanded to include using laser diodes, optical fibers, improved dispersive elements, and detector arrays. Optical diagnostics now in use include laser Raman spectroscopy, fluorescence spectroscopy, measurement of turbidity and solids distribution by light scattering, critical angle refractive index measurements, and combinations of these various techniques to name just a few. New areas will include video, microscopy, and others. These systems can be operated either at-line or remotely and are easily transportable. Optical fiber is key in the transport of information from the probe site to the analysis location (Figure 5.8).

However, current optical fibers are not robust in many chemical environments. Therefore, a serious barrier to the expanded use of optics-related process diagnostics is the development of a robust probe capable of operating in the adverse environments of chemical systems. Window performance is also a critical element, because an appropriate window material can allow the use of less robust probes by providing optical access to a chemical process. Windows must frequently operate

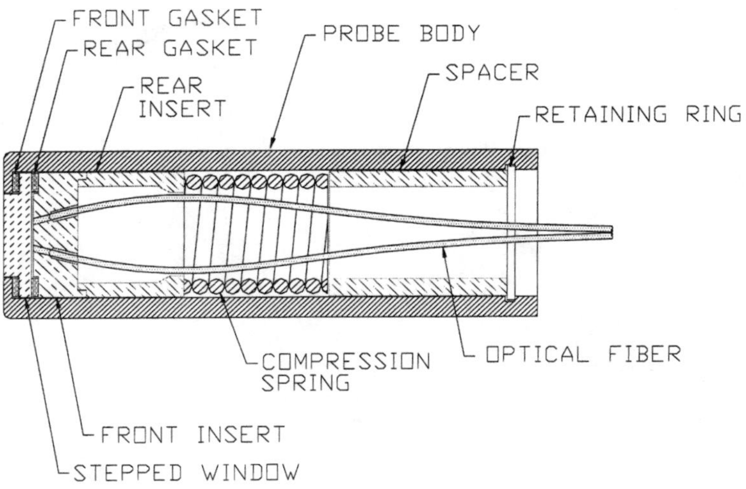

FIGURE 5.8 Design for a sealed fiber-optic probe. (Courtesy of R.S. Harner, Dow Chemical.)

in environments that not only are corrosive but also exhibit extreme conditions of temperature, pressure, flow (acoustics), mechanical forces, and electromagnetic interference.

Improvement to present technology will enhance productivity in the chemical industry. No advance in associated optical technology has been more important than the optical fiber, which both transports light energy into the system as a probe and transports it out of the system as a signal.

The implementation of optical sensors has been found in some instances to double the production output of an existing chemical plant. In other instances, chemical processes could not operate at all without real-time optical sensors. The use of such sensors is currently quite limited, however. The primary barrier to more extensive use is cost, both of the sensor or probe itself and of its installation into existing plants. Most applications require customized probe design since standardized optical probes are not commercially available.

Aircraft Manufacturing

Airplane manufacturers have similar needs featuring accurate location and alignment or layout of mating three-dimensional structures over large distances. Laser and other modern electro-optic systems are being effectively and routinely employed to save time and cost, while quality is maintained or improved relative to the older techniques such as photogrammetry that use theodolites or similar measurement instruments.

For example, three different modern optical instruments have been developed at Boeing. The Video Measurement System accurately positions large structural members as illustrated in Figure 5.9. Key fiduciary points are marked by retroreflectors. The illuminator features two to

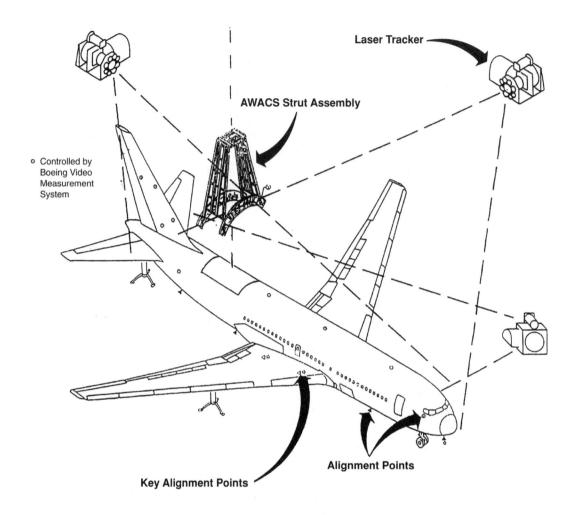

Laser Tracker

AWACS Strut Assembly

○ Controlled by
Boeing Video
Measurement
System

Alignment Points

Key Alignment Points

FIGURE 5.9 The Boeing Video Measurement System as used during installation of an AWACS strut assembly. (Courtesy of R. Withrington, Hughes Aircraft Company.)

four optical light sources and video detection to accurately position (~0.01 inch) the AWACS strut assembly on a 767 aircraft. The second instrument, the Laser Tracker Instrument, is a real-time coordinate measurement system for accurately mapping large structures. Typical operational range is more than 80 feet. The system uses amplitude modulators based on multiple-frequency semiconductor-lasers. Important uses of this instrument include verification of machine tool accuracy and profiling surfaces to archive engineering models. The third instrument, developed at Boeing, is a scanned laser template generator. Optical templates are used for locating and placing the cut edges of plys to produce ply dropoffs during composite component manufacturing. Planes of light, generated by a rotating laser head, are used for airplane body joint and wing alignment and for the generation of airplane interior reference planes during cabin outfitting. Crossed-fan laser beams are used to align a reference mark on a workpiece with a drill during machining.

It appears that aircraft manufacturing *problems* are the impetus for the use of specialized optical instruments specifically designed for an application as opposed to a particular existing optical technology or apparatus being adapted to this purpose. At McDonnell-Douglas and British Aerospace, new electronic theodolites have been very effective in equipment calibration with reported cost savings of more than 80%. Cost savings at Boeing are proprietary, but the installation of multiple units speaks to the success of the introduction of these optical techniques; active research and development on optical hole diameter measurements, surface profiling, wind tunnel instrumentation, and both cutting and welding indicate a considerable cost payoff in aircraft manufacturing.

The Construction Industry

New construction accounts for about 9% of the nation's gross domestic product each year. The industry includes the construction of residential and commercial buildings, factories, airports, tunnels, dams, landfills, environmental remediation systems, and a wide variety of other products. The annual market for construction is growing in the United States and worldwide, especially in the Pacific Rim and in Europe.

Construction projects typically require the acquisition and assessment of a lot of data. Optical techniques can make this process faster and cheaper by reducing the need for expensive labor or making it more efficient.

The use of optical methods in the construction industry is widespread but relatively straightforward. The techniques used fall into four categories: (1) optical systems incorporated into the final constructed product, (2) optical tools used in designing and building the product, (3) optical transducers that monitor activity or conditions at the construction site, and (4) optical elements that monitor the condition of finished structures over time.

The optical systems now being incorporated into some buildings and other constructed products can be quite sophisticated. Natural and artificial lighting systems are designed not only to illuminate the interior but also to control the structure's heat loss and gain. They incorporate optical coatings on lights and windows, lenses and mirrors, heliostats, light tubes, and other elements. Illuminating systems are discussed in more detail in Chapter 3.

Design and construction tools include optical image scanners, laser guidance systems for construction equipment (see Figure 5.10), geodetic measurements including fly-over and satellite-based mapping systems, and laser tools for precision cutting and welding of construction materials or monitoring of shifting structures and stresses. Laser guidance systems can be used to control the line and grade of tunnels, to control the blade elevation of grading equipment for site earthwork,

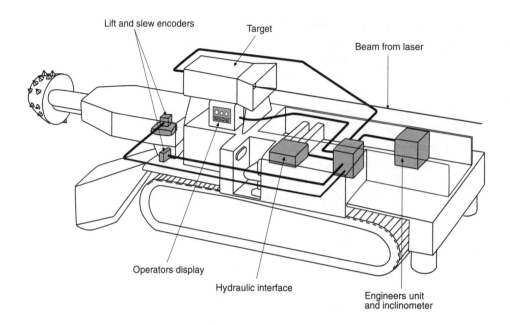

Lift and slew encoders

Target

Beam from laser

Operators display

Hydraulic interface

Engineers unit
and inclinometer

FIGURE 5.10 Laser-guided tunneling equipment. (Courtesy of B. Dorwart, Shannon and Wilson, Inc.)

and to control equipment for railroad track maintenance and highway repaving. Optical guiding can sometimes boost tunneling speed by as much as a factor of two, not only because fewer stops are needed for equipment realignment, but more importantly because straighter tunnels allow for faster and more efficient removal of waste material from the tunnel. Laser levels and targeting systems act as templates during construction by placing a spot or a line in an area to be excavated, eliminating the possibility that survey markers might be moved during excavation and often allowing replacement of an expensive survey crew by a less skilled person. Laser-guided tunneling has been accomplished from both tunnel ends, further reducing construction time.

At the construction site, optical transducers are used for gathering engineering data and for geodesy (surveying), including making distance measurements, measuring positions, and monitoring physical and chemical properties that can be detected from changes in optical properties. For example, in building a highway or a railroad, laser-based equipment may be used to measure displacements, grades, and loads. The alternatives, such as manual, mechanical, or sonic techniques, are often slower, more expensive, and less accurate. In an environmental remediation project, optical monitors may be used for groundwater, air quality, or stack emission measurements. Here the competing technology is usually manual sampling and analysis in a laboratory, which is often expensive and slow, especially for subsurface measurements such as groundwater, tunnels, or utility pipes. Video systems and stereo photogrammetry are becoming increasingly popular for documenting conditions at the construction site and their change over time.

What are some of the industry's other needs that can be met by further developments of optical equipment or employment? Cost and ruggedness are key. Conditions on a construction site are often extreme, and equipment must be able to tolerate heat, cold, dust, humidity, and vibration. Equipment that requires extremely careful handling or a highly trained operator is unlikely to be accepted for construction use. Survey equipment has to be portable, repairable, or replaceable while in the field, and able to operate 12 hours on a battery charge. Monitoring equipment must be accurate, reliable, low power, and remote-sensing. For soil or water characterization, the most useful parameters to measure include pH, total dissolved solids, identity and concentration of hydrocarbons and chlorinated hydrocarbons, and turbidity. Inexpensive and accurate continuous monitoring of water pressure (to within 0.1 pound per square inch) and distance (to 0.01 mm accuracy over distances up to 1 km) would also be helpful. The quality of the laser spot—including size, clarity, steadiness, and roundness—is important for guidance systems, marking of transfer tools, and welding.

Many of the barriers to more widespread use of optics in the construction industry are nontechnical. Equipment suppliers have difficulty finding qualified application engineers that understand the problems of the construction industry. Development has to be directed more closely to usable products, and education should be updated for modern equipment. Specific project goals are important drivers for the introduction and acceptance of optical adjuncts; for example, tunneling and surveying technology advanced tremendously as a result of the Superconducting Super Collider project.

The Printing Industry

The U.S. printing market, including such documents as periodicals, catalogs, newspapers, financial and legal documents, and greeting cards, represented about $7.5 billion in shipments in 1994, with real growth averaging about 4% per year. Compared to the 4% overall annual growth rate of the printing industry, digital production printing has been growing at 16.5% per year (Box 5.6). Manufacturers of commercial printing equipment are predominantly non-U.S. based.

• •

BOX 5.6 GROWTH OF MARKET FOR OPTICS IN PRINTING

The printing industry is large and growing steadily. The market for optical techniques in printing is growing at an annual rate four times that of the industry overall.

A strong drive toward shorter production times and just-in-time printing closer to the end user is forcing the printing industry to move from technologies such as traditional platemaking and printing to digital techniques, where the information to be printed is provided as digital input directly to the press. Figure 5.11 shows the current state and projected increase in use of digital technologies for a variety of printing applications. Digital platemaking, used for low-volume, high unit price applications, and digital printing, used for high-volume applications, are predominantly optical processes. Digital platemaking employs 25- to 100-W Nd:YAG, argon, or gallium aluminum arsenide lasers to expose traditional silver halide or a variety of photopolymers. Desktop digital publishing will continue to be dominated by inkjet printing and digital production printing by electrophotography.

For digital platemaking, photopolymers currently offer resolutions of 2,500 pixels per inch at a processing speed of 8 square inches per second. Requirements for the next 10 years call for an increase in processing speed to 50 square inches per second. Current electroplating techniques offer faster processing speed, about 20 square inches per second, but the resolution is only 1,200 pixels per inch. For digital printing, desktop applications are anticipated to increase from 30 to 100 square inches per second within 10 years, with production applications increasing from 200 to 1,200 square inches per second. Resolution for both techniques is predicted to increase from 600 to 1,200 pixels per inch.

From these figures, it can be seen that the primary opportunity to meet the projected requirements is an increase in pixel rates. However, the anticipation that the printing industry will be fully digital within 10 years offers a variety of opportunities for optics to contribute to the growth of digital printing technologies. Opportunities for optics include the development of higher-powered lasers for use with photopolymer plates, with emphasis on high-efficiency blue-green lasers, and imaging arrays to address the pixel limitations of current electronics.

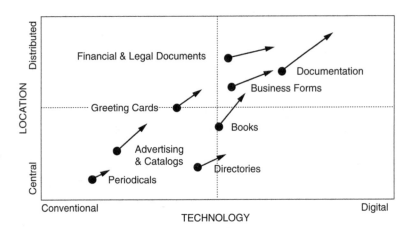

FIGURE 5.11 Market segments adopting digital print production technologies. (Courtesy of M. Fleming, Duplex Products, Inc.)

Increasing Use of Optics in Industrial Manufacturing

The incorporation of optics and optical systems into industrial manufacturing can be divided into two broad categories: (1) optical systems in applications that improve on current practice, and (2) optical systems that provide a new capability. For example, laser levels and targeting systems for tunnel construction provided an improvement over mechanical theodolites in the ability to sight and construct tunnels; three-dimensional rapid prototyping enabled an entirely new method of constructing models without requiring the development of tools to build individual components. For each new situation, developers and users balance the rewards offered by the new technique against the risks inherent in inserting new technology.

Because of the overriding importance of maintaining a controlled, reproducible process, incorporating new techniques and technologies into a manufacturing process is often avoided until cost or delivery time pressures from competitors compel manufacturers to change or until their current process is no longer capable of delivering the product with the performance needed. In the former case, manufacturers will often incorporate incremental changes into their process, as in the IC manufacturers' introduction of a chemically amplified photoresist in the 1980s to obtain smaller feature sizes with current photolithography equipment. However, the limits of the current generation of photolithography tools are approaching, and manufacturers are now contemplating the introduction of a new generation of equipment.

Over the next few years, several key advances are expected in the use of light to perform manufacturing. In the area of photolithography, a new generation of deep-ultraviolet and extreme ultraviolet photolithography equipment and processes will have to be introduced to produce features in tomorrow's 16-gigabit chips. The use of excimer lasers will make it necessary to develop entire new families of polymer materials for use as photoresists at wavelengths of 193 nm and shorter. Many exciting advances are anticipated in the field of laser materials processing, where new laser sources will provide shorter wavelengths, higher beam intensity, and sharper focus. An exciting possibility is the use of adaptive optics to achieve true diffraction-limited resolution on arbitrarily shaped workpieces in environments with poor optical quality. The use of three-dimensional solid modeling for rapid prototyping and manufacturing will continue to expand as new solid-state, high-power, cw ultraviolet laser sources are developed and improved optics for beam delivery makes it possible to achieve submicron root-mean-square accuracy for surface roughness.

FIGURE 5.12 Oxygen is introduced into molten steel in a furnace through the four large holes in the tip of this lance (top). Behind the smaller central hole is optical sensor equipment (bottom) for measuring position and temperature. (Courtesy of the American Iron and Steel Institute and B. Fuchs, Sandia National Laboratories.)

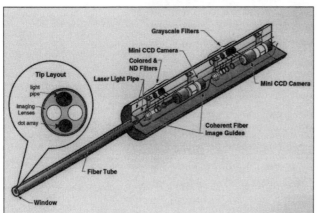

Major advances in the use of light to control manufacturing are also expected. Optical metrology should benefit from the development of smart sensors that incorporate data processing capability and from improved optical figure measurement techniques with 1-nm accuracy. Improvements in window and optical fiber materials will make it possible to use optical sensors to control manufacturing processes in increasingly hostile environments such as foundries. Figure 5.12 shows the recent successful use of optical sensors to measure temperature by submerging a probe in a bath of molten steel. Machine vision promises to increase its impact on manufacturing provided improved image processing and pattern recognition algorithms can be developed to make generic or plug-and-play solutions feasible.

Summary and Recommendations

Photolithography is the single most significant application of optics in industrial manufacturing. Submicron resolution, narrow-field-of-view photolithography is essential for the mass production of semiconductor integrated circuits, a major component of the U.S. and world economies.

The resolution achievable by photolithography will continue to be improved as far as possible. Dimensions as small as 0.18 μm will be achieved in the near future using 193-nm excimer laser sources. To achieve even higher resolutions, new technologies such as electron beam projection lithography and extreme-UV projection lithography are being developed.

Key technical barriers to be overcome for improved photolithography include the development of new families of photoresist materials for extreme-UV and electron beam sources, development of practical high-resolution masks or of maskless photolithography systems, and realization of all-reflective exposure tools for use at wavelengths of less than 193 nm.

Lasers perform a variety of materials processing operations as part of the manufacturing processes currently employed in a wide range of industries, including the semiconductor, aircraft, aerospace, construction, and automotive industries. These processes include cutting, welding, drilling, and surface hardening. Compared to conventional techniques, laser materials processing tools operate without physical contact, provide high localized energy density, and are truly versatile in their energy delivery.

The United States once dominated the production of specialized industrial lasers to perform materials processing functions but now has only a minority share of the market. To aid in reestablishing U.S. leadership, the Precision Laser Machining Consortium was formed through a partnership of government, industry, and academia.

Key technical barriers to be overcome for laser materials processing include improved opto-mechanics for beam delivery in adverse environments. An exciting possibility is the use of adaptive optics to correct for thermal and other aberrations in the beam path so that true diffraction-limited application and resolution can be achieved in manufacturing environments.

An emerging optical technique for performing manufacturing with great potential importance is laser-based rapid prototyping. Solid three-dimensional structures can be created in several different media by using laser irradiation to build a solid design directly from the information stored on a computer-aided design tape, layer by layer.

Among the techniques for using optics to control manufacturing, optical metrology is pre-eminent. Major uses of optical metrology systems include defect detection, inspection, measurement of product dimensions, monitoring manufacturing process conditions, providing real-time manufacturing process feedback control, alignment, and multidimensional measurement.

Optical sensors play a major role in many diverse industries. In the chemical industry, the implementation of robust, noninvasive optical sensors has been found in some instances to double the productivity of an existing chemical plant. In the semiconductor industry, optical sensors are used for contamination-free manufacturing, adaptive process control, and environmental safety and health. Optical sensors are employed in the construction industry to measure groundwater, air quality, and stack emissions as well as for geodesy.

Because of the overwhelming importance of photolithography to the U.S. economy, government agencies such as DARPA, in concert with commercial alliances such as SEMATECH, have been motivated to play an essential role in providing high-level oversight and coordination in the development of U.S. photolithography technology. There is a need to continue this role to guide future technical progress. A particular requirement is new families of photoresist materials, especially for extreme-UV and other new lithography technologies.

Advanced laser materials processing techniques are widely used in automobile assembly and other factories overseas, but less widely used in the United States. A factor that has undoubtedly aided the more rapid acceptance of laser materials processing in the foreign automobile industry is the greater level of emphasis that European and Japanese universities place on training engineers to be familiar with laser manufacturing techniques. The establishment of an application test facility in a service center setting in the United States would be particularly useful.

The use of three-dimensional laser-based or other rapid prototyping tools should be investigated for manufacturing limited quantities of actual working replacement parts. Such a capability would have important logistic benefits for military and other mobile or remotely sited applications.

Current problems in metrology for industrial manufacturing include the high-accuracy measurement of dimensions and position, the measurement of complex three-dimensional parts and surfaces, and the inspection of nanoparticle contamination on semiconductor wafers. Improved measurement standards and practices are needed for lithography systems, film thickness measurement, high-definition imaging systems, and colorimetry. More flexible applied optical metrology systems are necessary that do not require customization before each implementation.

In the manufacturing environment, the frequent goal of machine vision is replace human inspectors and allow automatic adjustment and optimization of the manufacturing process, quality control, and inspection. Two major limiting factors are the poor performance of presently available image processing and pattern recognition algorithms, and the need for custom algorithms and hardware configurations for each specific task. Nevertheless, tens of thousands of machine vision systems have been used.

For expanded utility in the chemical industry, optical sensors require improved optical configurations that can withstand extremely harsh environments involving corrosive materials at elevated temperature and pressure. Standardized optical probes are generally not commercially available. The special need of the construction industry is for portable sensors that are rugged enough to survive field use.

To preserve and enhance this critical technology base, coordinated government-industry-university activities are recommended in the following areas:

A multiagency-supported application and test facility should be established in a service center setting using the DARPA-sponsored Precision Laser Machining Consortium as a model for extension of laser materials processing and other optically assisted manufacturing techniques.

The National Institute of Standards and Technology should support development of optical metrology and machine vision systems with improved performance, with the ultimate objective of plug-and-play capability.

References

American Society for Metals. 1983. Pp. 647-671 in *Metals Handbook*, 9th ed., Vol. 6: *Welding, Brazing and Soldering*. Materials Park, Ohio: ASM International.

American Society for Metals. 1989. Pp. 572-576 in *Metals Handbook*, 9th ed., Vol. 16: *Machining*. Materials Park, Ohio: ASM International.

American Society for Metals. 1991. P. 265 in *Metals Handbook*, Vol. 4: *Heat Treating*. Materials Park, Ohio: ASM International.

Azer, M.A. 1995. Laser powder welding: A key to component production, refurbishment and salvage. *Photonics Spectra* 29(10):122-127.

Belforte, D. 1995. Presentation to the Committee on Optical Science and Engineering, October 12.

Belforte, D. 1997. Belforte Associates. Personal communication to the Committee on Optical Science and Engineering, June.

Bell, I., and N. Croxford. 1995. Fiber delivery gives YAGs an edge. *Photonics Spectra* 29(10):117-120.

Brueck, S. 1995. Optoelectronic diagnostics for semiconductor manufacturing. Presentation to the Committee on Optical Science and Engineering, October 12.

Dinda, S. 1996. Chrysler Corporation. Personal communication to the Committee on Optical Science and Engineering.

Hu, S.J., and S.M. Wu. 1992. Identifying root causes of variation in automotive body assembly using principal component analysis. *Trans. NAMRI* 20:311-316.

Industrial Laser Review. 1997. User profile: Successful solid-state sheet-metal cutting, Vol. 12, No. 5, pp. 15-17.

Levenson, M.D. 1993. Wavefront engineering for photolithography. *Phys. Today* 46:28.

Mak, C.A. 1996. Trends in optical lithography. *Opt. and Photonics News* (April):29.

Molian, P.A. 1986. Engineering applications and analysis of hardening data for laser heat treated ferrous alloys. *Surf. Eng.* 2:19-28.

Moore, G.E. 1975. Progress in digital integrated electronics. *IEDM Technical Digest* 11.

Photonics Spectra. 1995. Laser and light sources applications, Vol. 29, No. 10, p. 142.

SEMATECH. 1997. Critical level exposure technology potential solutions roadmap. Available online at <http://www.sematech.org/public/roadmap/doc/graphics/lithoro04.gif>. July 22.

Semiconductor Industry Association (SIA). 1994. *National Technology Roadmap for Semiconductors.* San Jose, Calif.: SIA.

Weiss, Stephanie A. 1995. Think small: Lasers compete in micromachining. *Photonics Spectra* 29(10):108-114.

6
....

Manufacturing Optical Components and Systems

Introduction

Since the early part of this century the manufacturing of optical components and systems has changed dramatically throughout the world, both in the types of products that are made and in the approach that is taken to making them. Once devoted entirely to passive image-forming components (such as lenses and mirrors) and to the instruments made from them, the industry now also manufactures a wide range of active elements such as lasers and optical sensors. Until recently, the industry depended heavily on a craftsman-style approach to manufacturing, with much of the work being carried out on an order-by-order basis by very small businesses. As new mass consumer markets have emerged that rely on optical technology—such as compact disk (CD) players and laptop computer displays—the implementation of high-volume mass-manufacturing techniques similar to those of the electronics industry has revolutionized this segment of the optics industry.

To take just one example of this new manufacturing technology, more than 100 million diode lasers are now produced each year, on highly automated production lines. The availability of these inexpensive diode lasers has revolutionized entertainment (in CD players), made high-quality printing affordable for small businesses and home users (in laser printers), and made possible numerous other new products that together account for hundreds of billions of dollars in global business revenue each year.

These changes in manufacturing are exciting, but they are reflected most prominently in the globalization of the optics industry, rather than in the domestic development of U.S. industry. Indeed, almost all mass production of optical components and systems now takes place outside the United States. There are only a handful of large U.S. optics companies engaged in the volume production of optical components, most of them in the plastic lens component business. This U.S. trend toward specialty products and small companies has been due in large part to the special technical needs of the Department of Defense, which has long been a vital customer for the industry. Government programs such as Small Business Innovative Research (SBIR) have also encouraged the formation of small, innovative optics companies. The main strength of the U.S. optics industry is now in high-precision manufacturing of low-volume specialty optical components and devices with high added value. This strategy has produced a strong industry based on the diverse activities of many small companies but lacking the manufacturing base required for expansion into mass consumer markets.

There are several thousand small optics and optics-related companies in the United States, with an average of 50 or 60 employees each. Together they account for more than 200,000 jobs and annual net revenues of about $30 billion for optical components and systems (excluding ophthalmics).[1] Yet even these impressive statistics do not adequately indicate the strength that small businesses provide to the U.S. optics industry as a whole, by making available a broad range of technical skills. A key finding of this report is that despite the optics industry's significant contribution to the U.S. economy, this contribution comes in so many small pieces that it is not usually fully recognized and understood.

The enabling character of optics, a repeated theme of this report, is an especially important consideration for the manufacture of optical components. The value of a component such as a laser diode or an aspheric lens is usually small compared with the value of the optical system that it enables. It is even smaller compared with the value of the resulting high-level application. Advances in the manufacturing of optical components are greatly magnified into improved capabilities and economic advantages at the systems and applications level. Advanced optical components cannot be considered commodity items.

This chapter addresses two distinct challenges. First, how can we maintain and strengthen the U.S. optics industry's leadership in high-precision manufacturing of low-volume specialty products? Second, how can we ensure the U.S. optics industry's ability to compete internationally in the increasingly important mass markets, especially the new mass markets that continue to emerge? Following a brief history of optics manufacturing in the United States and a short overview of the current state of the industry, the chapter divides into two main parts: (1) low-volume manufacturing of high-performance specialty products

---

[1] These numbers are based on a sample of the companies listed in the annual *Photonics Directory*.

and (2) high-volume manufacturing for the mass markets. The chapter ends with a discussion of some crosscutting issues, such as metrology and design, and the industry's composition, size, and growth.

A Brief History

Before about 1910, the U.S. optics industry consisted of just a few manufacturers of optical instruments such as binoculars and inspection equipment. Virtually all such products were imported from Europe. World War I stimulated demands for a domestic capability, and the need to provide components for these instruments was the basis for the U.S. optics industry's growth throughout the early part of the century.

The 1920s and 1930s supported several medium-to-large optical companies, such as Bausch and Lomb, American Optical, and Eastman Kodak—high-volume producers of both traditional and new optical instruments. A well-organized photographic industry provided almost all the cameras demanded by U.S. consumers. Most microscopes, binoculars, telescopes, and optical inspection equipment were also manufactured domestically.

The needs of the military during World War II placed significant demands on the industry's capabilities, and when military contracts ceased abruptly at the end of the war, most optics companies fell on hard times. Demand for cameras and other optical instruments for consumer and civilian uses grew, but Japanese and European competitors could satisfy this demand more cheaply than most U.S. companies, many of which succumbed to the competition. The remaining domestic camera and instrument manufacturers cut costs by turning to component suppliers in the Pacific Rim, first in Japan and more recently in China and Malaysia.

From the 1950s through the 1970s, the industry became increasingly divided, with overseas suppliers dominant in the high-volume markets and U.S. industry focused on assembly, systems building, and low-volume specialty components. Small companies came to dominate the U.S. optics industry.

In 1960, the invention of the laser spawned an entirely new segment of optics manufacturing, a segment that has grown astonishingly. Technologies developed to take advantage of the laser's capabilities have led to additional major markets for optical fibers, optical communications systems, optical sensors, and a broad range of other new applications. Mass U.S. markets for these applications have been based on aggressive growth in overseas manufacturing of the basic components.

An Overview of the Industry Today

The nature of the optics industry continues to change. Mass production techniques are used to manufacture components for an increasing number of high-volume optics applications. Among the products manufactured in this way are optical fiber for telecommunications and flat-panel displays for computers. Most of this type of manufacturing currently takes place overseas, not in the United States.

At the same time, demand remains strong for high-performance specialty products that are manufactured in small numbers. There are three main markets for these items: (1) the military, (2) other high-technology scientific and government programs, and (3) specialized industrial applications. Many high-performance military optical systems have very specialized capabilities but low production volumes. Some federal facilities for civilian research and development have similarly specialized needs. A key private-sector market for high-precision optical systems is the electronics industry, in which a relatively small market for photolithography systems enables the huge semiconductor business. The United States excels in this high-value, low-volume portion of the optics industry.

Most of the industry that serves the low-volume, high-accuracy component market remains dependent on very traditional fabrication methods, although it is increasingly facilitated by high-quality interferometric test equipment. This sector of the industry, made up mostly of small companies, faces increasing competition and must adapt to the new global marketplace. To maintain market share as overseas competitors improve their accuracy, domestic manufacturers will have to develop and use more deterministic fabrication methods that achieve the same results at lower cost with fewer high-skill workers.

For each of these types of manufacturer, an important element in the future growth of the industry is the growing integration of passive image-forming components with active sensors and light processors. The acceleration of this trend will mean a corresponding integration of the traditional optical component industry with the developers and suppliers of electrooptical materials and devices.

The challenges of the future will require new, faster, more flexible approaches to optical component fabrication, with less reliance on skill-intensive, iterative production methods. Some programs have already been established to promote this goal. For example, the Center for Optical Manufacturing has developed a series of computer-controlled generating machines that use diamond tools to produce accurate surfaces on glass elements. The Manufacturing Operations Development and Integration Laboratory (MODIL) has developed techniques for fabricating certain specialized laser components. Similar approaches are being implemented overseas. It is not clear, however, that such methods will be enough to revitalize U.S. production of high-volume general-purpose optical components, because most of the small shops that currently dominate the U.S. optics industry lack access to the investment capital necessary to upgrade their equipment.

Collaborative programs in optics manufacturing should include universities so that students are trained in the latest technical solutions to production problems.

The Department of Defense (DOD), the National Institute of Standards and Technology (NIST), and the Department of Energy (DOE) national laboratories should establish together a cooperative program that provides incentives and opportunities to develop new ideas into functioning methods for optics fabrication.

A critically important asset of the U.S. optics industry remains its strength in optical design. U.S. software companies set the world standard for optical design programs, although their products are of course widely used overseas. The development of sophisticated lens design programs, with good interaction with the designer, is remarkable. Programs that will run on a high-level personal computer now give any optical engineer access to modern design tools, and this easy availability has stimulated a widespread interest in optical design. There is as yet little integration of active components into the design process, however, and comprehensive software for physical optical design is still at a relatively rudimentary stage. U.S. accomplishments in optomechanical computer-assisted design (CAD) and thermal analysis software would be even more effective if fully integrated into optical design software.

Manufacturing of optical components and systems requires a large skilled and semiskilled workforce, and emerging new mass markets will increase the optics industry's need for trained workers. The quality and availability of optics training at the technician level is a widespread concern.

A key challenge for the future is the establishment of standards for the interchangeability of optical components, which is an important driver for cost-effective manufacturing. U.S. participation in international standards-setting activities lags far behind the activities of foreign organizations.

Low-Volume Manufacturing of Specialty Optics

There continues to be strong demand for high-performance specialty products that are manufactured in small numbers. For many of these products, the customer is the government, especially DOD, but certain high-value items are also vitally important in the commercial sector. Specialized high-value applications, such as lenses for photolithography, continue to be an area in which the U.S. optics industry can excel.

As described in Chapter 4, military optical systems tend to have high performance and specialized requirements but low production volumes (Joint Precision Optics Technical Group, 1987). For example, forward-looking infrared (FLIR) systems require expensive infrared-transmitting materials as well as environmentally resistant surfaces, coatings, and mountings. Ring laser gyroscopes require low-scatter surfaces and very high-precision optical components. High-performance aircraft and missiles require unusual aspheric components, conformal to the shape of the airflow. Affordability is becoming increasingly important to the Department of Defense, but despite its wish to use commercial products off the shelf where possible, DOD supports design and manufacturing development for a number of specialized optical technologies. The volume of demand for such items, even including the commercial applications, is often too small to ensure the necessary development of fabrication techniques by industry alone. **DOD should continue to maintain technology assets and critical skills in optics manufacturing in order to meet future needs.**

Some government projects require so many specialized optical components that they have a significant impact on the entire optics industry, despite the low volume for each of their individual components. Among these are the National Ignition Facility and the Atomic Vapor Laser Isotope Separation (AVLIS) program. These two DOE programs will consume thousands of medium-to-large optical components with high-precision surfaces and coatings resistant to high-power lasers. The overall size of these programs allows the private sector to plan some investments in improved machinery and processes.

Photolithography for manufacturing electronics is a key private-sector use of high-precision optical systems. The production of short-wavelength photolithography systems of ever-higher quality is essential for continued growth of the semiconductor industry. The Moore's law trend of increasing semiconductor miniaturization will drive photolithography through deep ultraviolet (UV) wavelengths and into the soft x-ray region by the turn of the century. At present, most imaging tools are produced overseas, but there are opportunities for U.S. industry to take the lead as systems move into the far UV, if economical methods can be found for producing moderate-sized aspheric surfaces with an accuracy better than 1 nm.

Specialized applications such as these incorporate a wide variety of traditional and modern optical technologies, each with its own manufacturing issues.

Spherical Lenses

The curved surfaces of a lens cause rays of light from a point on a distant object to come to a focus. A single lens with spherical surfaces, although quite economical to manufacture, forms an image that is not a perfect point (see Figure 6.1). Optical design has traditionally been a search for combinations of spherical-surfaced components, made of

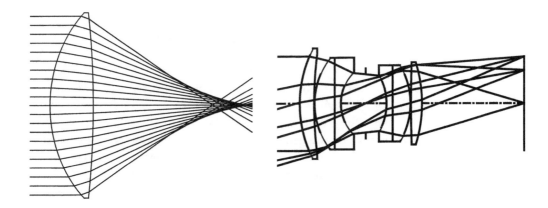

FIGURE 6.1 A single spherical-surfaced lens (left) forms an image that suffers from spherical aberration. To reduce this effect, a typical photographic or video lens (right) consists of many elements.

different types of optical glass, that result in nearly aberration-free images. In general, the wider the field of view or the more extended the spectral range required, the more elements will be needed.

The traditional approach to making spherical surfaces has been surface lapping, which can produce high-quality polished surfaces that deviate from the designer's specifications by as little as a few hundredths of a wavelength. This lapping or averaging method has been very successful in fabricating spherical and flat components, but it is by nature a time-consuming and craftsman-intensive activity. Improvements currently being investigated are directed toward deterministic fabrication, in which the accuracy of surface production is inherent in the machine carrying out the process rather than in the time-varying lapping of surfaces. Processes that are successful in finishing unusual materials, including optically active materials, have become more important.

There have been several attempts to improve and modernize the methods used for serial production. These approaches, however, such as high-speed surfacing, molding, and automated test and assembly machines, are usually directed at reducing the cost of a specific product. The improved production capability rarely extends to other products. A major improvement has been the very inexpensive production of plastic components. The use of plastic components is currently limited to systems of moderate and low quality, however, and to a limited range of environmental conditions. At least in the near term, most optical systems will continue to require glass components.

Currently, only one major domestic research and development program is directed toward the versatile production of economic spherical components. This is the Center for Optical Manufacturing (COM) at the University of Rochester, which has made significant progress in the development of high-speed machines to generate surfaces that require minimal polishing. COM has also developed a promising experimental polishing process called magnetorheological finishing. In magnetorheological finishing, the conventional rigid polishing lap is replaced by a suspension of magnetic particles and polishing abrasives. A magnetic field locally stiffens the fluid, creating a polishing spot that can be scanned over the part to polish it and correct its surface figure.

Although most spherical optical components are produced for use in imaging systems, other applications are also important for a healthy U.S. optics industry. Many specialized components are needed for laser systems, data storage systems, telecommunications equipment, a variety of analytical instruments, and endoscopes or other optical devices for minimally invasive surgery. The U.S. catalog houses that distribute stock produced overseas meet only part of this need.

Aspheres

If the surfaces of an image-forming component are allowed to be nonspherical, a major advance occurs. The addition of definable high-order curvature to the usual second-order spherical surface permits independent correction or balancing of spherical aberration. This leads to a reduction in the number of lens surfaces needed for aberration-corrected imagery and permits simpler, better optical systems (see Figure 6.2). Asymmetrical aspheres are also becoming important, especially in conformal applications, in which the outer surface of an optical component must conform to the aerodynamic shape of an aircraft or missile.

Among the diverse applications of high-precision aspheres are military aerospace systems, optical data storage, photolithography, and astronomy. Lower-precision aspheres have an even wider range of application, including photography and video imaging (especially zoom lenses), projection television, medical instruments such as endoscopes, telecommunications, and document scanners and printers. At the low end of the market, aspheres find use in such applications as condenser elements for illumination.

FIGURE 6.2 A single element with an aspheric surface can have significantly reduced aberration (compare with Figure 6.1).

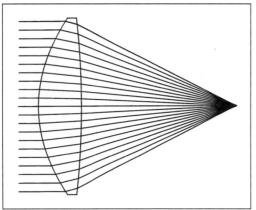

Some of these areas of application present valuable economic opportunities for the U.S. optics industry. The trend toward more compact optical systems with active or movable components—zoom lenses, for example—is increasingly driving designers toward aspheres. The future goals of the semiconductor industry (which will require asymmetrical reflective components for deep-UV photolithography), as well as other industrial and defense applications, cannot be achieved without the ability to produce high-quality aspheres cost-effectively.

Aspherical surfaces cannot be produced by the traditional methods used for manufacturing spherical-surfaced components. Three technologies are required to manufacture precision aspheres: machining, polishing, and metrology. Successfully generating, polishing, and testing aspheres all require considerable skill, and investment in all three is mandatory for successful production.

The two current options for machining defined aspheric surfaces are single-point diamond turning and computer-controlled (CNC) generation. Diamond turning is used to make aspheres in ductile materials such as metals, but it leaves an intrinsic surface roughness that must be removed by polishing if the component is to be used at visible or near-infrared wavelengths. CNC generators are used to grind brittle materials such as glass. Several countries have programs directed toward defined-shape generation of aspheric surfaces. These programs are beginning to produce numerically controlled machinery that makes the use of multiple aspheric surfaces feasible. In the United States, a major new program funded by the Defense Advanced Research Programs Agency (DARPA) is investigating the fabrication of conformal optical components.

Except in some infrared applications, machined aspheric surfaces usually require a computer-controlled polishing or finishing step. This step has two purposes: to correct the surface figure to the proper shape and to smooth its microroughness to reduce light scattering. The achievable accuracy of aspheric surface figuring has improved a hundredfold in less than a decade, from 0.5 μm [root mean square (rms)] in 1988 to about 5 nm (rms) in 1996. A variety of approaches have been developed based on loose abrasive polishing. These include passive and active flexible laps, deformable rigid laps, and small tool polishers. All of these approaches aim to solve the problem that a rigid lap, as is used in the conventional polishing of spherical elements, will not maintain good contact with an aspheric surface as it moves over the part. Magnetorheological finishing is also under development, as mentioned above. Ion figuring has proved to be a predictable method for final shape correction, but it does not reduce surface roughness.

The third requirement is metrology of the surface figure. Comparing surface measurements to the design of the component produces the data needed to drive computer-controlled polishers. Thus, the metrology system not only qualifies the finished part but also is essential during the polishing process. Contact and noncontact profilometers are used, but these systems are slow, and more importantly, they measure the surface only along widely separated one-dimensional traces. An interferometer coupled with a null optic (refractive, reflective, or diffractive) is preferable because it provides an accurate full-area test of the surface profile. The null lens must be calibrated before use. Asphere manufacturers find the use of null optics essential, but it is common for them to lament the high cost and long lead times.

All three of the above technologies need improvement, but the technology deficit is most severe in metrology, followed by polishing, and least severe in machining or generating. Also important for the future fabrication of precision aspheres will be the trend away from rotationally symmetric aspheres or their off-axis sections and toward generalized aspheres with little or no symmetry.

It is vital to the future of the U.S. optics industry that domestic production of aspherical optical components be made more cost-effective. Moderate-quality aspheres typically cost four to five times more than comparable spherical elements, so simply eliminating a spherical component from a design is rarely enough to make the economic case for using an asphere. Performance, size, or weight must be an additional driver. Thus, asphere fabrication technology has until now been driven by the needs of specific applications. More widespread application will require continued moves toward technology-based and computer-controlled manufacturing processes. These capabilities tend to lie in larger companies, and as a result, only a few U.S. manufacturers are currently able to produce aspheres economically. The ability to produce aspheres at a cost less than twice that of a spherical surface would open wider markets for such components. Without an incentive to develop a domestic capability, there is a real risk that precision asphere manufacturing will move offshore. **Government agencies should continue to support the activities necessary to introduce cost-effective precision aspheric components into both military and commercial products.**

Computer-Controlled Deterministic Grinding and Polishing

The primary reason high-volume manufacturing of image-forming components has left the United States is that current manufacturing techniques for these products are still operator resident, art driven, and based on labor-intensive machinery that often dates from the 1940s. As a result, the quality of the output depends on the skill and experience of the operator, rather than on computer controls and a scientific understanding of the manufacturing processes.

Revitalizing the industry will require a move toward transferable manufacturing processes that are based on smart operators and computer controls. This technology will be more capital intensive and less operator dependent than the current approach. Computer-controlled, deterministic processes will also lead to better consistency—all parts of identical quality and the first part as easy to produce as the last—and to significantly faster cycle times. A model to emulate is the metalworking industry, in which computer-controlled machinery is now the norm and handbooks are widely available that list feeds and speeds for various materials and configurations.

To begin addressing these needs, three universities and the American Precision Optics Manufacturers Association, with Department of Defense sponsorship, have jointly established the Center for Optics Manufacturing. COM's major thrusts are the integration of computers into optical component manufacturing, the development of computer-controlled optical component manufacturing equipment, the development of deterministic processes using this equipment, and the transfer of the resulting technology to industry. Some of the key challenges being addressed include improved inspection techniques for in-line process control, new machine geometries for making aspherical parts, and improved tools for optomechanical design.

This new deterministic approach creates some educational challenges. Advanced-degree programs must integrate mechanics and materials science into the traditional optics manufacturing curriculum. At the technician level, mechanical and computer training are required as well as training in optics, to ensure that machine operators' skills are appropriate for computer-controlled fabrication equipment.

Diffractive Elements

The use of diffractive optical elements, in which light is manipulated by a microscopic pattern on a surface rather than by its macroscopic curvature, is a fundamentally new approach with tremendous potential. Diffractive components exploit the wave nature of light to form images by the effect of a series of zones generated on a base surface. Some of the features of such components are their low weight even at large apertures, their ability to correct spherical and other aberrations, and their reduced need for exotic materials. Applications generally use diffractive elements in combination with refractive and reflective elements. The fundamental building blocks for diffractive technology are in place, and as a result, interest in this approach has surged recently.

Designs that combine aspheric lenses and mirrors with surface-relief diffractive lenses offer significant potential to improve performance and reduce size, weight, and cost. Among their applications are head-mounted displays, advanced sensor systems, laser and broadband imaging, optical interconnections for high-speed data transfer, optical data storage, optical correlators for target acquisition systems, consumer digital imaging applications, and precision testing of complicated optical systems.

Diffractive optical elements have significant market potential. Estimated at between $15 million and $20 million in 1995, the market is expected to grow to between $150 million and $200 million by the turn of the century. Almost one-quarter of the market will be in consumer products, with the remainder split roughly evenly between industrial and government customers. This projection represents 15 to 20% of the total optical component market in 2000.

To produce high-quality diffractive components, one must shape the surface profile within each diffraction zone with nanometer precision. There must be sharp transitions between adjacent zones, and the surface profile within each zone must be smooth. Several fabrication methods—including electron-beam and optical lithography, diamond turning, single-point laser pattern generation, and optical holography—have been used to create surface relief on "master elements." These master elements are then used, with varying degrees of success, to replicate diffractive components. An advantage of this approach is that the precision is built into the tooling and reproduced in each part.

Diffractive optics technology is at a critical stage of development. The United States has made significant investments in basic research and currently has a leadership position, but most of the basic research has been done and the need now is for development and commercialization. Federal R&D programs can play an important role in preserving U.S. leadership by focusing on the development of low-cost manufacturing processes for high-precision applications.

Optical Coatings

The last step in making an optical component, especially a high-value component, is often to apply a thin film of light-controlling or protective coating. These coatings range from simple metallic reflectors, to antireflection coatings, to multiple-layer filters, to high-efficiency dielectric reflection-enhancing bandpass coatings. The optical parameters of high-precision coatings must often be maintained to within a few percent. Large-area production coatings usually require less optical precision but may have special environmental or processing requirements.

Optical coatings are used for many light-control purposes. Coatings with as many as three layers are commonly used on photographic and video objectives to enhance transmission and reduce ghost images. Multilayer coatings are used on prisms, beam splitters, and windows to select spectral regions for enhanced transmission or reflection. Special large-area coatings are use to control the appearance of video displays and enhance the contrast of military cockpit instruments. Industrial and military laser systems depend on protective coatings to provide survivable surfaces for high-energy reflection or transmission. Special sharp-cutting filters are required to implement the wavelength-division multiplexing (WDM) technique for optical telecommunications. Government applications typically require high precision but relatively low production volumes. Many commercial applications also require small-batch orders and high precision, but some, such as coatings for projector lamps, require high-volume production at lower precision.

Several methods are used to apply optical coatings. Batch methods are typical for high-precision applications. Most infrared (IR) coatings for military and aerospace use are applied by evaporation. Visible or UV coatings, including some for commercial applications, can also be made by evaporation. Some coatings are densified by sputtering or ion-assisted evaporation to reduce porosity, which improves stability and environmental resistance. Moderate- and low-precision coatings with only a few layers can be applied using in-line flow machines, which process large areas efficiently. Roll coaters can be used for some low-precision applications on flexible substrates.

Among the technical issues for optical coatings are cost, durability, low body and surface scatter, high efficiency, production yield, and environmental stability. There are technical problems related to lengthening the evaporation source lifetime, acquiring pure materials, and ensuring the long-term reliability of equipment and processes. Interest in plastic substrates is increasing, but further process improvements are required in this area. The failure of an optical coating on a high-value substrate can be expensive for the user, but despite this, for specialized coatings the failure rate is sometimes as high as 50%. There is a growing market for the removal and replacement of degraded coatings. Some technical areas in which federal R&D support is most needed include the fundamentals of adhesion, defect avoidance, scattering and absorption reduction, stress minimization, improved uniformity for multilayer coatings, and the development of cost-effective high-throughput coating processes.

The U.S. market for optical coatings was about $635 million in 1996. It is projected to grow to $920 million by 2001. Worldwide, the market is $1.6 billion, with projected growth to $2.3 billion by 2001. These figures include optical materials, coating products, coating services, and coating equipment. Companies tend to specialize either in high-volume production applications or in low-volume specialty coatings. In most segments of the market, there are one or two large suppliers and a myriad of small ones. U.S. suppliers dominate the market, especially for high-quality coatings, but offshore competition is growing rapidly.[2]

Commercial coatings markets are growing by 15 to 30% per year. They offer long product lifetimes and attractive profit margins but require the coater to take on the risk of technology development. Some commercial markets have sufficiently high volume to justify significant investment in development; applications in office automation equipment and photocopiers, for example, can achieve very high volumes. The keys to success in the commercial coating business are holding down costs and meeting performance specifications.

---

[2]Market data are from Business Communications Company (1996).

The government coatings market, in contrast, is declining by about 10% per year. In addition, it tends to have low production volumes (typically a few hundred parts per order) and to require considerable paperwork for a limited profit margin. As a result, the focus of larger companies is increasingly on commercial work. Maintaining continuity and a critical mass of expertise is of major concern if future government needs for specialty coatings are to be met. There is federal support for optical coatings R&D, but it is inefficient to develop coating processes at government laboratories and subsequently transfer them to industry. Sponsored development in industry would be a better approach. It should emphasize manufacturing issues and be linked to follow-on production activities.

Optical Glasses, Polymers, and Specialty Materials

A wide variety of optical materials are used in image-forming components. In addition to their optical properties such as transparency and index of refraction, practical considerations such as isotropy, homogeneity, and environmental stability are usually also important. Mechanical and electrical properties must be appropriate for each application, and it must be possible to produce a material and fabricate it into an optical component at an acceptable cost. Thus, although there are many opportunities to develop materials with new optical properties, there are also many practical limitations. For some materials, only incomplete and sometimes inaccurate data on mechanical properties are available; optomechanical designers need such information to develop durable system designs.

The most used material is optical glass. A century of development has resulted in inexpensive, extremely homogeneous glasses with very well-defined optical properties. Catalogs list more than 300 varieties, although in fact, fewer than 50 are available without special melting and about 10 glass types account for 90% of the optical glass that is actually used. Environmental restrictions have encouraged the elimination of many glasses, and the current passion to eliminate lead from industrial processes will further reduce the available variety; for most applications, however, designers actually have little need for more than a handful of glass types, perhaps about 16. Physical limits on the dispersion parameters of glassy materials make it unlikely that new glasses will be developed that fall outside the ranges available today.

The vast majority of glass blanks are less than 1 cm in diameter and are manufactured using a continuous batch approach for just a few cents each. Even when the value added by finishing is included, the cost of a typical, completed spherical-surfaced glass element in volume production rarely exceeds five or ten dollars. Furthermore, for only a few glass types does annual production exceed a few thousand kilograms. Except in special cases, therefore, there is little incentive to invest in the development of new optical glass types.

The use of polymers in optical systems is increasing. Injection-molded plastic lenses have reached a quality that approaches that of glass lenses in many applications. Only a few useful optical polymers are known, however, and their use is limited by their low indices of refraction, relatively high dispersion, inhomogeneity, mechanical properties, and low tolerances to such environmental conditions as temperature and humidity. Great benefits could be obtained from the development of a few new good-quality optical polymers.

Certain specialized applications require other optical materials. For example, high-energy laser applications require reflective substrates that are resistant to thermal effects. A mirror that is reflecting a megawatt of power may have to dispose of as much as a kilowatt of heat. If absorbed in the substrate, this heat will lead to warping of the mirror's surface. One solution is to use a metal substrate below the mirror surface with channels for a circulating fluid coolant. Active cooling can be eliminated, simplifying both design and fabrication, by selecting substrates such as single-crystal silicon that are transparent at the wavelength of the laser radiation being used. This technique is used in space-based and airborne laser weapons, for example. The development of economic silicon boules for the semiconductor industry has permitted the manufacture of mirrors up to 30 cm in diameter for these applications. Other optical materials tailored for particular applications include ultrapure silica for optical fibers, erbium-doped glass for fiber amplifiers, and sol-gel materials for specialized filters and waveguide devices; some of these materials have become cornerstones of entire new industries.

Case Study: Photolithography Equipment

A key commercial application of high-value specialty optics is the equipment used for photolithography in the electronics industry. Advanced lenses for operation at 248 nm are crafted from pure fused silica, have 27 or more elements, weigh in excess of 1,000 pounds, and are a meter or more in length. Each element must be ground to exquisite tolerances in both figure and finish. The entire surface must be defect free, and the surface figure must be accurate to a few hundred angstroms. As lithography proceeds to even shorter wavelengths in order to achieve smaller feature sizes on the device, even more sophisticated optics will be required. There are many challenges in the design, manufacture, and implementation of these optical systems.

Lens Materials Issues

Conventional lenses that operate at I-line (365 nm) and g-line (424 nm) wavelengths have been the mainstay of optical lithography for more than two decades. Many materials are available with the correct refractive index, transmission, and other optical and mechanical properties, so materials were not a limitation in these lens designs. At deep-UV wavelengths such as 248 and 193 nm, however, most conventional optical glasses and coating materials are far too absorbing for use in lenses. Fused silica has become the standard for these applications. Current materials and material manufacturing methods simply cannot achieve the performance required if lithography tools are to produce devices with features smaller than 0.25 μm. Short-wavelength photolithography will increasingly depend on high-precision reflective aspheric components produced on stable fused-silica substrates. Among the many materials challenges at these shorter wavelengths are reducing absorption, increasing purity, and improving uniformity.

To yield a satisfactory lens, a glass must have almost perfect transmission. Photolithography lenses for operation at 248 nm or less need an optical path of a meter or more and must transmit at least 99% of the light; otherwise lens heating and defect generation become limiting factors. For 193-nm applications, inorganic materials such as CaF_2, MgF_2, and mixed-ion glasses are being studied, but these glasses are hydroscopic and difficult to polish because of their crystal structure.

Almost any impurity in a fused-silica structure will alter some property of the glass in an undesirable way. For example, most metals (including iron, sodium, cobalt, manganese, and magnesium) result in the generation of "color centers" after exposure to ultraviolet light. These color centers reduce the transmission of light and result in additional energy being absorbed in the bulk glass of the lens. Impurity levels must be kept at least below one part per billion and much lower than this for some elements, particularly iron and cobalt.

The refractive index and dispersion of the material drive the performance of the finished lens and must be uniform across the entire blank used to fabricate a lens element. This uniformity must be better than 10^{-7}! Other optical properties, such as polarizability and reflectivity, must be similarly uniform.

Challenges for Next-Generation Photolithography Equipment

Photolithography lenses operate at the diffraction limit. That is, the resolution achieved is equal to or exceeds the wavelength of the light used for imaging. To achieve such astonishing performance the lens must be fabricated and assembled to near-perfect tolerances. Important manufacturing issues for future systems include image and tolerance modeling, surface figure, surface finish, metrology, asphere production, mounting, and production of short-wavelength sources.

The design of these complex systems will require massive computer simulations that exceed current capabilities. Models must be accurate and robust. They must also be extremely fast since they will be used not only to design lenses but also to guide and control their manufacture.

Once the computer model tells the manufacturer the correct surface figure ("prescription") of each element, the glass must be ground and polished to the correct geometric shape, with an accuracy and precision of less than 1 nm. This precision is required even in elements that are often more than 20 cm in diameter and several centimeters thick! Removing or depositing material to such precision exceeds the capability of currently known methods.

Total reflectivity (summed over all surfaces) will have to be no more than 1 or 2%. Mechanical polishing will no longer be adequate to achieve this level of surface perfection, and plasma etching and ion milling techniques will be required. These techniques have not yet been developed to a degree that allows economical lens manufacture.

Metrology may be the single most difficult task. Material properties, lens figure, and surface finish must be measurable to at least 10 times better than the specification. Among the properties that must be measured are the following:

- Refractive index to better than 10^{-7} in all three dimensions;
- Phase interferometry of the projected wavefronts to better than $\lambda/80$ for both single elements and assembled lenses;
- Surface finish to better than 0.1 nm over the entire lens surface, possibly using atomic force microscopy (AFM);
- Glass purity and transmission; and
- Glass damage characteristics at the operating wavelength under long-term, high-dose conditions.

Some aspheric components will be required to minimize on- and off-axis aberrations and reduce the number of elements in the assembled lens. Aspherics are essential to the fabrication of the reflective lithographic lenses that will be required for next-generation systems operating at the 13-nm wavelength. Technology remains to be developed to produce these aspheres economically. The development of fabrication technology will require support from both equipment manufacturers and equipment users.

Wavelengths in the 13-nm region will likely be used to produce features smaller than 0.1 μm, and reliable new short-wavelength sources will be needed. Excimer lasers, high average power laser-plasma drivers, and synchrotron radiation sources are being developed. Leading the development and production of such sources will be a major opportunity for the U.S. optics industry.

Case Study: Optics for the National Ignition Facility

Government programs are another major customer for high-value specialty optics. A prominent example is the National Ignition Facility

(NIF) at DOE's Lawrence Livermore National Laboratory (LLNL). The NIF is a 192-beam laser at 351 nm, capable of focusing 1.8 MJ of energy onto a target. (The goals and technology of the NIF are discussed in more detail in Chapter 3). About 8,500 meter-class optical elements will be required, with production at a peak rate of about 2,500 elements per year between 1999 and 2002. The combination of tight specifications, high production rates, and low-cost goals is a significant manufacturing challenge.

Construction of the NIF will have a major impact on the entire U.S. optics industry. From development through production, approximately $200 million will be spent on optical components for the program. This is a significant fraction of the total U.S. market for large optical components, which is currently between $350 million and $700 million per year. Industry can meet the technical specifications, but the planned production schedule will be a challenge. Furthermore, in some areas such as laser glass, current manufacturing technology cannot meet NIF's cost goals.

A development program is in progress to improve industry's ability to meet these cost and schedule requirements. Some of the manufacturing technologies being developed include continuous melting for phosphate laser glass, high-speed polishing and deterministic figuring, rapid growth of large potassium dihydrogen phosphate (KDP) and deuterated potassium dihydrogen phosphate (KD*P) crystal boules, improved process design and control for mirror and polarizer coatings, meniscus coating technology for gratings and sol-gel multilayer mirrors, and improved process control and optimized boule geometry for low-cost fused silica. International Organization for Standardization (ISO) 10110 standards will be used for optics drawings. Metrology is a key concern, including 24-inch-diameter phase-shifting interferometry and characterization of mid-spatial-frequency errors by measurements of power spectral density.

As well as substantially reducing the cost of the NIF program, this development program will help the U.S. optics industry compete in the construction of large inertial-fusion lasers being considered in France, Japan, and Britain. The technology developed will also find a variety of other applications:

- Laser glass melting and forming technology can be applied to making a variety of UV- and IR-transmitting glasses for applications in ozone detection, laser surgery, and medical spectroscopy.
- Technology for the rapid growth of KDP and KD*P crystals can be adapted to improve the growth of other important crystals for applications such as frequency conversion and detectors for high-energy physics.

- Improved coating capabilities will be applicable to expanding markets such as computer displays and medical instruments. Meniscus coating technology already developed is being used to manufacture high-efficiency gratings for high-power lasers.
- Technology created to reduce the cost of transmitting fused-silica optical components for NIF will be applicable to the next generation of photolithography equipment.

Because of their size and technical sophistication, government programs such as NIF can have a major impact on the development of optics manufacturing, even though they do not themselves lead directly to large commercial markets.

Key Technical Challenges

Among the major challenges in the design and fabrication of low-volume optical components are the following:

- The cost-effective manufacture of general aspherics and conformal components and their use in a wide variety of instruments and devices;
- Deep-UV optics for microlithography to permit the continued reduction in feature size and enlargement of chip area that will take the semiconductor chip industry through several more cycles of economy and speed; and
- Low-cost, low-volume, surge-capable optical manufacturing, which is essential to maintain efficiency and support the continued development of military optical systems.

High-Volume Manufacturing of Optics

The high-volume mass production portion of the optics industry uses manufacturing methods quite different from those discussed above and faces its own technical and structural challenges. In some cases, there can be a natural progression from low-volume specialty capabilities to high-volume manufacturing of similar items. For example, it may be possible to produce glass lenses in bulk using computer-controlled deterministic grinding and polishing machines. In other cases such as optoelectronic components, mass markets require quite different manufacturing techniques. Often these techniques are closely related to those used in the electronics industry, with which the optics industry is becoming increasingly integrated in both markets and manufacturing methods.

Although high-volume optical components tend to generate low profit margins, the ability to manufacture an essential component may be key to maintaining a strong position in the profitable systems markets that the component enables. For example, a diode laser is quite inexpensive but is a key component in CD players and laser printers. The enabling character of optics is a theme found throughout this report.

The current U.S. strength in manufacturing molded glass and plastic image-forming components will probably continue, but prospects are doubtful for new ventures into mass production. There is currently little effort in the U.S. optics industry to develop mass production of integrated electrooptical systems with both active and passive components, although U.S. leadership in this area is a desirable goal. It is possible that investment in manufacturing techniques by DOD may help jumpstart the industry's efforts to enter high-volume markets.

A key high-volume user of optical technology is the telecommunications industry. The revolution in communications technology has been enabled by advances in optical materials: materials for optical fibers for the transmission medium and compound semiconductor materials for optical sources and detectors. (See Box 6.1.) Serious development of both technologies began in the late 1960s and early 1970s. The advent of the erbium-doped optical amplifier in the late 1980s revolutionized systems design by eliminating the need to regenerate signals electronically every few tens of kilometers to compensate for attenuation by the optical fiber. Even more importantly, the optical amplifier made WDM practical for the first time, by making it possible to route individual wavelengths separately in a point-to-multipoint

··

BOX 6.1 THE IMPORTANCE OF MATERIALS SCIENCE AND ENGINEERING IN OPTICS MANUFACTURING

Materials production and materials processing are essential enabling elements in the manufacture of optical devices and systems.

Efficient materials preparation techniques are the foundation for cost-effective production of optically pumped lasers, nonlinear optical modulators, and harmonic generators. The epitaxial deposition of single-crystal structures for III-V semiconductor lasers is a triumph of materials control and finesse; further improvements in the understanding and control of materials processing could increase yields for semiconductor lasers as much as tenfold. Optical communications relies on controlled vapor deposition of fibers with better than part-per-billion purity, controlled composition, and controlled profile.

Materials processing breakthroughs were key to the reduction of light losses in optical fibers. A landmark realization was that gas-phase oxidation of silicon tetrachloride (highly pure and readily available because it is used in the silicon semiconductor industry) could—under appropriate conditions of dopant and apparatus geometry, gas flow rates, and so on—produce high-purity, low-optical-loss dowel "preforms" from which fibers could be drawn. Additional materials challenges involved the invention and implementation of fiber drawing and coating processes capable of producing many kilometers of fiber with strength exceeding that of steel. Higher fiber draw speeds will require increased understanding of coating dynamics.

communications network. These new systems designs created a demand for new low-cost photonic components for multiplexing, filtering, routing, and monitoring. Further advances in optical telecommunications will depend on continued cost reduction and increased functionality and integration. Such advances will require the introduction of high-volume manufacturing technology that incorporates automated production lines and on-line process controls.

Optical Fiber, Fiber Devices, and Waveguides

In 1996, the world market for optical fiber was about 2×10^{10} meters and increasing at about 15% per year. Fiber is currently being installed at close to 1,000 miles per hour! The deregulation of telecommunications in the United States and elsewhere has increased fiber demand such that the current world supply is insufficient.

The U.S. industry has been a leader in fiber manufacturing from the beginning, and it continues to be so. Japan and Europe are also major players.

Silica Optical Fiber

The technology for manufacturing optical fiber is now a relatively mature three-stage process: preform fabrication, fiber draw, and cabling. At each stage the speed of fabrication is critical, because throughput is what ultimately determines cost. Other critical issues are uniformity in geometry and composition, high purity in the fiber core, and uniform application of the polymer coating that protects the fiber surface. All of these properties continue to improve, and optical fiber is currently the most reliable and reproducible component of all photonic technology.

The cost of silica fiber has fallen below that of copper wire, and continued cost reduction can be anticipated. The transparency of the glass used is now close to the theoretical limit, so further improvements in fiber technology will mostly be in the areas of fiber design and cable technology for specific applications. One potential concern is the limited world supply of germanium, which is used to increase the refractive index of the fiber core. Germanium prices have soared in recent years.

Nonsilica Optical Fiber

For certain applications, plastic optical fiber can reduce overall systems costs by permitting the use of low-cost molded plastic connectors. Plastic has considerably higher attenuation than silica, however, so plastic fiber is practical only for short-distance uses such as on-premises distribution and applications in cars and airplanes. Until recently, high losses in the near infrared prohibited applications in this spectral region, but a recent breakthrough with perfluorinated plastic fiber should now permit applications in the telecommunications window from 1.3 to 1.55 µm. A great deal of development is still needed, on both materials and draw technology, to make low-cost manufacturing of plastic fiber feasible.

Fluoride fibers have also been studied as an alternative to silica. Because the optical transparency of silica fiber is limited by Rayleigh scattering from fluctuations in composition, researchers have searched for alternate materials having lower fundamental loss. Although fluoride-based fibers have exhibited lower scattering losses than silica, the technology to achieve lower absorption loss has not been developed. With the advent of optical amplifiers there is little incentive to continue the development of ultralow-loss fibers for telecommunications, especially since taking full advantage of fluoride fiber's potential would mean operating at longer wavelengths where other components are not as mature. Thus, the principal application of fluoride fibers will be for optical power transmission for medical and sensor applications beyond 2 µm.

Fiber Devices

Early devices fabricated by the controlled fusion of fibers, such as splitters and couplers, offered limited functionality as purely passive devices; introduction of the erbium-doped fiber amplifier began an entirely new class of devices. High-power fiber lasers, Raman amplifiers, dispersion compensators, and fiber-grating filters and reflectors are now reaching the commercial marketplace. Such devices will be integral parts of broadband telecommunications networks as well as non-telecommunications applications such as optical sensing, optical gyroscopes, automobile collision avoidance, and medical applications. These markets will exceed $100 million by the year 2000.

All these devices draw on the technology already in place for manufacturing transmission fiber, but device applications require additional understanding of dopant properties and the optimized design of geometries and dopant concentrations for specific applications. Key technical goals for fiber amplifiers are high power output, high efficiency, and flat gain over a broad spectral region.

The principal manufacturing challenges in meeting these goals for fiber devices are making doping more reproducible and reducing cost. Specialized splicing techniques are often needed to connect these specialty fibers to transmission fiber, but there are few major obstacles to achieving all of these goals as the markets develop.

Planar Waveguide Devices

Two waveguide technologies are currently in use: lithium niobate-based devices for the integration of electrooptic devices such as switches and modulators, and silica-on-silicon devices for the integration of passive waveguides. Both of these technologies are based on the formation of planar lightguiding films by diffusion or deposition, followed by the definition of the waveguiding structures by photolithography. Manufacturing technologies are now quite well developed; lithium niobate modulators are in commercial use for cable television and high-end communications systems, and planar silica waveguides are being used commercially in telecommunications as splitters, combiners, and WDM routers. As the need for functionality increases, more photonic devices will be integrated into monolithic structures to avoid complex assembly of discrete photonic devices.

There are several challenges for the low-cost manufacture of these devices. First, because they are planar, connecting them to fiber transmission lines is complex. Demanding alignment tolerances have until now required active alignment of individual fibers or fiber ribbons. Passive alignment techniques are under development. It is important for most applications that the waveguides be polarization independent, which further complicates the design of two-dimensional structures.

A more serious issue is the small size of the current market. Since discrete devices satisfy most current needs, the volume demand for integrated devices remains low and the sizable capital investment required keeps costs high. Japanese manufacturers have been steadfast in their commitment to integration, in the expectation that as the technology matures, higher levels of integration will eventually reduce costs. In contrast, investment in the United States remains small.

Semiconductor-Based Optoelectronic Components

The semiconductor-based optoelectronic components market, excluding flat-panel displays, generated nearly $6 billion in revenue in 1995. Over the next decade, growth to $20 billion is anticipated. This growth is linked to the increasing demands of the global information infrastructure, which it will support and enable. The major manufacturing challenges in this area include the following:

- Reducing the cost of high-volume chip manufacturing, automated chip handling, measurement, testing, and package assembly;
- Developing CAD tools for photonic devices and packaging, like those used in the silicon industry, to reduce design cycles and cost—this is becoming increasingly important as integration begins to place more functionality on each chip;
- Improving laser performance (including linearity, speed, and tunability), laser arrays, low-cost packages, and higher-power lasers;
- Developing practical green, blue, and ultraviolet lasers; and
- Developing high-performance, low-cost imaging arrays.

For all photonic components, performance is critically dependent on materials quality, composition, and control (see Box 6.2).

BOX 6.2 PHOTONIC MATERIALS

The performance of all photonic components is critically dependent on the quality, composition, and dimensional control of the materials used in their manufacture. For instance, threading dislocations in the substrate wafer of a semiconductor device can propagate through the epitaxial film growth and adversely affect the yield of devices. Extremely small variations of epitaxial layer thickness can adversely affect the wavelength of operation of multilayer semiconductor structures.

Every stage of crystal growth, materials processing, and device fabrication requires stringent control of defects, interfaces, and layer deposition rates and a precise knowledge of the interrelationships among material quality, process parameters, and device performance.

Considerable progress has been made in recent years in improving the defect density of commercially important substrate crystals such as gallium arsenide, indium phosphide, and lithium niobate. For today's generation of small-area discrete devices, substrate quality is not typically a major limitation on device yield or performance. As materials processing improves and the integration of multiple photonic components on a chip becomes standard, however, continuous improvements will be required in substrate material quality. In the meantime, materials processing does remain the major limitation on the wafer yield of complex photonic devices, and improvement in all aspects of process control and equipment design will be essential in achieving the high yield and low cost required to make extensive photonic integration viable. For II-VI semiconductors, considerable improvements in crystal growth are required, since these materials do not yet enjoy the level of materials quality that is found in silicon. The growth of wide-bandgap semiconductors for blue lasers still requires a great deal of research on all aspects of materials growth and processing.

An issue of concern is the declining talent in the United States in the area of bulk crystal growth. For many years, there have been few university centers that train new students in this area; most of the current university focus is on epitaxial growth. As a result, in the next decade there will be a shortage of talent in this important discipline. Some of the most impressive new nonlinear optical materials for the next generation of devices are being developed in China, which has nurtured expertise in bulk growth.

••

Components for Long-Distance Transmission Systems

Long-distance telecommunications is dominated by optical fiber transmission systems that use high-performance 1.3- or 1.55-µm laser transmitters and p-i-n field-effect transistor (PIN/FET) receivers. (See Chapter 1 for details.) Many challenges remain to be overcome in manufacturing these components. The low-distortion requirements of analog cable-television systems place stringent demands on the linearity of lasers; the dominant factors in the price of these lasers are yield, packaging, and measurement and testing. Wavelength-division multiplexing systems require such stringent wavelength control that laser tunability or at least wavelength programmability will become necessary, but no broadly tunable communications lasers are yet commercially available. Full WDM functionality will require optical cross-connects and optical frequency converters, devices that are still in their infancy with no commercial suppliers. Meeting the demands of fiber to the home will require very inexpensive bidirectional transceivers; increased automation of chip fabrication, handling, testing, and assembly; and alignment tolerances better than about 1 μm.

Data Communications Links

The demand for optical data communications (datacom) links has grown rapidly over the last few years. Most such links are low-cost, multimode systems based on large-diameter plastic (1-mm) or glass (0.2-mm) fiber. Their low cost is attributable largely to the low cost of connectorizing such fibers and the undemanding alignment tolerance of the photonic components. As datacom transmission rates and communication spans increase, single-mode systems incorporating laser or detector arrays are becoming desirable. Vertical cavity lasers (VCSELs) are attractive for such applications. At present, however, 1.3-μm VCSELs are hard to fabricate with high mirror reflectivity.

Through-the-Air Infrared Links

Through-the-air links are becoming ubiquitous in portable products. Links that operate at 115 kilobits per second (kb/s) are already used to download files from handheld calculators and printers. In 1995, a standard of 4 megabits per second (Mb/s) was agreed on, and the rate is expected to reach 65 Mb/s soon. At 4 Mb/s the requirement for the gallium aluminum arsenide (GaAlAs) light-emitting diodes (LEDs) can barely be met; high-power VCSEL arrays are the likely alternative. The challenge will be to develop VCSEL arrays in the 0.2- to 0.5-W range for a cost less than $2 per unit.

Semiconductor Lasers

The cost of GaAlAs lasers is dropping to $100 per watt in the 800- to 900-nm wavelength range, with a conversion efficiency of 25 to 40%. At 650 to 690 nm, the conversion efficiency is 20 to 25%. A recent advance in the cladding-pumped fiber laser is the enormous (approximately a thousandfold) brightness conversion of GaAlAs laser arrays in an optical fiber. Single-mode lasers have been fabricated that emit several watts. As power continues to increase, such lasers may be able to replace traditional gas and solid-state lasers in medical and industrial applications.

In CD players and CD-ROM applications, 780-nm GaAlAs lasers dominate. Increased storage density requires shorter wavelengths in the green and blue. Prototype ZnSe lasers have been demonstrated, but their lifetime is only a few hours. The GaInN system looks promising, but no prototype lasers have yet been demonstrated. As the brightness of GaAlAs lasers increases, up-conversion to blue wavelengths may become practical.

Visible LEDs

Visible LEDs account for the largest segment of the indicators and display portion of the optoelectronic components industry. The challenges for future growth are threefold: higher efficiency, expanding the green and blue range, and lower cost per lumen. Visible emitters have considerably lower internal quantum efficiency (10 to 40%) than do IR emitters (approximately 100%). A better understanding of internal loss is key to progress.

The recent breakthrough in blue and green LEDs based on GaInN on a sapphire substrate, which has 3% external quantum efficiency, suggests that the entire visible spectrum may soon be covered by LEDs. The manufacturing challenge is formidable, however. The development of cost-effective, lattice-matched substrates (SiC, GaN, or AlN) may be required to improve performance.

Imaging Arrays

Imaging arrays are dominated by charge-coupled device (CCD) technology. On the high end, 2000×2000 pixel arrays are used in low-volume civilian and military applications that cost $4,000 to $40,000 per array. At the low end, video cameras with 300 to 400 lines sell for $35. The major challenge is to develop high-performance arrays with photographic resolution, since that would revolutionize amateur photography.

A recent challenge to CCDs, based on complementary metal oxide semiconductor (CMOS) technology, includes electronic circuitry integrated with the array that can greatly enhance imaging functionality. CMOS arrays can be manufactured on conventional production lines.

Laser and Waveguide Packaging

The history of laser and optical waveguide packaging has largely been one of adapting methods from electronics packaging (a much larger industry) to meet the needs of fragile and incompliant photonic devices. The most troublesome issue has no counterpoint in electronics, however. The issue is coupling an external optical fiber connection to an optical waveguide with submicron tolerances. Elaborate schemes for fiber attachment have gradually been replaced by simplified designs that use a few high-precision mass-produced parts, usually fabricated by photolithography. Market leadership in this area is determined largely by manufacturing improvements that reduce costs; the U.S. industry currently lags behind Japan.

The ideal package is easily assembled from inexpensive parts, provides a secure thermal and chemical environment for the device, maintains appropriately stable optical and electrical connections, dissipates heat and various overstresses, and conforms to applicable standards. The design issues involved in meeting these goals fall into four categories: (1) optical design, including coupling, mode matching, minimization of reflections, polarization management, connectorization, and stability to submicron tolerances; (2) physical design, including support of the fragile photonic chip materials, stress, ease of assembly, and standards; (3) electrical design, including electrostatic discharge protection, radio-frequency (RF) supply, and power; and (4) environmental design, including heat dissipation and hermeticity.

Although packaging costs contribute at least half of the cost of final assembly, investment in packaging remains relatively small compared to investment in chip fabrication. Packaging R&D is relatively unglamorous, yet increased investment here has become essential for photonics component manufacturing. More attention to passive fiber attachment and low-cost packaging of high-end components, together with improved yield and automated device testing, are essential components of cost reduction and the ubiquitous use of photonic components. Only a few universities have attempted programs that emphasize packaging; in part because the technology changes so rapidly, skills in this area are typically acquired in the manufacturing environment.

At present, essentially all photonic components are discrete devices in individual packages. In this respect, photonics is in its infancy, much as silicon electronics was four decades ago. Improvements in yield and reductions in cost will bring increased integration. This trend has already begun. For example, the recently introduced integrated electroabsorption-modulated laser contains both a distributed-feedback laser and an external modulator, monolithically integrated on a single InP substrate. Similarly, optical transceiver packages based on silicon optical bench technology are now integrating lasers, WDM splitters, and detectors on a silicon platform. This hybrid integration within the package simplifies assembly and packaging.

Many of the challenges for photonics integration are closely related to those of silicon electronics. Because different photonic components are often based on different materials systems, integration relies on advanced photolithography methods. As in electronics, techniques to increase the level of integration and reduce device size typically exacerbate the interconnection problem; breakthrough technologies for interconnect design will be necessary to realize integration's full potential. Because photonic components are some three orders of magnitude larger than silicon transistors, however, photonic integration will not be done on the same scale as electronics.

Trends in telecommunications, especially the move toward WDM, will increase the demand for integration. As the use of optical amplifiers based on erbium-doped fiber increases, so does the importance of other photonic components, including splitters and combiners, multiplexers, attenuator filters, dispersion compensators, and fiber lasers for amplifier pumps. These fiber devices are readily packaged in either capillaries or fiber reels and can be spliced directly into the fiber transmission line, but as waveguide packaging and integration advance, many of them will be integrated directly into the waveguide to reduce size and interconnection costs. In the long term, optical multichip modules will further simplify interconnection and integration with silicon electronics. Surface-emitting laser technology simplifies the packaging and testing of components, and successful commercialization of InP- or GaN-based VCSELs in addition to GaAs-based devices will radically change laser and optical waveguide packaging and interconnection. Indeed, the next generation of lasers will incorporate features to simplify packaging and interconnection into the semiconductor chip design itself.

Key Technical Challenges

Among the major challenges in the design and fabrication of high-volume optical components are the following:

- Integrated design, fabrication, test, and assembly methods, including active optical components, which will permit the smooth advance of the hybrid optical devices of today into the fully integrated optical systems required for miniaturization and high performance in future products; and
- Development of an optical foundry for active and passive optical components, similar to the foundries of the silicon world, which will permit a new generation of integrated optoelectronic and optomechanical systems in the future.

Crosscutting Issues

There are some issues that cut across the diversity of optics manufacturing, including both the low-volume specialized applications and the high-volume mass markets. This section discusses two technical issues of this type, optical design and the role of metrology, and two structural concerns, the question of standards and the difficulty of identifying and characterizing the "optics industry."

Optical Design and the Impact of Increased Computer Power

The development of effective and comprehensive optical design programs that are capable of handling a wide variety of optical components is a success story of the U.S. optics industry. These programs have become world standards for use in the development of new optical systems and devices and have contributed to U.S. leadership in developing innovative optical systems. The programs are remarkably capable with passive optical components, although there is as yet little integration of active optical components into the design process. The dramatic increase in computing speed and the simultaneous dramatic fall in computing cost have greatly influenced the optical design process.

For several decades, software based on exact ray tracing has been the primary tool of the optical design engineer. The performance of such software is determined by how fast the computer can trace the path of light rays through the system being analyzed. Speed is measured in ray-surfaces per second (i.e., the number of refractions or reflections of a ray at a surface computed per second). As shown in Figure 6.3, from the 1950s through the 1980s, optical design required large mainframe computers that cost millions of dollars and could trace no more than about 10,000 ray-surfaces per second. Today, desktop computers that cost only a few thousand dollars can trace 300,000 ray-surfaces per second. There have been at least four major ramifications of this revolution in speed and cost: freedom from mainframes, wider access to design capabilities, full analysis of designs, and global optimization.

First, because optical designers are no longer tied to mainframe computers, they are no longer tied to the owners of those computers, typically large companies or other large organizations. Today, small firms and even individual consultants have access to full-featured optical design systems.

A second and related result is that many engineers with little or no formal training or experience in optical design can now afford the capability to design their own optical systems. Paradoxically, many optical component manufacturers cite this trend as a major problem. Manufacturers are often presented with designs that have not been fully

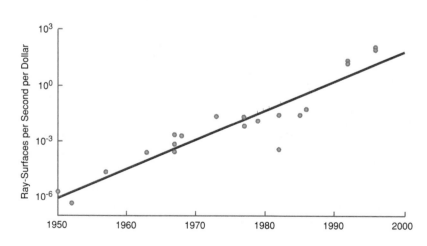

FIGURE 6.3 There has been a dramatic revolution in both the speed of ray tracing and the capital cost of the computers required. (note the logarithmic scale).

evaluated for tolerances, so they do not work when assembled or are impossible to build. There are less than a thousand expert optical design engineers in the United States, but many more people have access to design software. Experience plays a significant role in this field; it takes about 3 to 5 years of full-time effort for a competent designer to become comfortable with a wide range of applications.

Third, optical systems can now be fully analyzed for very little cost. The number of rays a designer can afford to trace no longer limits his or her abilities. The total design process begins with a meeting of the designer and the customer to define the goals of the system, including such factors as performance, cost, package, and delivery. About 25% of the design effort is usually spent in communication with the customer to ensure that the product is feasible, that the estimated cost is compatible with the budget, and that the design is indeed what the customer wants. Examination of manufacturability should start in midproject. This includes tolerancing limits of the design and checking compatibility with mechanical constraints, mounting, and environmental issues such as temperature range. The design is then fine-tuned by taking into account all known issues. The total design time can range from a few weeks to 6 months depending on a number of factors including the complexity of the system and the cost sensitivity of the manufactured product. Surprisingly, despite the availability of tolerancing and manufacturability tools in design packages, many designers fail to examine these issues, concentrating only on analyzing and optimizing the design's image and aberration. In most cases, design software can model, specify, and predict the performance of complex surfaces that are beyond what can be accurately fabricated, tested, or aligned. Designers must be aware of these limitations. Better tolerancing is a challenge for the optical design community. Areas that need improvement include communication of the fabrication and testing limitations of the optics shop, communication of the tolerancing set and compensators, better understanding of the manufacturing cost breakpoints, and more enthusiasm for the tolerancing process. New tools for computer simulation of multiple-step assemblies will result in better correlation between computer-predicted and as-built performance.

The fourth major ramification of the revolution in design software is the possibility of global optimization, a major breakthrough that allows multiple local aberration minima to be traversed in a single computer run. Historically, local minima have been a common problem facing optical designers. The system they have designed may be the best of all similar systems for the given application, but there is no guarantee that other classes of design (using more or fewer elements, other types of glass, and so on) would not produce superior results. Real problems with many variables and many constraints—such as designing a lens that fits within a volume or length constraint—can now be analyzed overnight on a desktop computer that generates a family of alternate solutions of different types. Global optimization is one of the most significant innovations in optical design in the last decade.

Despite these changes, no foreseeable advances in design software will eliminate the need for human intervention. Incremental designs, minor improvements, or small configurational changes in existing systems can result from optimization techniques built into a software package. Nonincremental designs represent entirely new configurations of optical elements and cannot be obtained automatically. Truly innovative nonincremental designs represent new technology and require invention by an expert optical designer. By one estimate, only about 3% of design jobs result in nonincremental improvements.

New optical technologies continue to pose new challenges for the optical design community. Optical systems now range from surgical endoscopes a few millimeters in diameter to space-based telescopes 25 m wide. Systems containing adaptive elements force new optimization guidelines on the designer. Improvements in software are necessary for the design and optimization of nonimaging components. Optical design tools must advance to handle integrated optical and mechanical design; for example, standards for representing precision surfaces are needed that are compatible with mechanical design programs. In the future, as computation becomes ever cheaper, end-use image simulation will be employed instead of ray fans for system analysis and optimization. Resources must be found to incorporate all of these changes into optical design software packages. The optical design community is small, and optical design is taught formally at only a few universities, which has limited the number of people dedicated to developing design software. Rarely are students given more than introductory training in optomechanical design. The backlog of potential productivity enhancements from improved designs is measured in years.

Role of Metrology

Advances in optical metrology over the past decade have opened up vast new areas in optical fabrication. There is an old saying that "you can't make it if you can't measure it." A corollary to this has now been well established: "If you can measure it, you can indeed make it." Advances in fabrication technology have sometimes outstripped our measurement capabilities. A lack of routine, cost-effective, timely metrology solutions is often the bottleneck in the manufacturing process.

Much of the advance in metrology is due to advanced light sources, improved sensors, and the tremendous increase in computing power and reduction in computing cost. The availability of a variety of light sources at useful wavelengths, the development of two-dimensional high-density detector arrays, and the computational and storage capability of today's machines have made practical advances that previously were possible only in theory. Continued progress will lead the way in the development of new metrology systems. Its strength in computing puts the United States in a very strong position internationally for optical metrology.

Previously unprecedented surface finishes have been made routine by commercially available noncontact surface profilers and scanning-probe microscopes that measure to the angstrom (Å) level. Even for large-scale optical components, such as mirrors for the Advanced X-Ray Astrophysics Facility (AXAF), x-ray test data have established that the fabrication metrology has an accuracy of better than 250 Å over a 1-m scan.

Future developments in optical metrology must embrace conflicting requirements. There is an ever-increasing desire for higher spatial resolution at the same time as larger and more complete area coverage. The ability to manufacture diffractive optical components with high throughput requires the ability to sculpt profiles on submicron structures to accuracies on the order of tens of angstroms. The advancement of microlithography requires concomitant advances in metrology. Even on large scales, optical components such as those required for space-based observatories are affected by angstrom-level height variations over spatial scales in the hundreds of microns.

The need to maintain special environmental conditions for high-precision measurements is a huge driver of system costs, and the final measurement accuracy is often limited by how well these environmental conditions are maintained. Consequently, the development of a metrology system insensitive to environmental conditions would be of tremendous benefit.

Some other areas for future technology development include these:

- Rapid and inexpensive measurement of aspheres and anamorphic elements;
- Nondestructive evaluation of subsurface damage;
- Measurements of material homogeneity on large blanks in an unfinished or raw state;
- In-process metrology for material removal rates, dimensional changes of the blank, surface figure, subsurface damage, and surface finish;
- Metrology for surface feature generation in microlithography;
- In-process metrology for the deposition of well-defined multilayer coatings over large areas;
- Figure measurement for meter-sized parts with absolute accuracy on a scale of tens of angstroms;
- Metrology for diffractive optics;

- Rapid remote and direct measurement and quantification of both particulate and molecular contamination; and
- Development of sophisticated computer-generated null lenses.

The current level of R&D support in the United States in these areas, whether by industry or government, is a strong ground for concern about continued U.S. technological leadership.

There are no recognized calibration standards for surface roughness, scattering, or cosmetic defects. In addition there are no standard techniques for certification or calibration of holograms or binary optics to be used for optical testing. The quantification of surface defects and imperfections that affect system performance (scratches and digs) should be redefined along with other areas in an international standard. It is necessary to adopt such international standards in order to remain competitive globally.

The increasing sophistication of metrology has made it difficult for many engineers and technicians to understand the meaning of data provided by current instruments. Results are often misinterpreted, and this problem will be compounded in the future as even more complex instruments are developed and the amount of data processing required makes the results still more remote from most users' intuition. Therefore, it is necessary to develop and incorporate practice-oriented instruction aimed at creating an in-depth understanding of state-of-the-art metrology as it relates to what is being measured and to the limitations of its use. Also, because software algorithms, which are not accessible to the user, are so important in advanced measurement instrumentation, instrumentation manufacturers must continue to provide instrument-specific training and user-friendly software. Metrology training should be done at the undergraduate and the graduate levels. Similarly, technician-oriented courses should be developed for technical schools and two-year degree programs.

Standards

The issue of standards has been mentioned several times in this chapter as a key challenge for the future. For example, standards for physical design (e.g., support, stress, and ease of assembly) are an issue in the packaging of photonic components. Photonic integration will require well-defined standards for form, fit, and function. Integrating optical design with mechanical design will require standards for the representation of precision surfaces. In the area of metrology, there are no recognized calibration standards for surface roughness, scattering, or cosmetic defects. There are no standard techniques for certification or calibration of holograms or binary optics to be used for optical testing. There is a need for redefined standards for the quantification of scratches and digs. Standards for the interchangeability of optical components are an important driver of cost-effective manufacturing.

Successful competition in international markets depends on establishing workable international standards in areas such as these. Unfortunately, U.S. participation in international standards-setting activities lags far behind the activities of foreign organizations. Virtually all new optics standards have hitherto been developed overseas, with no support from U.S. industry or the U.S. government. As a result, the U.S. optics industry has been a follower, not a leader, in adapting to new international opportunities. Active government and industrial participation in the setting of strong international standards for optical components is especially important because of the diversity of the U.S. optics industry.

Government agencies and the optics community should recognize the importance of optics standards, especially their significance in international trade. The U.S. government should participate actively in the setting of such standards. NIST should be given the funding necessary to take the lead in this area.

Size and Composition of the Optics Industry

There is no satisfactory comprehensive source of data on the optics industry. No single professional or trade organization represents the industry as a whole, and the industrial data collected by government agencies are of limited use because their classification scheme does not clearly identify optical products. Nevertheless, certain facts are clear.

As mentioned in the introduction to this chapter, large companies do not dominate the U.S. optics industry. Some large companies, such as AT&T, focus on integrating optics at the systems level, whereas others, such as Kodak, focus on systems and components. Small entrepreneurial companies play a critical role in developing the component technology (Council on Competitiveness, 1996).

Optoelectronics appears to be the most rapidly growing segment of the optics industry. The most comprehensive source of recent data on optoelectronics manufacturers is a survey of 106 U.S. companies conducted between July 1992 and February 1993 by the Bureau of Export Administration (BEA, an agency of the Department of Commerce). Of the 106 companies in the BEA survey, 77 were primarily manufacturers and 17 performed both R&D and manufacturing. (To protect proprietary information, BEA did not identify individual companies.) For the purposes of the survey, BEA defined optoelectronics as all systems, equipment, and components that emit, modulate, transmit, and/or sense light or are dependent on the combination of optical and electronic devices (see Table 6.1) (U.S. BEA, 1994).

According to the survey, in 1989 these 106 companies had net sales of $4.4 billion. Respondents projected that by 1995 this figure would

TABLE 6.1 Composition of the Optoelectronics Industry

Category	Products
Fiber-optic communications	Transmission, amplifiers, cable television distribution, optical modulators, switches, fibers, muliplexers, connectors, transmit and receive modules
Fiber-optic information equipment	Optical processing units, memory or storage devices, bar-code readers, printers, image processing, interconnects, faxes, displays
Industrial/medical equipment	Machine vision, optical test and measurement, night vision and surveillance, laser processing equipment, nonlaser medical equipment, lasers
Nonmilitary transportation equipment	Automotive interior displays, traffic control systems, fly-by-light, cockpit displays, lidar for sensing turbulence, optical gyroscopes
Military equipment	Fiber-optic ground and satellite communications, lidar, optical gyroscopes, FLIR, night vision, munitions guidance, laser weapons
Consumer equipment	Televisions, video cameras, CD players, home faxes, appliance displays
Subsystems/components	Photo detectors, semiconductor light sources, hybrid optical devices

TABLE 6.2 Sales of 106 Surveyed Companies in Optoelectronics Sector of Optics Industry

	1989	1990	1991	1992	1993 E	1994 E	1995 E
Net sales ($ million)	4,412	4,974	5,576	5,873	6,430	7,134	7,921
Change from previous year (%)	—	12.8	12.1	5.3	9.5	10.9	11.0

NOTE: E = estimated. Source: U.S. Bureau of Export Administration (1994), Table IV-4.

approach $8 billion, which represents an average annual increase of 10% (see Table 6.2). This growth rate is probably representative of the optoelectronics industry as a whole, but the survey was not all-inclusive, and so the exact size and composition of the industry are not known (U.S. BEA, 1994, p. iii-2).

Another source of data is the annual survey of manufacturers (ASM) conducted by the Bureau of the Census. This survey is one of the primary sources of information regarding the performance of the U.S. manufacturing industry. The most recent ASM, based on a sampling of approximately 58,000 manufacturing establishments, estimates the value of 1995 shipments for approximately 1,750 classes of manufactured products (U.S. Bureau of the Census, 1997). Two of these product classes are *fiber optic cable* and *laser systems and equipment*; each shows an average annual growth rate of about 15% in recent years (see Figure 6.4).

Recent information in the trade press indicates that worldwide 1996 sales of industrial laser systems were $1.5 billion, a 29% increase from

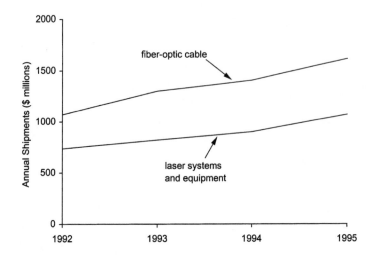

FIGURE 6.4 Shipments of fiber-optic cable and laser systems and equipment by U.S. manufacturers. (Source: U.S. Bureau of the Census, 1997.)

1995. A further 23% increase is expected in 1997. U.S manufacturers have a 35% share of this market by revenue—about $400 million in 1995 (Belforte, 1997). These growth rates are about twice those derived from the 1995 ASM, whereas the U.S. market size estimate is substantially smaller than the Census Bureau's estimate.

The Optoelectronics Industry Development Association (OIDA) estimates that in 1995, world production of optoelectronics was $64 billion, of which U.S. production was $12.5 billion (see Table 6.3). OIDA predicts that the optoelectronics market will continue to grow by about 10% per year, resulting in a world market in 2013 that approaches $500 billion.

TABLE 6.3 Optoelectronics Market Segments and Their 1995 Production ($ billion)

	World	U.S.
Cathode-ray tube displays	16.3	1.8
Data storage (media only)	13.0	1.0
Flat-panel displays	11.5	0.2
Sensors (including imaging)	7.0	1.0
Lightwave transmission, interconnects, etc.	6.0	3.5
Fiber, passive	6.0	2.0
Military	2.5	2.0
Lighting	1.5	1.0
Solar cells	0.1	0.0
Total	63.9	12.5

Source: A. Bergh, Optoelectronics Industry Development Association.

BOX 6.3 INDUSTRIAL CLASSIFICATION CODES

The basis for collecting industrial statistics such as those described in this section is a system of product classes or groupings of the individual products of an industry. Since 1939, product classes have been designated by a system known as the Standard Industrial Classification (SIC), which was established by the Department of Commerce in concert with industry representatives. Each SIC code has four digits. Special surveys often add a fifth digit or letter to better identify specific groups of products.

A new six-digit North American Industry Classification System (NAICS) is being introduced beginning in 1998. Its primary goal is improved relevance to current economic activity, in response to the rapid creation of new industries and the need to better characterize existing industries. Additional goals are to include Canada and Mexico in a unified system and to facilitate comparison of economic activity in North America with activity in the rest of the world. Classification under NAICS is based primarily on production methods; that is, products that use identical or similar production processes will be grouped together (*Manufacturing News*, 1997).

· ·

A critical issue for all these data sources is that the standard system of industrial classification (see Box 6.3) does not align well with the various segments of the optics industry. This makes it extremely difficult to compare data from different sources, draw inferences about trends, or assess the effects of policies. Even though it produces billions of dollars in revenues each year, there is no statistically well-defined "optics industry." Many important and growing components of the optics industry described in this report cannot be found in the Standard Industrial Classification (SIC) database. For example, the category *laser systems and equipment, except communications* (SIC code 36992) is included only as part of *electrical machinery, equipment, and supplies not elsewhere classified* (SIC code 3699), which also includes Christmas tree lights, electric insect lamps, automatic garage door openers, and outboard electric motors. There is no way, except through a special survey, to sort out the data for laser systems, and the special survey that identifies *laser systems and equipment, except communications* is performed only every 5 years.

Despite recent efforts by the Department of Commerce to provide some much-needed increased visibility to the rapidly growing high-technology segments of the U.S. industrial base, much of the optics industry appears to have been overlooked. For example, the new North American Industry Classification System (NAICS) appears to have dropped the category *laser systems* altogether! **The Bureau of the Census should involve representatives of the optics industry in the next revision of the NAICS codes.**

TABLE 6.4 Some NAICS and SIC Codes for Optics-related Products

NAICS	Description	SIC	Description
333314	Optical instruments and lens manufacturing	3827	Optical instruments and lenses
333315	Photography and photocopy equipment manufacturing	3861	Photography and photocopy equipment
334413	Semiconductor and related device manufacturing	3674	Semiconductor and related devices
334613	Magnetic and optical recording media manufacturing	3695	Magnetic and optical recording media
33511	Electric lamp bulb and part manufacturing	3641	Electric lamp bulbs and parts
335921	Fiber-optic cable manufacturing	3357	Nonferrous wire drawing and insulated wire and cable
336321	Vehicular lighting equipment manufacturing	3647	Vehicular lighting equipment
339112	Surgical and medical instrument manufacturing	3841	Surgical and medical instruments
339117	Eyeglasses and contact lens manufacturing	5995	Eyeglasses and contact lenses

Some optics-related NAICS and SIC codes can be found in Table 6.4. In 1997 the major optics-related professional and trade organizations have come together to form the Coalition for Photonics and Optics (CPO). The goal of the CPO is to better represent the character of the optical industry as a unified whole. As the CPO develops, it is to be hoped that more (and more consistent) details about the character of the industry will become available.

Summary and Recommendations

Small companies, generally quite specialized, make up much of the U.S. optics industry. Thus, although the industry makes a significant contribution to the economy, this contribution comes in so many small pieces that it is hard to fully recognize and understand. Annual revenues are in the tens of billions of dollars, but precise characterization is difficult because the optics industry does not align well with the government's standard statistical categories. Furthermore, no single professional or trade organization represents the entire industry, although in recent months several major organizations have come together to form the Coalition for Photonics and Optics.

New technologies are presenting new manufacturing challenges. Aspheric and diffractive elements offer new options for design and packaging, but design capabilities in this area currently far exceed capabilities for economical manufacturing. Driven by the growth of broadband telecommunications and other applications, fiber devices are developing rapidly and demanding lower-cost manufacturing techniques. New opportunities are emerging for semiconductor devices as their cost per watt falls. The photonic materials InP, GaAs, and $LiNbO_3$ are maturing, but for future devices, CdTe, SiC, GaN, and AlN are still in need of improvements in crystal growth and material quality.

Optical design is a key strength of the U.S. optics industry. Design and analysis capabilities have become dramatically more powerful as a result of advances in computing power. For photonic components, however, CAD tools like those used in the silicon industry are not available. Such tools are becoming essential for shortening design cycles and increasing functionality.

The Department of Defense has historically been a key customer of the U.S. optics industry, and its special needs have strongly influenced the industry's development. Despite the trend toward increased reliance on commercial, off-the-shelf technology, continued DOD support of optics manufacturing technology is vital. There are many specialized optical technologies with important military applications whose commercial markets are insufficient to motivate development by the private sector alone.

Photolithography equipment is a key comercial application of high-value specialty optics. Achieving the ever-smaller feature sizes demanded by the semiconductor industry will require industry support for improved fabrication of deep-UV aspheric elements. Metrology may be the single most difficult challenge in manufacturing the next generation of photolithography equipment.

Photonics applications in information technology are a major driver of the optics industry's low-cost mass markets. Today, most photonic devices consist of discrete components. Advances will depend on lower cost and increased functionality, and these requirements will drive integration. Also required will be high-volume automated manufacturing lines; improvements in assembly and process control; and well-defined standards for form, fit, and function. Packaging, testing, and fiber connection account for the majority of the cost of photonic components; skills in these areas are underemphasized in the U.S. optics community. The shrinking pool of U.S. talent in bulk crystal growth is also an area of concern.

Successful competition in international markets depends on establishing workable international standards. Virtually all new optics standards have hitherto been developed overseas, with no support from U.S. industry or the U.S. government. As a result, the U.S. optics industry has been a follower, not a leader, in adapting to new international opportunities. Active government and industrial participation in setting strong international standards for optical components is especially important because of the diversity of the U.S. optics industry.

As discussed earlier in this chapter, the committee recommends the following actions:

Government agencies and the optics community should recognize the importance of optics standards, especially their significance in international trade. The U.S. government should participate actively in the setting of such standards. NIST should be given the funding necessary to take the lead in this area.

Government agencies should continue to support the activities necessary to introduce cost-effective precision aspheric components into both military and commercial products.

DOD should continue to maintain technology assets and critical skills in optics manufacturing in order to meet future needs.

The Bureau of the Census should involve representatives of the optics industry in the next revision of the NAICS codes.

Collaborative programs in optics manufacturing should include universities so that students are trained in the latest technical solutions to production problems.

DOD, NIST, and the DOE national laboratories should establish together a cooperative program that provides incentives and opportunities to develop new ideas into functioning methods for optics fabrication.

References

Belforte, D. 1997. Annual market review. *Industrial Laser Review* (January):17-21.

Business Communications Company. 1996. *The Future for Optical Coatings*, RGB-187. Norwalk, Conn.: Business Communications.

Council on Competitiveness. 1996. *Endless Frontier, Limited Resources.* Washington, D.C.

Joint Precision Optics Technical Group. 1987. *Precision Optics Study.* Washington, D.C.: Pentagon Joint Group on the Industrial Base.

Manufacturing News. 1997. Goodbye SIC, Hello NAICS. January, p. 1.

The Photonics Directory. 1997. Pittsfield, Mass.: Laurin Publishing Company.

U.S. Bureau of the Census. 1997. *1995 Annual Survey of Manufactures*, M95(AS)-2. Washington, D.C.: U.S. Government Printing Office.

U.S. Bureau of Export Administration (BEA). 1994. *Critical Technology Assessment of the U.S. Optoelectronics Industry*, NTIS PB93-192425. Washington, D.C.

7

....

Optics Research and Education

Research in optics has a long and distinguished history, dating back even further than the work of Galileo and Newton. In recent decades, optics research has blossomed with the invention of the laser, an increasing interaction between optics and electronics, the development of new materials with unique optical properties, and other extraordinary advances. The first part of this chapter highlights some examples of research areas that hold special promise for further discoveries. This is a time of great excitement for all optics researchers, whether in universities, industry, or government laboratories.

The second part of the chapter discusses the state of optics education. Combining this report's discussion of research and education issues in a single chapter is only appropriate. The creation of research universities, which combine research and education endeavors to create a synergy between the discovery of knowledge and the education of students, has been one of the key institutional developments of the past century for research and education in optics as in other fields.

Introduction

Both basic and applied research are motivated by a deeply rooted human instinct: the desire to know and understand nature. Over hundreds of years, humans have gradually learned to explore nature in a systematic way known as the scientific method—the use of experiments and observations, guided initially by intuition and hypothesis and later by theory. The resulting understanding of nature is the hallmark of the scientific process. Roger Bacon, a multitalented thirteenth-century scientist and philosopher whose work in optics included a description of a telescope (more than 300 years before anyone actually built one) and the first application of geometry to the study of lenses and mirrors, is often considered the forerunner of experimental science and the scientific method.

Applied research, which is often indistinguishable from basic research, is motivated by the need to understand *how* as well as the need to understand *why*. Louis Victor de Broglie, recipient of the 1929 Nobel Prize in physics, explained that "the two aspects of science [pure and applied] correspond to the two principal activities of man: thought and action. They are inseparable if human science is to progress as a whole and fulfill with increasing success its high and twofold task." Decades of basic and applied research are often needed to lay the foundations for key discoveries, such as the invention of the laser. The most important discoveries often arise at boundaries between established fields—in the case of the laser, at the interface between physics and electrical engineering.

The ultimate impact of research is rarely predictable. For example, as the earlier chapters of this report demonstrate repeatedly, the invention of the laser has had a major economic and social impact, with remarkable applications in areas that range from communications to the environment to medicine. Yet Arthur Schawlow, the laser's coinventor, once commented that "if I had set out to invent a way to improve eye surgery, I certainly would not have invented the laser."

Development, often guided by understanding gained through basic and applied research, is motivated by a need to make something that works and is of commercial value. However, development—especially the development of commercial products—usually requires a much larger financial investment than most research. Such resources are often unavailable at universities; therefore, in the United States, development activities are generally concentrated in the commercial sector. The exception is in certain areas of special interest to the nation, such as defense, space, and the environment. Military needs in particular have historically motivated substantial support for long-term optics R&D by the Department of Defense—support that has been very important over the years but that is being reduced, leading to concern about a gap emerging between basic conceptual research and commercial hardware development. Development often involves multiple scientific and engineering disciplines, multiple approaches to problem solving, and the unique techniques and resources of large public and private organizations.

Education at all levels is critical to the future of optics. Optics is a natural tool for a visual approach to the education of K-12 students in many aspects of science and mathematics. Post-high school education, including 2-year (associate) and 4-year degree programs, is important for meeting the future needs of this rapidly growing area of knowledge. There is a need for graduate education, as discussed below, as well as special courses of study in optical engineering provided by universities with an historical focus on optics. Jobs are available for those trained in optics.

It is often asked whether optics constitutes a separate discipline, and this question requires critical attention. Optics is certainly recognized as an integrated field of knowledge. Its teaching extends across disciplinary boundaries and includes science departments, engineering departments, and departments in schools of medicine.

Because the field of optics is not sharply defined by a job title or membership in a single professional society, it can be difficult to develop a quantitative picture of the size and breadth of the optics research community. One indicator is participation in professional conferences. About 6,000 scientists and engineers usually attend the annual U.S. Conference on Lasers and Electrooptics (CLEO). The Optical Fiber Conference attracts more than 7,000 participants each year. Other major conferences in optics include the annual meeting of the Quantum Electronics and Lasers Society, the International Quantum Electronics Conference, the annual meeting of the Optical Society of America, the Lasers and Electrooptics Society annual meeting, and conferences organized throughout the year by SPIE, such as Photonics West and Photonics East. Altogether, the field involves at least 30,000 active scientists and engineers worldwide, and many areas of the optics industry are growing rapidly.

Another indicator of the strength and growth of the field is its impact on the economy. Optoelectronics is now a major component of U.S. import and export trade. The rate of formation of optics-related businesses has grown rapidly during the 1990s.

The need for greater investment in research and education in optics was recognized in a 1994 National Research Council (NRC) study, *Atomic, Molecular, and Optical Science* (the FAMOS study). Although this study focused on a narrow segment of the field of optical science and did not address the engineering aspects, it nevertheless found that this field is ripe for breakthoughs, growing rapidly, and having an impact on sciences beyond optics, including biology, chemistry, and medicine. Today, the United States invests approximately 0.5% of its gross domestic product (GDP) on research in its research universities. This is the lowest rate of such investment by any major industrialized country.

As a result of a workshop in May 1994,[1] the National Science Foundation (NSF) in 1995 announced a multidisciplinary research initiative in optical science and engineering. The program attracted more than 600 pre-proposals and 70 full proposals. Ultimately only

---

[1] The workshop produced a report; see NSF (1994).

18 projects could be funded. These numbers indicate both the vitality and the competitive nature of optics research and the strong interest nationwide in the opportunities it presents.

In recent years an important feature of optics research has been the growing interaction between optical physics and electrical engineering. The first hint of this link was the discovery by Heinrich Hertz in 1883 that radio waves and light waves are described by the same theory of electromagnetism and are distinguished only by their different frequencies. Marconi soon demonstrated the transmission of information by pulsed radio waves, and by the 1920s there were more than 30 million radios in use in the United States alone. From the development of electrical devices for radio and other applications came a new research discipline, electrical engineering. In this same period, physicists were exploring newly discovered properties of light. Max Planck found that light is quantized in units called photons. Einstein used this discovery to explain the emission and absorption of light quanta by atoms and predicted that it should be possible to use light to stimulate an atom to emit more light. Other physicists—most of them engaged in basic research with little thought about potential applications—explored electrical discharges, the properties of atoms in solids, the properties of semiconductors, and atomic and molecular spectroscopy. The two communities—physicists, with their emphasis on quantum mechanics and the basic understanding of nature, and electrical engineers, with their understanding of electronic circuits and wave propagation—were brought together in 1960 by the invention and demonstration of the laser (named for light amplification by stimulated emission of radiation). Today, work at the interface of physics and electrical engineering is a key element of optics research, producing such important developments as the semiconductor diode laser.

Optics is a multidisciplinary field that cuts across many of the traditional academic disciplines. The funding of initiatives in such crosscutting areas is often hindered by the structure and organization of the federal agencies that support research and development. In an effort to overcome these difficulties, the government has created the National Science and Technology Council (NSTC) and charged it with addressing the following goals: to maintain leadership across the frontiers of scientfic knowledge; to enhance connections between fundamental research and national goals; to stimulate partnerships that promote investments in fundamental science and engineering and in the effective use of physical, human, and financial resources; to produce the finest scientists and engineers for the twenty-first century; and to raise the scientific and technology literacy of all Americans. These goals serve as guideposts for NSTC in making recommendations and setting priorities for investment in new crosscutting research and development opportunities that involve multiagency and multidiscipline collaboration and support. NSTC works to identify important crosscutting science and technology programs and works with multiple agencies to fund new initiatives. For example, an initiative on high-performance computing and communications involves nine agencies and an investment of $1 billion (National Science Board, 1996). As demonstrated throughout this report, the broad area of optics is multidisciplinary and growing rapidly. It is ripe for support as a crosscutting initiative at a level comparable to the earlier investment made in high-performance computing and communications.

Multiple government agencies should form a working group to collaborate in the support of optics, in a crosscutting initiative similar to the earlier one for high-performance computing and communications systems.

NSF should develop an agency-wide, separately funded initiative to support multidisciplinary research and education in optics.

The Department of Commerce should explicitly recognize optics as an integrated area of knowledge, technology, and industry and should structure its job and patent databases accordingly.

Research Opportunities

Opportunities for research in optics have the potential for significant benefit to society in the next decade. Many new developments, with a high leverage for return on the investment of increasingly scarce research dollars, have been identified. However, research by its nature returns benefits on a portfolio of many possible avenues of investigation. A few examples of the many areas of research in optics that show promise and offer high leverage for future return are presented in this section. It is impossible to cover all areas of research in the extensive and multidisciplinary areas of optics. A more inclusive set of research opportunities in optics is presented each year at large annual meetings such as CLEO.

Areas of optics research selected for discussion include control of atoms by light, fundamental quantum limits of measurements, and light in biology. Recent advances in optical microscopy are highlighted. Femtosecond laser technology and its application to ultrafast physics, chemistry, and engineering is identified as a particularly promising opportunity. Advanced laser sources and frequency conversion of lasers using nonlinear optical devices offer significant potential for applications in semiconductor processing, reprographics, and image display. Semiconductor lasers and solid-state lasers are promising sources of coherent optical radiation at ever-decreasing cost. Advances in optical materials are recognized to play an important role in the next generation of laser sources, nonlinear frequency converters, and new optical elements such as gradient index optics and media for optical storage of information. Finally, progress in the generation of coherent radiation progressed from radio frequencies in 1900 to microwave frequencies in 1930 and to optical frequencies in 1960. We are now entering the era of extreme ultraviolet (UV) and, in the future, coherent x-ray sources with applications that extend from lithography to advanced x-ray imaging.

Quantum, Atomic, and Biological Optics

Control of Atoms by Light

Light continues to be the principal method of probing matter. Powerful spectroscopic techniques continue to be developed as light sources extend into new regimes of the electromagnetic spectrum and as optical sources with extremely short pulses such as femtosecond lasers, synchrotron sources, and free electron lasers become more widely used.

In the past decade, light has begun to be used to detect and to *control* matter, particularly the position and velocity of atoms, molecules, and small particles. Currently, atoms in the gas phase can be laser-cooled to microkelvin temperatures where their velocities are on the order of a centimeter per second. Once cold, atoms can be manipulated with relative ease. Atoms tossed upwards in a ballistic trajectory as in a fountain form the basis of a new generation of atomic clocks. Prototype atomic fountain clocks already have short-term stability an order of magnitude better than atomic clocks based on thermal beams of atoms. Alternatively, ions trapped in magnetic and electric fields could lead to atomic clocks based on optical transitions in which the stability would be improved another thousandfold. (Since ion clocks will have a low signal-to-noise ratio, they will need stable "flywheel" frequency references, such as ultrastable hydrogen masers or cryogenic microwave oscillators, to take full advantage of this potential for superior accuracy.) Atom traps have also been used to store radioactive isotopes for nuclear physics studies and for tests of fundamental symmetries of nature such as parity and time.

Atom optics has emerged as a new field. Atom lenses were used to deposit 0.05-μm lines and dots on a surface. These structures can be written over large areas and may be used to pattern a surface for high-density optical storage. One promising approach uses the optically guided atoms to react chemically with a resist with better than 0.1-μm spatial resolution. Atom beam splitters, mirrors, and diffraction gratings have been used to construct atom interferometers. Atom interferometers have already proven to be sensitive accelerometers and gyroscopes.

BOX 7.1 BOSE CONDENSATION

In 1925, Einstein made a dramatic prediction. If the atomic density n increases to $n\lambda_{DeB}^3 = 2.612$, where λ_{DeB} is the DeBroglie wavelength, a sizable fraction of the atoms will condense into a single quantum state. Examples of Bose condensation media include superfluid helium, paired electrons in a superconductor, and paired atoms in superfluid ^{3}He.

In 1995, 70 years after Einstein's prediction, research groups at the University of Colorado and the Massachusetts Institute of Technology achieved Bose condensation of a dilute gas of rubidium and sodium atoms by using a laser to cool the atoms. A year later, a group from Rice University was able to show that ^{7}Li atoms also form a Bose condensate.

Figure 7.1 shows the momentum distribution of a gas of (Rb or Na) atoms in the vicinity of the Bose condensation threshold. Just above the required phase space density (*left*), the atomic energy is distributed in all directions equally among the many occupied quantum states, in accord with the principles of statistical physics. As the threshold condition is crossed (*center*), a central peak at $v = 0$ begins to form, signifying the onset of a condensation. With further cooling (*right*), most of the atoms condense into the ground quantum state of the system.

The properties of dilute Bose gases are being explored, and it is too early to predict how this field will develop. Perhaps new physics will be discovered. Perhaps Bose condensates will be used as intensely bright sources of atoms analogous to photons from a laser. The one certain prediction is that the future is hard to predict: The inventors of the laser had no idea of the myriad of ways in which their invention would contribute to science and technology.

· ·

Freely falling atoms have been used to measure the acceleration of gravity with a precision of one part in a billion after 100 seconds of averaging time, and an atom interferometer gyroscope has exceeded the sensitivity of the best commercial laser gyroscopes. Atom gravity gradiometers currently being developed may eventually be used for oil exploration.

The control of atoms permits the creation of new states of matter. For example, standing wave interference patterns caused by interfering laser beams create periodic potential wells in which atoms can be trapped. These so-called optical lattices offer a unique model system of "condensed matter" concepts in which the characteristics of either fermionic or bosonic particles can be probed. Atoms have been cooled with sufficient density that they have condensed into a single quantum state known as Bose-Einstein condensation (Box 7.1), predicted by Einstein (see Figure 7.1). Once the atoms are in a single quantum state, the concept of temperature is no longer valid. The properties of this new state of matter are now being studied with great fervor, and important applications will likely result from these studies. Photons, which share the same quantum statistical-mechanical properties as atomic bosons, have been made to "condense" into laser light. As the atom analogue to the

FIGURE 7.1 *The momentum distribution of atoms during Bose condensation of a laser- cooled dilute gas. (Courtesy of M. Matthews, JILA.)*

laser, the Bose condensate may also revolutionize the way atoms are used just as the advent of the laser revolutionized the way light is used.

Fundamental Quantum Limits of Measurement

In addition to providing some of the most refined measurement tools, optics is and will continue to be the primary means of studying the fundamental limitations of any measurement. The measurement of a physical quantity is limited by the size of the signal relative to the noise in the measurement. Quantum mechanics defines the fundamental noise floor. For example, the uncertainty principle tells us that the simultaneous measurement of position and momentum with uncertainties Δx and Δp is limited to an accuracy $\Delta x \Delta p \geq h$, where h is Planck's constant. Similarly, the simultaneous measurement of the amplitude and phase of an electromagnetic wave is also limited.

Although these fundamental limitations were known since the early days of quantum mechanics, we have only recently shown that the partition of quantum noise can be altered. For example, in a laser interferometer measurement of distance, the phase of the light in the interferometer is the quantity of interest. The phase noise due to the vacuum can now be "squeezed" into the amplitude sector, allowing more precise measurement of the phase of the interferometer. Before this refined measurement sees wide application, the methods of altering noise at the quantum limit have to be made easier.

Quantum mechanics also tells us that any measurement of a physical system necessarily alters the system. This effect of the measurement on the system being studied can be tailored to influence different aspects of the system. To measure the intensity of a pulse of light, it is now possible to make a "nondemolition" measurement in which the number of photons in the pulse is not altered; the unavoidable perturbation of the light pulse manifests itself with a change in the phase of the optical pulse.

The quantum dynamics of strongly coupled quantum systems allows precise coherent control of both the internal state and the external dynamics of an atom. This type of quantum state engineering has already been pursued in order to control chemical dynamics. More recently, it was shown that it may be possible to engineer correlated quantum states where the sensitivity of an optical interferometer could be improved by the square root of the number of photons used in the measurement. Since the number of photons used in a measurement can be very large, the gain in sensitivity could be staggering.

The study of highly correlated quantum states may also offer fundamentally new ways of transmitting information and computing. Unlike a classical computer based on irreversible, dissipative transitions between distinct states such as 0 and 1, a quantum computer based on the reversible quantum evolution of linear superpositions of quantum states could in principle greatly reduce the time needed to solve certain problems. For example, on a conventional computer the time required to determine a number's prime factors (a problem with important applications in cryptography) grows exponentially with the size of the number, so any computer can easily be overloaded. On the other hand, a factorization algorithm based on the logic of quantum computing necessitates only a polynomial growth in computing time. A practical quantum computer may never be realized because of the difficulty of maintaining the phase coherence of a system with many degrees of freedom, but this area of research will no doubt lead to greater facility in handling complex quantum systems and eventually to new technologies. The laser, the transistor, and magnetic resonance imaging are examples of how a deeper appreciation of quantum mechanics has led to technological revolutions. Optics will continue to provide some of the best tools for the exploration of quantum mechanics and quantum systems.

Light in Biology

Not only are optical techniques driving a revolution in the manipulation of atoms and the refinement of physical measurements, but they are also generating a resurgence in biological fields. Even the venerable optical microscope, invented in 1625 and "perfected" at the beginning of this century, is having a renaissance because of the confluence of several technological developments (see Figure 7.2).

Fluorescence methods made possible with high-performance interference filters, laser illumination, and the synthesis of new dyes resistant to photobleaching have contributed to this resurgence. The fluorescence from single molecules can now be detected with detectors such as microchannel plate image intensifiers or avalanche photodiodes.

BOX 7.2 ADVANCES IN OPTICAL MICROSCOPY

Technical innovations in the optical microscope have often been followed by major advances in biology. The invention of the first compound microscope in 1625 was followed quickly by the discovery of microscopic organisms such as yeast, algae, and protozoa. The construction of good-quality, high numerical aperture microscope objectives, the development of a method of achieving uniform sample illumination (Kohler illumination), and the introduction of cell-staining methods elevated microscopy to a general-purpose diagnostic tool in medicine. The introduction of phase contrast methods for observing unstained biological samples opened up the possibility of observing live cells.

We are in the midst of a renaissance in the development of the microscope, generated by the integration of other technologies such as video cameras, image intensifiers with single-photon counting sensitivity, and computer pattern recognition and image deconvolution. Research into laser trapping techniques has led to "optical tweezers" that allow one to micromanipulate samples with light while simultaneously viewing them. The scanning confocal microscope, coupled to a computer, gives three-dimensional images of samples. Near-field optical microscopes have achieved molecular resolution. Molecular biology techniques now permit the staining of specific proteins expressed in living cells.

What are the current technological challenges? Near-field optical techniques have to be improved so that molecular resolution can be routinely achieved for biological samples in water. Fluorescent tags with greater resistance to photobleaching and additional colored protein markers to complement the green protein marker will allow biologists to track protein-protein interactions in living cells. The most important ingredient is the invention of methodologies that will exploit these new tools. History has shown repeatedly that new discoveries in basic science have always followed significant technological advances.

Proteins manufactured in cells can be tagged with a green fluorescent protein (GFP) by adding instructions to the protein DNA that result in a protein-GFP molecule. The fluorescent protein can then be seen as it moves in a live cell. The development of alternate color tags (such as red and blue dyes) will enable the observation of protein-protein interactions in a live cell in real time. Two-photon excitation from inexpensive diode-pumped femtosecond lasers will permit the observation of the intrinsic fluorescence of biological molecules and molecules in the interior of tissues that could not otherwise be detected. Fluorescent dyes can also be used to monitor their local biological environment since the fluorescence is affected by conditions such as temperature, electric fields, and pH.

New laser methods such as confocal microscopy for three-dimensional imaging and near-field optical microscopy with nanometer resolution are adding to the set of powerful, noninvasive optical diagnostic tools (Box 7.2). Lasers have also been used to control micron-sized

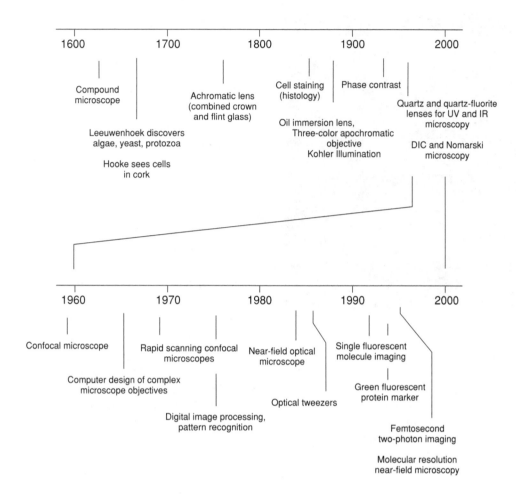

1600   1700   1800   1900   2000

Compound
microscope

Achromatic lens
(combined crown
and flint glass)

Cell staining
(histology)

Phase contrast

Quartz and quartz-fluorite
lenses for UV and IR
microscopy

Leeuwenhoek discovers
algae, yeast, protozoa

Oil immersion lens,
Three-color apochromatic
objective
Kohler Illumination

DIC and Nomarski
microscopy

Hooke sees cells
in cork

1960   1970   1980   1990   2000

Confocal microscope

Rapid scanning confocal
microscopes

Near-field optical
microscope

Single fluorescent
molecule imaging

Computer design of complex
microscope objectives

Green fluorescent
protein marker

Optical tweezers

Digital image processing,
pattern recognition

Femtosecond
two-photon imaging

Molecular resolution
near-field microscopy

objects at room temperature in aqueous solution. These so-called optical tweezers have given us the ability to manipulate individual cells, organelles within the cell, and even molecules in an optical microscope. For example, muscular contraction at the molecular level has been reduced to the interaction of a myosin molecule on an actin filament. Optical tweezers have been used to measure, in real time, the force and displacement of the myosin molecule during hydrolysis of a single ATP (adenosine triphosphate) molecule. Just as powerful new insights have come from the study of single, isolated quantum systems, the study of biology at the level of a single molecular event will undoubtedly reveal unexpected behavior.

The National Science Foundation should recognize the dramatic new opportunities in fundamental research in atomic, molecular, and quantum optics and should encourage support for research in these areas.

FIGURE 7.2 Four hundred years of progress in the history of the optical microscope. Rapid progress over the past 40 years is illustrated on the expanded scale.

C h a p t e r   7

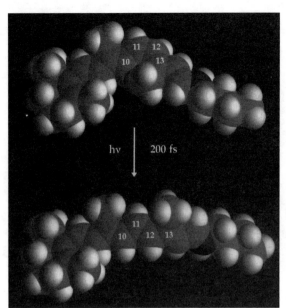

FIGURE 7.3 Ultrafast laser techniques have shown that the first step in vision, a rapid structural change in the rhodopsin molecule, occurs in only 200 fs.

Femtosecond Optics

Throughout history, short flashes of light have provided a means of capturing and studying high-speed events. High-speed electronic flashes can do this with accuracies of nanoseconds. With lasers, the potential quickness and precision of measurement has been sharpened by almost a factor of a million. We now have the capability to observe, on a time scale of femtoseconds, some of the most fundamental interactions among the atoms, electrons, and molecules that compose materials (Figure 7.3). These interactions determine the effectiveness of important chemical and biological processes as well as the ultimate speeds of electronics and opto-electronics. Observing previously unseen phenomena and understanding them better will have a major impact on the development of a wide range of technologies.

In addition, the use of femtosecond pulses is expected to go beyond fundamental measurement to new ways of controlling and changing matter. Femtosecond pulses (see Figure 7.4) capture the full bandwidth of optics coherently. Thus, they can be sculpted by frequency-domain methods to produce the unique optical waveforms required to manipulate selected quantum states of atoms and molecules. The short duration of femtosecond pulses also makes it possible to amplify them to extremely high powers with relatively modest energy. Today's power levels (which produce electric fields 100 times greater than those holding the hydrogen atom together) are already leading to the discovery of new phenomena and the creation of effects similar to those in high-energy accelerators.

High-Intensity Laser-Matter Interactions

At optical intensities greater than 10^{13} W/cm^2, the electric field of the focused light beam exceeds that found in the interior of atoms by many orders of magnitude, and atoms fly apart or ionize during one optical cycle. In this regime, the ponderomotive wiggle energy of the emitted electron as it oscillates in the laser field becomes comparable to the photon energy, and atomic energy levels and widths can no longer be calculated with perturbation theory. The subject of high-intensity laser-matter interactions has important applications, which range from the understanding of laser-induced inertial confinement fusion to the investigation of laser-plasma instabilities at high field strengths. Many new phenomena have already been uncovered, such as above-threshold ionization (ATI), new pathways to direct multiple ionization, fine intensity-dependent resonances due to large ac (alternating current) Stark shifts in the bound-state spectrum, and bond softening and above-threshold dissociation in molecules. Yet the intensity range over which these phenomena occur is only a small fraction of what will be achievable with amplified femtosecond solid-state lasers. Several lasers are now capable of focused intensities a million times higher than the threshold for atom ionization. At these intensities, relativistic effects are important. Two laboratories have recently reported generating relativistic electrons from low-density plasmas using high peak power lasers. Further exciting developments can be expected during the next decade, including laser acceleration and laser-driven inertial confinement fusion.

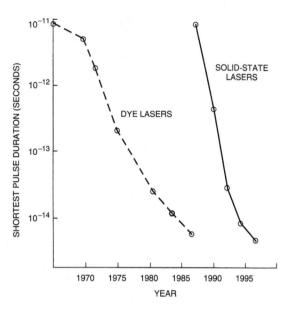

FIGURE 7.4 The femtosecond time domain was pioneered in the 1970s and 1980s with dye laser technology. The development in the 1990s of femtosecond solid-state lasers is opening up a wide range of new applications.

Intense femtosecond pulses have also been found to have some dramatic advantages in materials processing. By optimizing pulse duration for the specific material being processed, one can greatly improve the reliability and localization of the desired damage or ablation. One important application is laser surgery in the transparent vitreous fluid of the eye without damage to the retina. Micromachining in solid materials with a resolution of better than 1 μm could open up new possibilities for high-technology fabrication.

Lasers in Chemistry

With the rapidly improving ability to control all aspects of ultrashort laser pulse shapes, optics is poised to have a dramatic impact on progress in chemistry and biochemistry. Developing an improved, predictive understanding of chemical reaction dynamics has always been a major goal of research in physical chemistry. New excitement has been generated in this area recently by novel laser-based capabilities for studies of transition states by photoelectron spectroscopy and femtochemistry, for control of reaction pathways, and for bond-selective chemistry. The potential for optical control of chemistry in particular is made possible by the combination of advances in femtosecond lasers, sophisticated pulse shaping methods, theory, and computational capability. Experimental demonstrations of the control of reaction products, vibrational dynamics, and population transfer have already been demonstrated. This research is generating new knowledge and has significant potential for experimental discovery.

A new frontier also exists in efforts to understand, at a microscopic level, processes that occur in the solution phase. Most chemical reactions, especially those of commercial significance, occur in solutions. Yet only recently have femtosecond laser developments facilitated direct studies of these important reactions. Some of the most exciting recent results concern reactions that have coherent excited states as products. The key to chemistry lies in understanding the influence of the solvent on the dynamics of these transition states. Small molecules are permitting the first clear observations of such reactions in solution, but similar phenomena are also beginning to be seen in complex biological molecules. Emerging from these experiments are new paradigms of condensed-phase reactivity. Here, in particular, is where advances in optical science are the rate-limiting step in advancing the understanding of chemistry. A further step could be direct visualization of the nuclear dynamics during such reactions by means of x-ray spectroscopy, and optics would play an enabling role in that as well. The first femtosecond x-ray pulses have recently been produced by Thomson scattering of femtosecond laser pulses by an accelerated electron beam. Efforts to apply these pulses to questions of microstructural dynamics are under way. Other possible applications include the study of protein folding using time-resolved x-ray scattering.

Technology of Femtosecond Lasers

The current rapid advance in femtosecond technology is made possible by recently invented methods for ultrashort pulse generation with solid-state lasers. Early pioneering work with dye lasers opened the femtosecond time domain to science and laid many of the foundations for present technological advances. Solid-state technology now brings greater reliability, shorter pulses, and higher powers to scientific applications. It also opens the door for more widespread commercial application by making femtosecond sources more efficient, more compact, and less expensive.

Key to the all-solid-state femtosecond revolution has been the invention of nonlinear optical pulse-forming methods and devices, such as additive pulse modelocking (APM), Kerr lens modelocking (KLM), soliton modelocking, antiresonant Fabry-Perot saturable absorbers (AFPSAs), and saturable Bragg reflectors (SBRs). Each is important for a somewhat different type of application. What they have in common, which sets them apart from previous technologies, is that they are no longer simply determined by the properties of basic materials. They can be optimized artificially, engineered, and enhanced by synthetic design techniques. Opportunity still exists for the discovery of new enabling

BOX 7.3 LASER-DRIVEN PARTICLE ACCELERATORS

The first particle accelerators, developed in the 1930s, explored the properties of the atomic nucleus. The accelerator is basically a transformer, in which electrons gain energy as they traverse the voltage applied across the structure. A key breakthrough in accelerator design was the use of voltage at radio frequencies in the cyclotron to accelerate electrons. This was followed by the use of microwave frequencies in a linear accelerator to reach electron energies of billions of electron volts. Today most accelerators operating in the world are driven by microwave radiation. Conceptual designs for the Next Linear Collider, based on high-frequency microwave radiation, call for a 20-km linear collider with 500-GeV colliding particle energy.

Livingston studied the evolution of particle accelerator technology and noted that each new step toward high energy involved a new design approach or a new source of high peak power electromagnetic radiation. One new approach to accelerator technology is to use the laser to generate future tera-electron volt (TeV) accelerators. Recent success in experimental laser acceleration has been achieved by a group at the University of California, Los Angeles, using a plasma structure (Joshi and Corkum, 1995). However, not all researchers think that this approach will be practical.

Laser-driven TeV accelerators, if constructed, would allow probing of fundamental particles that were formed in the earliest few moments at the birth of the universe. Why the universe is formed of matter and why we are here are questions that could be studied with the next generation of high-energy particle accelerators.

Today, terawatt (10^{12} W) peak power lasers operate on a table top. In the future, these lasers would operate at high efficiency and high average power to drive kilometer-long TeV linear accelerators.

•••

mechanisms, but now there are at least some direct pathways for technology development by improved design and manufacturing.

Research into the optical materials for these ultrafast lasers and into devices for femtosecond pulse manipulation continues to play a crucial role in the development of this emerging technology. New laser crystals for different wavelengths, improved nonlinear optical materials, novel active and nonlinear fiber devices, and a better understanding of light-matter interactions are all still needed to nurture the technology's success and widen its impact.

Technology Applications

To date, the principal beneficiaries of developments in femtosecond technology have been scientific applications, but the prospect of compact, practical, cost-effective femtosecond laser sources is now creating opportunities for a variety of mainstream commercial applications. Some take direct advantage of the ultrahigh-speed aspect of the technology. Others derive benefit from related aspects such as broad spectral width, high focused peak intensities, phase coherence, improved detectability, and precise repetition rates. Optical communication systems have already benefited from advances in ultrafast optics, and they remain a fertile domain for new applications based on ultrashort pulses, such as soliton transmission, all-optical switching and networking, wideband wavelength-division multiplexing (WDM), ultrahigh-rate data processing, and clock distribution.

Femtosecond laser sources with high peak and average power open the possibility of using a laser to accelerate electrons in place of the microwave source traditionally used for this application. Laser sources with higher peak power and shorter wavelengths (Box 7.3) offer the possibility of achieving acceleration gradients of about 1 GeV per meter. If successfully developed over the next two decades, the laser-driven particle accelerator may open new avenues for studies of the physics of fundamental constituents of matter at the TeV energy scale (Figure 7.5).

Measurement technologies such as device and circuit testing, optical ranging, surface monitoring, and microscopy can take advantage of the peak powers and low temporal coherence of femtosecond pulses as well as their short durations. Low-coherence confocal microscopy and two-photon microscopy are examples of rapidly developing applications, as are biomedical imaging by optical coherence tomography and laser surgery. The feasibility of combining ultrafast microscopy with the measurement of ultrafast optical response will expand the list of possible applications.

A particularly novel and unexpected application of femtosecond light pulses is the generation of terahertz radiation, or T-rays. T-rays in the wavelength range of 0.05 to 1 mm provide a new technology for imaging hidden items and some invisible material properties (Figure 7.6).

FIGURE 7.5 (a) Evolution of laser peak power and focused intensity versus year and the physical phenomena that can be explored. Over the last 35 years, peak power has increased by 11 orders of magnitude. (Courtesy of P. Pronko, University of Michigan.) (b) A proposed dielectric-based laser accelerator structure with 1/3-mm repeat distance per stage. The average acceleration gradient is calculated to be 0.7 GeV/m.

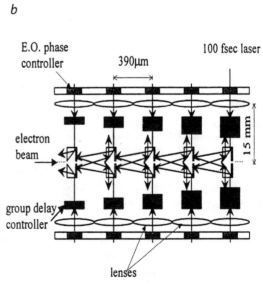

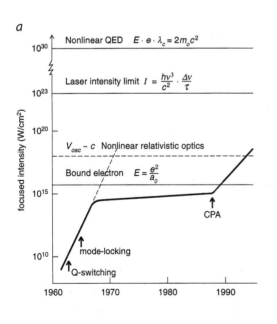

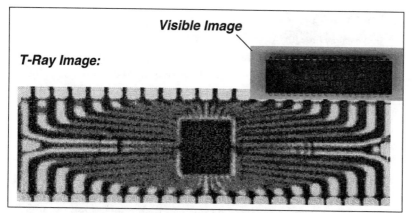

T-Ray Image:

Visible Image

FIGURE 7.6 *T-rays with wavelengths between 0.05 and 1 mm provide a clear image of this electronic circuit, still inside its manufacturer's packaging. (Courtesy of M.C. Nuss, Lucent Technologies.)*

Successful development of this and other emerging applications will depend on the availability of cheap and versatile optical pulse power supplies. The availability of such supplies can also be expected to stimulate still other applications, as yet unforeseen.

Femtosecond optics and sources offer an opportunity for dramatic impact on science and technology. Agencies should focus attention on this opportunity and encourage innovative work in this cutting-edge field.

Semiconductor and Advanced Solid-State Lasers

Lasers are now essential to the national economy, enabling applications that extend from CD-ROM to the fiber information network; from the processing and fabrication of semiconductors to the cutting of cloth, plastic, and industrial materials; and from laser vision correction to theraputic medicine. Laser sources have made a transition from the gas discharge tubes associated with the ubiquitous red helium-neon lasers used in supermarket scanners and the blue-green argon ion lasers familiar in laser light shows to solid-state laser diodes (LDs) and advanced solid-state lasers. This transition to the solid state will have a profound impact on the future growth of laser markets. Laser diodes and solid-state lasers can be produced at low cost and operate at high electrical efficiency with high reliability. One consequence of the transition from discharge tubes to the solid state is lower production costs because of the increasing volume of production. Moore's law for the exponential increase in the number of transistors on a wafer, a consequence of the lithographic production techniques applied to silicon, is for the first time applicable to lasers. We can expect that laser diodes and solid-state lasers will be produced at ever-increasing volume and ever-decreasing cost as the markets continue to expand. This in turn will lead to a continued growth of many new laser applications. The markets for laser applications will grow from $3 billion today to tens of billions of dollars in the next decade.

Laser Sources

Semiconductor Laser Diodes

Lasers diodes based on the semiconductor gallium arsenide (GaAs), an extension of silicon technology, are widely used 30 years after the invention of the laser. GaAs laser diodes operate from the red to the near infrared (IR) at power levels from milliwatts to multiple watts. Red laser diodes are now used in compact disk (CD) players, and more than 200 million of them are produced each year at a cost of less than $1 each. The GaAs material technology is more complex than silicon-based material technology, but lithography-based planar-processing technology developed for the silicon industry is applicable to GaAs laser diode production, so laser diodes can be manufactured in large volumes at low cost.

The cost of laser diode sources decreases rapidly with production volume. An annual market volume of $30 million is adequate to bring the production cost to less than $1. This low cost, along with reliable operation and small size (a grain of sand is larger than the diode laser source), opens many new markets for laser diode devices.

The market for optoelectronic equipment is expected to grow from $139 million in 1998 to $463 million in 2013. Applications of diode lasers include data storage, optical displays (Box 7.4), telecommunications, local area network communications, sensors, and medicine. Worldwide laser diode sales in 1996 were $1.92 billion. The worldwide market for all laser sources was $2.82 billion. Of course, the laser source is a critical and enabling element in a larger system. For example, the diode laser in the compact disk player enables a market for CD players that now exceeds $10 billion per year. The growth and market size of laser-based systems continue to track the growth and market size of computer-based systems with a delay of about 15 years.

....

BOX 7.4 ORGANIC AND SILICON-BASED LED DISPLAYS

Research into organic-based LEDs is now under way in many laboratories. Organic LEDs, manufactured from polymer films that are patterned and electrically excited directly, offer the promise of very low-cost optical displays that are like paper in flexibility, cost, and ease of use. Progress in research toward brighter, longer-life, organic-based LEDs continues to show promise (Kido et al., 1995).

Porous silicon emits light that is broadband in nature and rather weak in output power. Nevertheless, silicon-based LEDs offer promise as inexpensive sources of light for speciality applications such as alphanumeric indicators directly on the silicon circuit based electronic component (Collins et al., 1997).

Laser diodes based on GaAs are the key to the future growth of the laser markets. GaAs refers to not just one material system but a complex set of material systems based on extensions of GaAs. The progress in wavelength extension of diode lasers from the near IR to the blue depends critically on the development and understanding of the III-V semiconductor material systems. Today, commercially available laser diode products operate from the red to the near infrared. Power levels range from 10 mW for reading compact disks to more than 20 W for pumping solid-state lasers. These lasers operate at greater than 40% electrical efficiency, which makes them among the most efficient sources of light.

The 45% operating efficiency of laser diodes is far greater than that of the tungsten bulb (~1%) and even greater than that of fluorescent bulbs based on mercury discharge (~25%). In the future, the application of laser diode sources to general lighting has the potential for saving significant electrical power usage in the country since 19% of generated electricity goes for lighting (see Chapter 3).

The laser diode is more efficient and can operate at higher output power than the light-emitting diode (LED; Box 7.5). Further, the laser diode has a coherent output beam that can be focused into an optical fiber or onto the surface of a CD. Thus, LDs complement LEDs, and both will have widespread applications in the future. Like LEDs, laser diodes are being improved and extended in wavelength. Today, the critical area of research is to extend the laser diode wavelength into the blue. The first realization of a blue laser diode based on GaN (gallium nitride), a material closely related to GaAs, was announced by the Japanese company Nichia in 1996. The GaN material system is not as well developed as that of GaAs. It took 10 years from the first demonstration of a room-temperature GaAs red laser diode to the introduction of commercial laser diode products. A similar time might be required before blue laser diodes are available commercially.

When introduced as a product, the blue laser diode will open major application areas including information storage and optical displays. Other applications include fluorescence microscopy for biological and medical applications. There is significant economic leverage for the development of blue laser diodes for application to the optical storage systems of the future.

Today companies in the United States lead the world in the manufacture of high-power LD sources. However, investments in R&D by companies in Europe and in Japan are challenging that lead. High-power laser diodes are used directly for medical and industrial applications and to pump advanced solid-state lasers. The leading producers of laser diodes for information storage, such as CD players, are located in Japan as is the first company to demonstrate the GaN blue laser diode.

BOX 7.5 THE LIGHT-EMITTING DIODE

Light-emitting diode (LED) light sources, the incoherent versions of laser diodes, operate at near 10% efficiency and milliwatts of output power. LEDs are less efficient and less powerful than laser diodes but are also much less expensive. LEDs are familiar as indicator lights in electronics and as red brake lights in some automobiles. However, they will soon be found in signal lights, road signs, exit signs, and lighting displays (see Chapter 3). The market for LEDs is larger than the laser market and is growing rapidly because of the small size, long operating life, high visual brightness, and low cost of LEDs. As blue LEDs become commercially available, three-color displays will become possible and will add to the growth of the multibillion-dollar LED market.

Blue LEDs have been introduced to complement the already available green and red LEDs. Applications for blue LEDs include color displays, specialty lighting, and medicine and biology where blue LEDs induce fluorescence in molecules for detection. For example, an array of 100 blue LEDs has been used for research on the treatment of jaundice in newborn babies.

Visible laser diodes are used for information storage, bar-code readers, and laser alignment in construction.

High-power laser diodes are used for materials processing, soldering, xerographic image generation, and medical applications. Laser diodes coupled to optical fibers enable direct processing of materials through cutting, welding, and annealing. High-power laser diode arrays are also widely used today to pump advanced solid-state laser sources.

Advanced Solid-State Lasers

Advanced solid-state lasers play a unique, significant, and growing role in scientific and industrial applications. Solid-state lasers offer a combination of performance characteristics that lend themselves to a wide variety of tasks, from the frontiers of research to applications in reprographics, medicine, and defense. High-power solid-state lasers enable high-value manufacturing processes in the automotive and aerospace industries.

Some of the favorable characteristics of solid-state lasers are the ability to store optical energy for periods up to 1 ms and to extract the stored energy in a short time, yielding a high peak power pulse of a few-nanosecond duration (Q-switching); the ability to operate with a large gain bandwidth for ultrashort pulse generation by modelocking; and the ability to emit optical radiation in a single spectral and spatial mode with extreme spectral purity and stability. When extended in wavelength by nonlinear optical frequency converters, solid-state lasers offer the prospect of high peak and average power across a wide wavelength range from the ultraviolet to the infrared.

In 1995 the sales of solid-state lasers reached $394 million. Sales were projected to reach $500 million in 1997. The fastest-growing segment of this technology is diode-laser-pumped solid-state lasers (DPSSLs). Here advantage is taken of the high operating efficiency of the laser diode compared with that of the traditional flashlamp pump source. The DPSSL allows the power from multiple laser diodes to be summed and extracted in a single high-power beam (Anderson, 1997).

Figure 7.7 shows a schematic of a high-power laser-diode-pumped solid-state laser. Using this approach, solid-state lasers now operate at power levels of 1,000 W, and 5,000-W lasers are under development. The electrical operating efficiency of DPSSLs is greater than 10%, which is a significant improvement relative to traditional lamp-pumped solid-state lasers. Further, reliability is improved from the 200-hour lifetime of lamps to the greater than 7,000-hour lifetime for laser diodes. These factors, coupled with the small size of the DPSSL, have opened new areas of application in reprographics and medicine and have extended earlier applications in laser radar, remote sensing, semiconductor processing, and industrial materials processing such as welding, cutting, and surface hardening.

The United States has recognized the importance of DPSSLs to industry. A major program is now under way involving a consortium of companies to develop and utilize kilowatt-class DPSSLs for industrial materials processing applications in the automobile and aircraft industry. If continued, this program should enable the introduction of this new technology into industrial processing applications in the United States.

Internationally, both Germany and Japan have explicitly recognized the potential and the value of DPSSLs to industry and have established assertive national programs to capture the projected $10 billion market for this technology. Germany's Laser 2000 program emphasizes

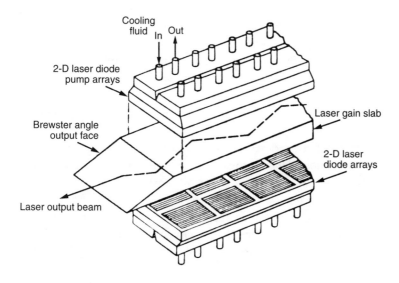

FIGURE 7.7 Schematic of a laser-diode-pumped solid-state laser that operates at output power levels in excess of 1 kW.

BOX 7.6 LASER INTERFEROMETER GRAVITATIONAL-WAVE OBSERVATORY (LIGO) PROJECT

Einstein predicted in his general theory of relativity that accelerating massive objects radiate gravitational waves. Taylor and Hulse earned the Nobel Prize in 1994 for their precise measurements of the period of a pulsar that indicated a slow loss of energy by gravitational radiation. This was the first indirect evidence of the existence of this extremely weak form of radiation. The goal of LIGO and related projects is to detect gravitational waves directly by laser interferometry.

The LIGO project (for more information, visit the Web site at www.ligo.caltech.edu) was established in the United States as a research project to detect and study the very weak gravitational waves emitted by co-orbiting neutron stars, by binary black holes, or perhaps by exploding supernova stars. Similar projects have been initiated in France and Italy (VIRGO), in Germany and Scotland (GEO), and in Japan (TAMA). Detection of the extremely weak gravitational waves requires the highest degree of sensitivity. The detector is a 4-km-long Michelson/Fabry-Perot interferometer housed in an ultrahigh vacuum and illuminated by a 10-W Nd:YAG solid-state laser. Gravitational waves propagating through space and matter (including Earth) cause space to "warp" a very small amount. The change in space-time induced by gravitational waves is detected by a very small change in the optical path length of the 4-km interferometer—one part in 10^{21}. This is equivalent to measuring the diameter of an atom at the distance of the Sun.

The interferometer detectors for LIGO are being constructed at two sites: Hanford, Washington, and Livingston, Louisiana. First measurements are expected to be initiated in 2002.

DPSSLs for manufacturing applications and is funded at more than $20 million per year for the next 5 years. Japan's larger Advanced Photon Research Program is sponsored by the Science and Technology Agency (STA) and is being carried out by the Japan Atomic Energy Research Institute (JAERI) as the lead organization. A new laboratory is being constructed in the Kansei region near Kyoto. Complementary programs are being carried out by MITI and by the Ministry of Education at Osaka to develop high peak power class lasers for inertial-fusion energy generation. Collectively these programs have been funded at a level of $500 million over 5 years. The program was initiated in April 1997.

Several big science projects are in progress in the United States that depend on advances in solid-state laser technology. These projects serve as grand challenges for the coming decade and will drive the performance of laser sources. The Laser Interferometer Gravitational-Wave Observatory (LIGO) project (Box 7.6), a research project supported by NSF, requires a solid-state laser with extreme frequency stability, high average power, and extended operational lifetimes (Figure 7.8).

The DOE National Ignition Facility (NIF) will use advanced glass lasers in an array of 192 beam lines to focus 1.8 MJ of energy in approximately 3 ns onto a millimeter-size target containing hydrogen

FIGURE 7.8 The LIGO site at Hanford, Washington, showing the 4-km arms stretching from the main building. The interferometer mirrors, or test masses, are isolated from ground vibrations to allow extremely sensitive measurements for the detection of gravitational waves. (Courtesy of LIGO.)

isotopes. The target will be compressed and heated to temperatures that may ignite a laboratory-scale fusion reaction, thus releasing more energy than used to ignite the target (Box 7.7). In the future, DPSSLs may provide the 10 MJ of energy at the required 100-MW average power for the generation of electricity using the laser-driven inertial confinement fusion process.

The prospect for scaling the output power of advanced solid-state lasers from 5 to 250 kW is now being explored for possible defense applications. The recently announced Airborne Laser project is based on a chemical laser source with solid-state lasers used for target location and tracking. However, the compact size, overall electrical efficiency, ease of long-term storage, and supply of prime power to the solid-state laser make it a possible candidate for future megawatt-power lasers. One application of such a laser is to destroy short-range missiles in flight, as demonstrated in February 1996 in a test at White Sands, New Mexico (see www.boeing.com/airborne laser).

Given the substantial progress in advanced DPSSLs in the past decade and the rich prospects for continued advances with the potential

FIGURE 7.9 (a) A Hubble Space Telescope image of supernova SN1987A. (b) A simulation of the plasma instabilities expected to be found in supernovas. (c) A two-dimensional experiment on the NOVA laser showing plasma instabilities and a three-dimensional computer simulation of plasma instabilities.

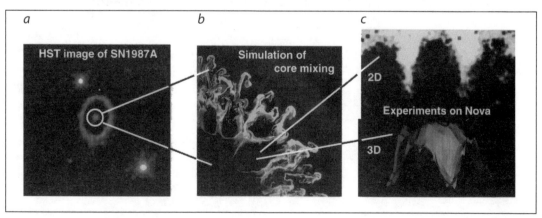

BOX 7.7 ASTROPHYSICS ON THE NOVA AND NIF LASERS

The existing NOVA laser and the next-generation National Ignition Facility (NIF) lasers are capable of irradiating matter to achieve densities and temperatures equal to those found in stars. At these extreme temperatures and densities, matter exists as a plasma. Further, matter at high density and temperatures can undergo fusion such that hydrogen isotopes burn to create helium and in doing so liberate fusion energy. The NIF laser is designed to reach the temperatures and densities required to ignite a fusion burn in the laboratory.

However, these plasmas exhibit instabilities that are representative of the instabilities found in exploding stars or supernova. A striking example is the Hubble Space Telescope image of supernova SN1987A in the Large Magellanic Cloud, shown in Figure 7.9 (a). A simulation of the instabilities expected to be observed is shown in (b). Experiments on the NOVA laser show a striking resemblance to the instabilities thought to exist in stars. Thus, the high peak power NOVA and, in the future, NIF lasers will allow astrophyics studies to be done in the laboratory.

The next generation NIF laser will enhance our ability to study and to understand the astrophysical plasmas exhibited by stars in space. The same laser will enable the grand challenge of demonstrating fusion in the laboratory to be realized. With further research and development, fusion in the laboratory may lead to a source of fusion energy in the future.

* * *

for enormous impact in the coming decade, continued and even increased funding of R&D in this area is justified. This conclusion was stated previously in the FAMOS study (NRC, 1994), which recommended that "the third (and final) priority is to promote research that promises new and improved lasers and other advanced sources of light for a broad range of applications and for furthering studies of the properties of light and its interation with atoms and molecules."

R&D and applications of solid-state lasers are cross-disciplinary and should be supported by a special initiative involving multiple agencies.

Advanced Materials for the Generation and Control of Light

Materials for Nonlinear Frequency Conversion

A few well-developed lasers have made an impact in the commercial market. These include the helium-neon, argon ion, and carbon dioxide gas discharge lasers, which operate in the red, blue-green, and infrared ranges. Solid-state lasers of commercial importance are the semiconductor laser diodes and the Nd:YAG solid-state laser. The difficulty and the cost of inventing and developing laser systems that operate at new wavelengths makes it important to extend existing well-developed lasers to new wavelengths by nonlinear frequency conversion.

Nonlinear frequency conversion was first demonstrated in 1962 shortly after the invention of the laser. In less than a decade, hundreds of new nonlinear materials were identified, tested, and demonstrated for converting lasers to new frequencies. Thirty years after this early research, which earned N. Bloembergen of Harvard the Nobel Prize, there are less than a dozen commercially available nonlinear crystals. These crystals are produced in small volumes and require hand polishing; they are therefore expensive, costing more than $1,000 per crystal. The 1980s saw the introduction of new nonlinear crystals based on extensive research programs under way in China and Japan. These new crystals such as barium borate (BBO) and lithium triborate (LBO) opened the ultraviolet wavelength region where there are important applications such as UV lithography, CD master production, and materials processing by laser ablation.

To be of commercial interest, nonlinear crystals must meet the physical requirements of adequate nonlinear coefficient, good optical quality and transparency, ease of growth and fabrication, and chemical stability. In addition, they must meet demanding cost objectives. For research applications the cost must be less than $1,000 per crystal. For original equipment manufacturing (OEM) applications, the cost must be less than $200. For widespread consumer applications, the cost must be less than $1.

Recently, a breakthrough in the technology of nonlinear materials has led to a reduction in cost by a factor of 1,000. By applying lithographic planar processing techniques developed for the mass production of silicon integrated circuits, it is now possible to fabricate nonlinear crystal "chips." Since thousands of nonlinear crystal chips can be fabricated from a standard 4-inch-diameter wafer of lithium niobate costing $300, the cost per chip is less than $1, thus meeting the cost objectives for widespread consumer applications. Further, the nonlinear interaction can be "engineered" to optimize conversion efficiency by using the spatial modulation of the ferroelectric domains to achieve quasi phase matching (QPM). QPM has allowed the generation of wavelengths from the ultraviolet to the infrared. For example, a Nd:YAG pumped tunable optical parametric oscillator can generate output from 1.4 to 4.5 µm in the infrared. In the latter case, more than 3 W of continuous-wave (cw) output power was demonstrated at the important 3.5-µm wavelength, a wavelength where hydrocarbon molecules absorb and can be detected remotely in the atmosphere. The conversion efficiency of the tunable parametric oscillator exceeded 90% and approached the theoretical maximum of 100%. This promising work requires continued support to make the transition from the research laboratory to commercial and defense application areas.

Chapter 7

BOX 7.8 ENGINEERED NONLINEAR FREQUENCY CONVERSION BY QUASI PHASE MATCHING

In 1962, in their seminal paper on nonlinear frequency conversion, Bloembergen and colleagues suggested that efficient nonlinear frequency conversion might be accomplished by periodically changing the sign of the nonlinear response to reset the phase (Armstrong et al., 1962). This approach to phase velocity matching is called quasi phase matching (QPM). In 1968, Bloembergen patented the idea.

Twenty-five years later, in 1988, after the patent had expired, the first engineered QPM interaction was demonstrated in the ferroelectric crystal lithium niobate using spatially modulated chemical diffusion and crystal growth to achieve periodic inversion of the ferroelectric domains and periodic change in the sign of the nonlinear coefficient. In 1993, room-temperature electric-field poling was used to invert the ferroelectric domains, and by 1996, wafer-scale processing of 3-inch wafers of lithium niobate was demonstrated. Today this technology is being transferred to industry for use in commercial applications of nonlinear frequency conversion. Interest in this technology stems from the tremendous cost reduction possible by using the lithography and planar processing technology developed for silicon. Nonlinear crystals are now being fabricated by lithographically printing patterns on the wafer, electric-field poling, and then dicing the wafer into chips. Waveguides can be implanted on the chip for enhanced nonlinear conversion at lower power levels.

· ·

The technological know-how for processing nonlinear crystal chips is being transferred to industry (Box 7.8). In the United States a number of companies are pursuing product development using this breakthrough. However, the importance of this technology has been recognized internationally. Research to take advantage of this breakthough is also under way in Europe, Japan, and China. Continued investments are necessary to extend our knowledge and to maintain our technology lead.

Future applications of nonlinear crystal chips may include the frequency doubling of laser diodes to generate blue radiation for information storage. Prototype products are being investigated at this time. This approach directly competes with the GaN-based blue laser diode source that is now a subject of extensive research. Other applications include the generation of red, green, and blue output for laser color projection displays. The laser-based projection display provides a brighter, crisper image than existing technology and may be the future choice for the projection of high-definition color images in theaters.

Future applications of tunable parametric oscillator technology include remote sensing, coherent laser radar, and medical applications that require control of the wavelength. The tunable solid-state laser is a direct replacement for the tunable free-electron laser source that has been under research and development for medical applications for more than a decade.

Over the long term, nonlinear materials will enable applications such as optical switching of fiber-to-the-home wideband communication (see Chapter 1). In these applications, the nonlinear devices will amplify and channel-shift or band-shift the optical radiation. It is likely that new materials will be needed for these applications, such as polymer-based nonlinear materials. A significant challenge remains to develop these new materials, especially in the United States, where research funding is short-term and is not consistent with the decade time frame required to develop a new material to the degree necessary to make the transition to commercial products.

Semiconductor Quantum-Well Materials

The ability to modulate the properties of a semiconductor at periods of tens of nanometers allows control of the quantum states of the conduction electrons and holes. In particular, if the bandgap of the semiconductor is modulated periodically, electrons are found in well-defined potential wells in the solid and the electron is confined to these quantum wells. Quantum wells are used to engineer the density of states of semiconductor diode lasers to provide both higher gain and wavelength control. Most laser diodes, and especially high-power laser diodes, are constructed by the use of quantum wells to control the electron states.

In laser diodes, electrons and holes combine and emit a photon at the bandgap energy. The output wavelength of the laser diode is thus controlled by the bandgap energy. In 1995 a new type of quantum-well laser was demonstrated for the first time, the quantum-cascade laser. In this laser the electrons remain in the conduction band and cascade down a quantum-well ladder emitting photons at a high rate along the way. The quantum-cascade laser emits radiation in the transparent region of the semiconductor at energies determined by the cascade's quantum energy levels. Quantum-cascade lasers open the midinfrared to laser diode radiation and are a promising new type of laser source.

Recently, voltage controlled quantum-well absorption has been used in conjunction with a Fabry-Perot interferometer to provide a high-speed optical switch. A small change in the absorption within the quantum-well band induces a large change in the reflectance or the transmittance of the interferometer.

The ability to control the depth of a quantum well allows control of the quantum-well states and the nonlinear response of the quantum-well semiconductor. Progress in quantum-well nonlinear materials (exhibiting a nonlinear response that is 1,000 times greater than the typical nonlinear response of the bulk semiconductor crystal) has been impressive. The large nonlinear response has allowed IR generation by frequency mixing and by harmonic conversion. In the future, the combination of guided-wave structures with the quantum-well enhanced nonlinear response should allow efficient nonlinear conversion at very low power levels across the infrared region of the spectrum.

Photonic Bandgap Materials

Artificially structured materials are already used effectively in many areas of optical science and technology. Multilayer dielectric coatings provide the ultrahigh reflectances needed for laser resonators and eliminate unwanted surface reflections. Semiconductor quantum wells and other nanostructures make a variety of novel optoelectronics devices possible. For example, photolithographically produced distributed feedback corrugations control the frequency of advanced semiconductor laser diodes for use in communications.

Recently, a new concept in dielectric structuring has emerged that could lead to an even more widespread revolution in optical device functionality and practicality. Theory and microwave model experiments have shown that strong three-dimensional structuring of dielectric materials can be used to control light (photons) in a manner similar to that used by semiconductors to control electrons. Emission of light can be suppressed or enhanced, greatly improving the efficiency of lasers and LEDs; new devices created from optical waveguide circuits can be concentrated at structural defects to produce miniature wavelength filters and create novel nonlinear optical behavior. The dielectric structures with these novel optical properties are referred to as photonic bandgap materials.

To realize practical photonic bandgap devices will require breakthroughs in device technology. New concepts in materials processing of three-dimensional structures must be sought, with pattern dimensions of the order of one-tenth of an optical wavelength. The materials processing challenges are great, but so is the payoff if photonic bandgap crystals can be realized.

Materials for Shaping and Focusing Optical Radiation

Lenses have been important in the development of optical devices and other technologies for centuries. However, new approaches to focusing and controlling light have emerged that offer revolutionary potential. These new approaches are based on materials advances and on the application of planar processing technologies developed for the silicon industry.

The gradient index (GRIN) lens is an example of an advance in materials that impacts approaches to the control and focusing of optical radiation. GRIN lenses are formed by diffusion of ions into glass, thus altering the index of refraction. One-dimensional arrays of GRIN lenses are now an important technology that has enabled low-cost fax and photocopying machines. This lens market alone is approximately $100 million per year and enables a market worth in excess of $10 billion per year.

Diffractive optical elements, or holographic optical elements (HOEs), are another example of a new approach to controlling optical radiation. HOEs are now widely used in optical systems along with standard lenses to focus laser radiation. There is a significant leverage in supporting research to improve diffractive optical elements for applications that include telescopes, cameras, and medical imaging devices.

It has been known for more than a century that the use of aspheric surfaces in optical systems would provide better imaging quality. Only in the past decade has it become possible to produce aspheric surfaces by pressing glass or plastic into a mold (see Chapter 6). Virtually all video cameras sold today use pressed aspheric plastic lenses. Thus, this technology offers tremendous leverage in opening new markets to lower-cost but high-performance optical systems. For large-scale manufacturing of aspheric optics, the use of the new technique magnetorheological polishing is promising. This technique is early in the development stage and offers an opportunity for leverage if supported by R&D funding.

Materials Research and Development Opportunities

Research opportunities in lasers and engineered optical devices that offer significant leverage for research investment are based often on a better understanding and control of materials (Box 7.9). In laser diode sources, progress in the understanding and synthesis of III-V semiconductor materials will enable new laser diodes and blue laser diodes. In solid-state lasers, the development of laser host crystals with low optical loss and high optical quality will enable more efficient, high-output-power solid-state laser sources. In optics, advanced materials will enable improved optical storage devices, possibly leading to the reinvention of the printing press as an optical disk printer the size of a small copying machine.

The challenge is to sustain research and development through the decade time frame that it takes to fully develop a new material. Research and development activities in the United States are often biased toward the immediate payoff. If devices are not demonstrated in a short time, the research effort is redirected. Often it takes more than 20 years for a new material to mature and gain wide acceptance. In the early 1960s, the Advanced Research Projects Agency (ARPA) took the bold step of establishing 13 interdisciplinary centers for materials research. (The program was transferred to NSF in 1972 as the Materials Research Laboratories program. Since 1994, the centers have been part of NSF's Materials Research Science and Engineering Centers program.) During the next 25 years, these centers produced a body of materials knowledge that is sustaining our industry today. However, they are now reduced in size and are being closed under the stress of reduced research funding. The challenge is to renew our commitment to long-range materials research that will enable new devices and new industries in the future.

BOX 7.9 MATERIALS RESEARCH AS AN ENABLER FOR OPTICS

The discovery and optimization of new materials is vital to the exploitation of optical phenomena and the development of optical devices and systems. Consider the role that materials research has played in the development of three optical technologies of vital economic importance: semiconductor lasers, optical fibers, and optical amplifiers. See Chapter 1 for more details about the applications of these devices.

SEMICONDUCTOR LASERS

The process that makes lasers possible was first postulated as early as 1917, but the first working laser, based on a single-crystal synthetic ruby rod, was demonstrated only in 1960. By 1962, helium-neon gas lasers and gallium arsenide (GaAs) semiconductor lasers had joined the ruby laser in this new class of device.

The first semiconductor lasers suffered badly from overheating. It was soon suggested that confining the laser action to a thin active layer—creating a semiconductor "sandwich" with the active layer set between layers of another material—would reduce the required current and keep the heat output manageable. In 1967, researchers proposed to add small amounts of aluminum to GaAs, growing confining layers of AlGaAs on either side of the GaAs active layer. Adding aluminum would change the atomic spacing of the crystal by less than one part in 1,000. Research on layer-by-layer growth of high-purity crystals, on defects and dopants, and on the effects of heat on the stability of compounds led to the demonstration in 1969 of an AlGaAs laser able to operate continuously at room temperature. Semiconductor lasers are now ubiquitous in optical communications and optical data storage.

OPTICAL FIBERS

As early as 1910, the equations had been determined that govern the theory of light transmission along what were then known as glass wires. By 1964, hair-thin strands of glass were carrying light over short distances, but up to 99% of the light was lost when it passed through as little as 30 feet

....

Progress in materials science and engineering is critical to progress in optics. The committee recommends that the Defense Advanced Research Projects Agency (DARPA) coordinate and invest in research on new optical materials and materials processing methods with the goal of achieving breakthrough capability through engineered semiconductor, dielectric, and nonlinear optical materials.

Extreme Ultraviolet and X-Ray Optics

Our common experiences involve visualizing the world around us with natural light, or optical radiation, to which our eyes respond. Visible light covers the wavelength range from blue (450 nm) to yellow (590 nm) to red (650 nm). To the shorter-wavelength side of the visible spectrum lie the ultraviolet, deep ultraviolet, extreme ultraviolet (EUV), of fiber. In 1966, a landmark theoretical paper asserted that these high losses were caused by minute impurities in the glass—primarily water and metals—rather than any intrinsic limits of glass itself. The paper predicted that eliminating impurities could reduce light loss from 1,000 decibels per kilometer of fiber to less than 20. This would allow signal-boosting amplifiers to be spaced at intervals of miles rather than yards, intervals comparable to those between the repeaters that amplified weak electrical signals along conventional telephone lines. Such predictions spurred a prodigious effort in materials research, especially in the area of materials processing, resting heavily on a knowledge of chemical thermodynamics, kinetics, gas dynamics, and polymer science. By 1980 the best fibers were so transparent that a signal could pass through 150 miles of fiber before becoming too weak to detect.

OPTICAL AMPLIFIERS

Despite these extraordinary advances in transparency, periodic amplification is still necessary. By the early 1980s, the transmission capacity of optical fiber communications systems was becoming limited by the capabilities of the amplifiers. The amplifiers of the day converted an optical signal to an electrical current, electronically amplified the current, and then used this to drive a laser that re-created the optical signal with greater intensity. The transmission capacity of these electronic amplifiers was considerably less than that of the lasers and optical fibers to which they were connected and thus limited the overall capacity of the system.

It was discovered that a short strand of erbium-doped glass, spliced into the main fiber and supplied with energy from an external source, could act as a laser in its own right, amplifying a weak optical signal without electronics. Key to turning this discovery into practical and effective optical amplifiers was mastering the materials chemistry of erbium doping, especially the appropriate reactions for volatilization. All-optical systems have now been demonstrated with transmission capacities 100 times that of any system built using electronic amplifiers.

. . . .

and vacuum ultraviolet (VUV, so called because the radiation is not transmitted by air). Beyond the VUV are the soft x-ray and finally the x-ray wavelength region. It is x rays that are used for medical and dental x-ray imaging. The spectrum continues to gamma radiation at very short wavelengths or very high photon energies.

The history of progress in the study of the electromagnetic spectrum has been one of steady progress in the generation of radiation beginning 100 years ago at radio wavelengths. Coherent sources were extended to the microwave region and, with the invention of the laser, to the infrared and visible wavelengths. This progress continues today with research aimed toward the generation of ever shorter wavelengths or higher frequencies. In recent work, coherent laser-like sources have been demonstrated in the deep ultraviolet, extreme ultraviolet, and soft x-ray spectral regions. The development of synchrotron sources has opened both the soft x-ray and the x-ray spectral regions for science and applications. Although synchrotron radiation is very intense, it is largely incoherent in nature. Research continues toward the possibility of generating coherent, laser-like, x-ray sources.

X-Ray Microscopy

The infrared and visible spectral region is the domain of light interactions with molecules and atoms. It is also the region where optical instruments, such as the microscope, allow us to extend our vision to very small objects. With microscopy, we are constantly striving to resolve ever smaller structures. It is well known that the limit of resolution of a microscope is set by the wavelength of the light used to produce the image. Thus, there is a great interest in microscopes that operate in the UV, EUV, and even x-ray regions for enhanced image resolution.

Recent research has led to the first images using soft x-ray microscopy. For example, soft x-ray images of a malaria-infected red blood cell show finer details than could be observed using a light microscope (see Figure 7.10). Microscopy in the EUV and x-ray regions offers the ability to tune the wavelength of the radiation to the atomic resonances of various elements—for example, carbon or oxygen—so that these elements, if present in the object, can be highlighted in the image. Although so far only preliminary, EUV and x-ray microscopic images give a glimpse of the future potential for higher-resolution microscopy with significant advances in areas such as nanometer materials science and subcellular structure of intact biological material.

EUV and X-Ray Lithography

The limiting resolution of the microscope also applies to writing features using advanced lithography. The smallest feature that can be written by lithography is determined by the illumination wavelength and the numerical aperture of the optics. Lithography is now moving from conventional mercury discharge lamps with output in the ultraviolet region, to excimer lasers with output in the UV and the deep UV.

FIGURE 7.10 X-ray microscope images of a red blood cell show features of a malaria infection site observed at a resolution not obtainable with an optical microscope. The left image shows a new infection. The right image was taken after 36 hours. (Courtesy of C. Magowan and W. Meyer-Ilse, Lawrence Berkeley National Laboratory.)

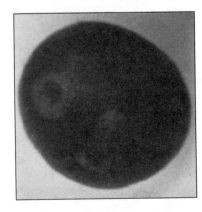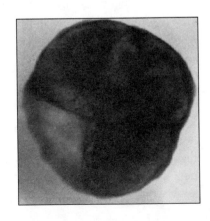

These lithography systems are under development and are just now moving to the production line.

The requirements for yet higher resolution, to meet the demands of semiconductor devices beyond the turn of the century, will force the move to even shorter-wavelength lithography. Research is under way to define approaches to both EUV and x-ray lithography, with potential feature sizes of the order of hundreds of atoms. Progress has been steady but certainly not rapid since research was initiated 15 years ago. Research for new sources and for a practical EUV and x-ray lithography process is driven by the enormous market demand for the next generation of semiconductor devices. Although no consensus has yet been reached as to what steps lie beyond current lithography based on UV excimer lasers, it is clear that shorter wavelengths must be used to obtain the required feature sizes. Shorter wavelengths demand that new EUV and x-ray sources and optics be developed. For example, an industrial consortium has been formed to invest $200 million in the development of EUV lithography. One promising approach with research under way in the United States, Japan, and Europe is to use high peak and average power solid-state lasers to generate a laser-produced plasma that radiates EUV and x-ray output. This is an active area of research and one that demands attention both for the laser drivers and for the control and focusing of the generated radiation.

X-Ray Optics and Sources

Plans are well under way to extend lithographic capability to the EUV and eventually to the soft x-ray region. Here research is needed to develop not only new sources of radiation but also optics and reflective coatings for the optical elements. This work is breaking new ground because of the lack of fundamental knowledge of material parameters in the EUV and x-ray regions. Sources of EUV and soft x-radiation include synchrotron sources, laser-produced plasmas, and laser-pumped soft x-ray lasers. Synchrotron sources are the most advanced but require a large facility and thus are located only in national laboratories and large corporations. There is a need for "granular" EUV and x-ray sources that can be accommodated in research and production laboratories and are of modest cost and size.

The progress in x-ray optics has been significant over the past decade. Today mirrors can be produced with greater than 60% reflectivity using high-Z and low-Z coatings such as molybdenum-silicon layers (see Figure 7.11). An alternative approach is to use diffractive optics such as a Fresnel zone

FIGURE 7.11 An EUV mirror with a multilayer coating of molybdenum-silicon. (Courtesy of J. Underwood, Lawrence Berkeley National Laboratory.)

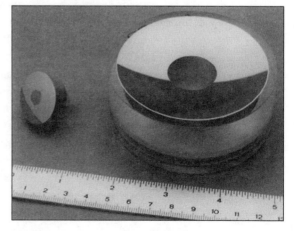

plate (see Figure 7.12) to focus the x rays. The short x-ray wavelength demands very high precision in the fabrication of optical elements. Fortunately, metrologies are now available with accuracies measured in fractions of a nanometer, which permit fabrication of x-ray components with greatly reduced wavefront distortion. Considerable research remains to be done to fully develop high-quality x-ray optics suitable for lithography, microscopy, and other emerging applications.

FIGURE 7.12 Diffractive-optic x-ray zone plate lens used to focus x-ray radiation. (Courtesy of E. Anderson, Lawrence Berkeley National Laboratory.)

Multiple agencies with interest in the crosscutting science and technology of EUV and soft x-ray optics and techniques should encourage research in this area because of the substantial potential economic payback in the near future.

Education in Optics

Formal university-level education in optics started in the United States in 1927, after considerable national debate. The debate arose from the experience of World War I when "in all the allied countries, requirements of the armed forces had revealed woefully inadequate production facilities and an abysmal lack of trained engineers to design and build optical devices of every kind" (Kingslake and Kingslake, 1970). As a result, optics centers and institutes were founded in France, England, and eventually, the United States.

The debate at the time was in some ways similar to discussions that continue today. Was there no tradition of optics in the United States before World War I? Indeed there was. Jackson's *An Elementary Treatise on Optics* was published in 1848 and was followed by a whole series of books, pamphlets, and handbooks. However, the debate, then as now, was about recognizing the existence of an integrated body of optics knowledge, that can and should be taught in a unified way. The recognition of optics as an integrated body of knowledge is the premise for this section's discussion of education. (This does not necessarily mean its recognition as a "discipline," however, or the creation of separate university departments. The field's diversity and multidisciplinarity are great strengths.)

The broad picture for science and engineering education is reviewed periodically by the National Science Board and published in *Science and Engineering Indicators* (1996): The 500,000 degrees earned in science and engineering in 1993 included 25,000 doctorates, 86,000 master's degrees, 366,000 bachelor's degrees, and 23,000 associate's degrees; in that year, 2.2 million people were employed in science and engineering jobs. Models of future employment demand are famously uncertain, of course, but only in the NSB's most aggressive growth scenario for the period 1994-2005, in which science and engineering employment expands by 62%, is there a strong demand for new scientists and engineers. The supply side has flexibility to respond to increased demand in a variety of ways, including the actions of individuals who have choices about education, retraining, postponed retirement, and moving to other technical fields. Small shifts in the career paths of employed scientists and engineers can have major effects on the supply and demand of scientists and engineers across fields.

Yet because optics is a new field that is undergoing rapid growth, demand may not be met by such adjustments and additional students may be needed. This is already the case for many subfields of optics. In some subfields, the demand for trained employees already exceeds the supply.

U.S. Optics Education Programs

U.S. colleges and universities offer a wide variety of educational programs in optical science, technology, and engineering. Some of these programs have broad curricula that span the field of optics. Others have more limited curricula that are devoted to specific subfields. Over the years, there have been several attempts to compile directories of these programs. Two major guides are produced by the professional societies. The most recent is *Optics Education 1996: Annual Guide to Optics Programs Worldwide*, produced by the International Society for Optical Engineering (SPIE). The Optical Society of America (OSA) published a three-volume guide to optics courses and programs in 1992.

Gathering such data is complicated, however, by the lack of a universally agreed-upon definition of the field. It should be noted that the major fields of engineering are recognized, in part, by the accreditation standards used by the Accreditation Board on Engineering and Technology (ABET) in evaluating university engineering departments. An accreditation program for optical engineering—similar to the ABET program for electrical engineering, for example—would provide several benefits, including an official government employment classification. This would help to identify prospective employees with specific experience and expertise; it would also help to provide a clear picture of the field's status by improving the basis for gathering statistics. The topic of accreditation was studied a few years ago by committees of OSA and SPIE. Their studies included discussions with ABET, but at the time, ABET concluded that there were too few formal programs to merit its attention and both societies decided against seeking an accreditation program. Continued rapid growth in the field would be likely to change this conclusion.

The *1996 Annual Guide* lists some 114 U.S. institutions that offer programs in optics. Of these, 93 offer one or more degree programs at the bachelor's level or above. A majority (55%) of these 93 institutions offer all three degrees: B.S., M.S., and Ph.D. About 14% offer the B.S. and M.S. only, and about 25% the M.S. and Ph.D. only. A few offer only a single degree (this is not to say that a stand-alone bachelor's program cannot be successful; there are excellent examples to the contrary). It is important to note, however, that although the *1996 Annual Guide* is quite useful, it is the result of self-reporting and should be evaluated carefully with this fact in mind. **The professional societies should evaluate educational programs in optics and jointly produce an annual guide.**

Because of the strong correlation between education and scholarly publishing, the major optics research journals are useful indicators of optics education, especially when comparing the level of activity in different countries. For example, the 450 papers published in *Optical Engineering* in 1995 involved 1,333 authors, 64% of whom were associated with academic institutions. A total of 33 countries were represented, corresponding closely with the 31 countries known to have educational programs in optics; 155 of the 450 papers (and 432 of the 1,333 authors) were from the United States. Other optics journals (of which there are more than 50, less than one-third of them published in the United States) appear to present similar statistics. The high representation of academic institutions in these journals attests to the quality of academic scholarship, which is of course tied closely to the quality of educational programs, particularly at the graduate level (Thompson, 1996, 1997).

Approaches to Academic Programs in Optics

The term optics, as used this report, encompasses an extraordinarily wide range of study, including such subfields as imaging and image processing, photonics, electro-optics, acousto-optics, fiber optics, optoelectronics, and lasers. The fundamentals of geometrical optics, physical optics, and quantum optics are the foundations of many of the teaching programs in these areas. Major educational institutions here and abroad offer structured programs that cover fundamentals, applied topics, and engineering applications from components to systems. Thus, unlike traditional departmental disciplines, optics cuts across traditional educational boundaries, involving the full range from very basic academic science to applied science to technology to systems engineering. For example, Chapter 1 notes the importance of education in systems issues for employment in the communications, display, and information storage industries. Chapter 6 notes that students educated in optics manufacturing are in high demand.

One reason optics educational programs achieve success in teaching across these boundaries is that educators in optics have an excellent network to help them with their work. Key to this network are the two major professional societies, OSA and SPIE, and a number of other societies that have sections devoted to optics or have optics as a parallel track to their main thrust. Recognizing the breadth of optics, the societies have formed the Coalition for Photonics and Optics to serve the broader community of optics professionals and to inform government agencies, educational institutions, and industry about the future importance of optics. **The professional societies should work to strengthen optics as a recognized crosscutting area of science and technology through the recently established Coalition for Photonics and Optics.** The society structure is supplemented by a number of international conferences on all areas of optics, including optics education.

There are two recognized approaches to the teaching of optics. One is a broad program of study, at an institute that recognizes optics as a separate discipline, covering the field with components in geometrical, physical, and quantum optics. The other model is an extended set of course sequences that bridge departments and schools and offer study in subfields of optics, such as lasers, image processing, electronic imaging, or electro-optics. The multidisciplinarity of the latter approach is a great strength, but the narrowness of some programs can be a disadvantage if graduates think they have been educated in optics when in fact they have been exposed only to certain subfields. No physics department would limit its program of study to solid-state physics or to quantum mechanics, and no electrical engineering department would teach courses only in circuits. **Universities should encourage multidisciplinarity in optics education, cutting across departmental boundaries, and should provide research opportunities at all levels, from the bachelor of science to the doctorate and from basic science to applied technology.**

These diverse optics programs are sufficient to meet current needs and probably also have the capacity and flexibility to meet future needs. The continued evolution of the field and increasing demand by industry may result in a need to extend optics education at the bachelor's or associate's degree level. There is no apparent need to create new programs of teaching in optics at the advanced level. The state of education and training in optics is generally healthy at present.

It has been traditional for many Ph.D. graduates in optics to go on to work in industry and for the government, unlike some other disciplines in which the "normal" track for a Ph.D. graduate is in academia. Thus, optics education is already meeting the goals stated in a recent report of the Committee on Science, Engineering, and Public Policy (COSEPUP, 1995).

The same can be said of most optics master's programs. These programs are styled after engineering programs but with some of the attributes that promote "adaptability and versatility as well as technical proficiency," as called for by COSEPUP. Master's programs in optics are independent programs with specific curricular goals, not just a set of courses on the way to a Ph.D. Thus, the field is also meeting many of the goals suggested in a recent article "Mastering Engineering" (Fitzgerald, 1996).

Continuing Education

The field is fortunate to have an excellent set of continuing education opportunities provided by academic institutions and national societies. Courses in continuing education offered by the OSA, SPIE, and other societies cover a wide spectrum from basic science to engineering to technology. They also include professional development courses in such subjects as business, management, ethics, marketing, and patents. Courses are offered in conjunction with national meetings and are also available by videotape and broadcast. They are supplemented by excellent written materials prepared through the societies' publication programs. **The professional societies should continue to expand their commitment to professional education in optics.**

Summary and Recommendations

Broad Issues

Research in optics cuts across disciplinary boundaries and enables advances in many other areas of research as well as commercial applications that are important for the nation's future growth and prosperity. As indicated by the response to the NSF optical science and engineering initiative in 1995, there is high demand and strong competition for optics research funding. However, optics often has difficulty finding research support in the federal agencies, despite the scientific excitement in the field and its significant and growing economic importance, because of its multidisciplinary and enabling character, which contrasts with the agencies' structure by discipline or by mission. Optics involves multiple agencies and multidisciplinary collaboration and support. Optics is a rapidly growing crosscutting area of science and technology that is ripe for investment at a level comparable to the support of high-performance computing and communications.

In the United States the combination of research and education provides the joint benefit of the discovery of knowledge and the education of students. Support for research in the nation's universities is 0.5% of the GDP, the lowest investment rate of advanced industrialized nations. Fundamental scientific and engineering research laid the basis of understanding required for the discovery of the laser in 1960, a discovery that has ultimately led to significant economic and social benefits for the nation. Continued investments in research and development have led to widespread applications of lasers and related technologies in basic science, engineering, medicine, communications, information storage, information display, and entertainment.

Multiple government agencies should form a working group to collaborate in the support of optics, in a crosscutting initiative similar to the earlier one for high-performance computing and communications systems.

The National Science Foundation should develop an agency-wide, separately funded initiative to support multidisciplinary research and education in optics.

The Department of Commerce should explicitly recognize optics as an integrated area of knowledge, technology, and industry and should structure its job and patent databases accordingly.

Research Opportunities

Quantum, Atomic, and Biological Optics

Light has historically been used to probe matter via spectroscopy. Important advances in spectroscopy techniques continue to be made. In addition, light is now used to manipulate atoms, molecules, and submicron particles. This control opens up possibilities for applications such as atomic-scale lithography and information storage. Laser-cooled atoms are being used to develop the next generation of atom clocks. Atom mirrors, beam splitters, and lenses have led to some of the most sensitive inertial sensors such as accelerometers and gyroscopes. Atoms laser-cooled to nanokelvin temperatures have formed a new state of matter, the Bose-Einstein condensate. Lasers have been used to create a form of light for which the fundamental noise limits of the Heisenberg uncertainty principle can be manipulated to enable more precise measurements. The unavoidable, quantum-limited perturbation of a physical system during a measurement can be modified. The disturbance of the quantum variable of interest can be minimized by shifting the effect of the measurement to another quantum degree of freedom. Quantum states with several degrees of freedom can be engineered so that the quantum phase of the entire system can be preserved. Although practical general-purpose quantum computers are unlikely to be developed, this research is at the forefront of the exploration of more complex quantum systems. Recombinant DNA techniques have been used to instruct cells to manufacture proteins with fluorescent tags. The intercellular motion of these labeled proteins can be tracked with an optical microscope. The optical microscope and new variations such as the confocal microscope and the near-field scanning microscope are undergoing rapid technological development.

This progress is further aided by the introduction of sensitive photon detectors and computer processing of images. Laser-based optical tweezers can manipulate submicron-sized objects such as organelles within live cells. This tool has also been used to manipulate single biological molecules; molecular forces and displacements during an isolated molecular event can now be measured.

Light is now used to control atoms, which opens up possibilities for new applications from laser-cooled atomic clocks to gyroscopes to gravitometers. Laser control of atoms has created a new form of matter, the Bose-Einstein condensate, in which all atoms occupy the same atomic state. Light control of atoms allows quantum states to be engineered, which may open up new possibilities for applications of complex quantum systems. The new fluorescent dyes and dye-labeling techniques allow the detection of a single molecule, which has important applications in chemistry, biology, and medicine. With these techniques, coupled with recent advances in microscopy and sensitive light-based techniques for detection and control, measurement and control at the single-molecule level appear possible.

NSF should recognize the dramatic new opportunities in fundamental research in atomic, molecular, and quantum optics and should encourage support for research in these areas.

Femtosecond Optics

Optics has opened the femtosecond time domain to science and technology through the use of lasers that produce ultrashort pulses of light. Extremely intense optical fields, made possible by the amplification and compression of femtosecond laser pulses, are creating new opportunities in physics and medicine. Femtosecond sources of high average power offer dramatic advantages for laser processing of materials. Femtosecond laser pulses, because they are coherent optical waveforms, offer new and unique capabilities for manipulating molecules to produce new chemistry and biochemistry. Femtosecond lasers open the possibility of generating subfemtosecond x rays for time-resolved studies of biological structures. The rapid advance of femtosecond technology is being driven by the invention of new techniques for pulse generation and new all-solid-state laser systems. Low-cost femtosecond sources are key to the rapid application of these unique optical sources. Optical fiber communication technology is already a major beneficiary of short-pulse laser research and it should benefit further from rapid progress in the generation of ultrashort pulses. Many new applications for femtosecond lasers are based on properties other than ultrahigh speed. Laser acceleration benefits from the high peak power of focused femtosecond lasers. Terahertz imaging and optical coherence tomography (see Chapter 2) are two examples of new applications opened by femtosecond lasers.

Femtosecond optics is an important developing science and technology that is opening new opportunities for advances in physics, chemistry, biology, and medicine. New, unexpected commercial applications enabled by femtosecond lasers are beginning to emerge. These applications and others could be greatly stimulated by the development of highly reliable, low-cost femtosecond sources.

Femtosecond optics and sources offer an opportunity for dramatic impact on science and technology. Agencies should focus attention on this opportunity and encourage innovative work in this cutting-edge field.

Semiconductor and Advanced Solid-State Lasers

The growth and market size of lasers and applications track the growth and market size of computers with a 15-year delay, and the market for lasers and applications is expected to grow from about $3 billion per year to greater than $20 billion per year in the coming decade. Laser diodes are efficient, reliable sources of coherent light that are mass produced. Increasing production volumes with decreasing costs per laser diode will lead to expanding markets. Light-emitting diodes are lower in power and efficiency than laser diodes, but their very low cost enables applications in markets that exceed $3 billion. Organic and porous silicon-based LEDs offer the promise of a low-cost, paper-like display that can be addressed by a pocket-sized optical-based information storage unit. Companies in the United States lead the world as commercial suppliers of high-power laser diodes. However, R&D investments in Europe and in Japan are challenging that lead. The current market for solid-state lasers is $394 million and is growing at more than 20% per year. Advanced solid-state lasers now operate at power levels in excess of 1 kW and are finding applications in cutting, welding, hole drilling, surface hardening, reprographics, and medicine. Japan and Germany have initiated R&D programs for advanced solid-state lasers funded at levels of $500 million and $100 million over the next 5 years to capture the multibillion-dollar market projected for the next decade. Major science projects in the United States, such as the LIGO and NIF projects, are technology drivers for advanced solid-state lasers. Research on new and improved lasers is a priority recommended in the 1994 FAMOS report (NRC, 1994).

Semiconductor laser diodes, LEDs, and advanced solid-state lasers are important technologies that enable applications critical to the nation. Light sources based on solid-state technologies represent an important opportunity for R&D investment with very high leverage for return on investment. Investment in this technology has declined in the United States, whereas it has accelerated in both Germany and Japan. R&D investment in semiconductor and advanced solid-state lasers is essential if the United States is to retain technological leadership.

R&D and applications of solid-state lasers are cross-disciplinary and should be supported by a special initiative involving multiple agencies.

Advanced Materials for the Generation and Control of Light

Frequency extension of laser sources by nonlinear crystals is critical to many applications that range from laser radar, to remote atmospheric sensing, to laser projection displays, to lasers for medical applications. The recent advances in engineered nonlinear materials and the breakthrough in cost reduction by a factor of 1,000 to less than $1 per non-linear crystal "chip" open the possibility of commercial applications of frequency-shifted solid-state lasers. Structured materials have created new opportunities for advanced optical technologies. Future realization of structures with photonic bandgap properties would have major impact on optics and optoelectronics industries. New fabrication techniques allow the control of light using lenses that are mass produced at low cost using gradient index materials, diffractive optics, and aspheric lenses. New materials are critical to the future progress in the extension of laser capability by frequency conversion. Materials research often takes a decade or longer before the material is developed to the degree required for applications. The long lead time for materials research has to be recognized in the support of research programs.

Advances in materials enable new approaches to devices and open new application possibilities. Engineered nonlinear materials, structured dielectrics, and nanostructured quantum-well semiconductors are advanced materials capabilities that have significant impact on optical science and engineering.

Progress in materials science and engineering is critical to progress in optics. The committee recommends that DARPA coordinate and invest in research on new optical materials and materials processing methods with the goal of achieving breakthrough capability though engineered semiconductor, dielectric, and nonlinear optical materials.

Extreme Ultraviolet and X-Ray Optics

There is considerable interest in developing EUV and x-ray microscopes for higher-resolution imaging of biological and physical structures. X-ray microscopes allow atomic species to be detected by tuning the wavelength to the atomic resonance. Lithography is moving from the UV to the EUV and soft x-ray wavelengths to meet the demand for smaller feature size in future semiconductor devices. There is a need for EUV and x-ray sources that are laboratory scale for research and manufacturing activities.

EUV and soft x-ray optics are providing new opportunities in broadly based science and technology, with potential impact for the manufacture of turn-of-the-century semiconductor electronics devices through the application of EUV and soft x-ray lithography.

Multiple agencies with interest in the crosscutting science and technology of EUV and soft x-ray optics and techniques should encourage research in this area because of the substantial potential economic payback in the near future.

Education in Optics

There is a recognized integrated body of knowledge that constitutes optics. Educational institutions have recognized that this body of knowledge must be taught in an integrated and unified way. Professional societies have recognized that optics is a broad area of knowledge and practice by forming a Coalition for Photonics and Optics. Unlike traditional departmental disciplines in science and engineering, optics is involved in both science and engineering and hence cuts across traditional educational boundaries from very basic academic sciences to applied sciences to technology to engineering systems in an integrated manner. Academic educational programs thrive and are effective when students are exposed to and involved with research activity. This is true at the bachelor's level but vital at the master's and doctoral levels. Research activity in optics covers the full spectrum from basic to applied science to engineering to technology.

Students educated and trained in optics are in demand at this time in many fields of optics. The diversity of the U.S. educational system, coupled with the flexibility of scientists and engineers to change direction, allows the supply of students educated in optics to meet projected demands in the foreseeable future. The quality and diversity of continuing education courses offered by professional societies is first rate and should be continued. Excellent mainstream optics programs are in place that have the ability and flexibility to meet perceived national needs. A national program to establish new teaching centers is not required, but support for the existing structure should not be permitted to erode.

NSF should develop an agency-wide, separately funded initiative to support multidisciplinary research and education in optics. Opportunities include fundamental research in atomic, molecular, and quantum optics; femtosecond optics, sources, and applications; solid-state laser sources and applications; and EUV and soft x-ray optics.

The professional societies should work to strengthen optics as a recognized crosscutting area of science and technology through the recently established Coalition for Photonics and Optics. They should evaluate educational programs in optics and jointly produce an annual guide. The professional societies should continue to expand their commitment to professional education in optics.

Universities should encourage multidisciplinarity in optics education, cutting across departmental boundaries, and should provide research opportunities at all levels, from the bachelor of science to the doctorate and from basic science to applied technology.

References

Anderson, S.G. 1997. Review and forecast of laser market. *Laser Focus World* 33(1):72.

Armstrong, J.A., N. Bloembergen, J. Ducuing, and P.S. Pershan. 1962. Interactions between light waves in a nonlinear dielectric. *Phys. Rev.* 127:1918.

Collins, R.T., P.M. Fauchet, and M.A. Tischler. 1997. Porous silicon: From luminescence to LEDs. *Phys. Today* (January):24.

Committee on Science, Engineering, and Public Policy. 1995. *Reshaping the Graduate Education of Scientists and Engineers.* Washington, D.C.: National Academy Press.

Fitzgerald, N. 1996. Mastering engineering. *ASEE Prism* 5:25-28.

Jackson, I.W. 1848. *An Elementary Treatise on Optics.* New York: A.S. Barnes and Company.

Joshi, C.J., and P.B. Corkum. 1995. Interactions of ultra-intense laser light with matter. *Phys. Today* (January):36-43.

Kido, J., M. Kimura, and K. Nagai. 1995. Multilayer white light organic electroluminescent device. *Science* 267:1332.

Kingslake, R., and H.G. Kingslake. 1970. A history of the Institute of Optics. *Appl. Opt.* 9(4):789-796.

National Research Council. 1994. *Atomic, Molecular, and Optical Science: An Investment in the Future.* Washington, D.C.: National Academy Press.

National Science Board. 1996. *Science and Engineering Indicators 1996,* NSB 96-21. Washington, D.C.: U.S. Government Printing Office.

National Science Foundation. 1994. *Optical Science and Engineering: New Directions and Opportunities in Research and Education,* NSF 95-34. Washington, D.C.

Optical Society of America. 1992. *International Guide to Optics Courses and Programs, 1992,* 3 vols. Washington, D.C.

SPIE. 1996. *Optics Education 1996: Annual Guide to Optics Programs Worldwide.* Bellingham, Wash.: SPIE Publications.

Thompson, B.J. 1996. 1995 in review. *Opt. Eng.* 35(2):325-326.

Thompson, B.J. 1997. 1996 in review. *Opt. Eng.* 36(2):297-299.

Appendixes

A
....
Collected Recommendations

This appendix collects all the major recommendation of the report together in one place for quick reference. Please refer to the body of the report for further details and explanation. The recommendations are listed twice, first in the order in which they are presented in the body of the report and second, sorted by intended actor.

Major Recommendations Sorted by Chapter

Optics in Information Technology and Telecommunications

- Congress should challenge industry and its regulatory agencies to ensure the rapid development and deployment of a cost-effective broadband fiber-to-the-home information infrastructure.
- To push compact disk (CD) and digital video disk (DVD) technologies to higher effective storage densities and performance levels, U.S. industry should develop multilayer storage media; low-cost optical systems for writing and reading data; and efficient, low-cost techniques for mass replication and assembly of multilayer disks.
- To retain the U.S. technological edge in three-dimensional recording, industry and universities should nurture and accelerate the development of advanced three-dimensional recording media, the design of low-cost optical systems, and the study of systems integration and architectures. It is imperative that these activities be coordinated among university and industrial researchers.

- The Defense Advanced Research Projects Agency (DARPA) should establish a program to seek new paradigms in optical storage that will reach toward the theoretical storage density limit of about 1.0 terabyte per cubic centimeter (TB/cm^3), with fast [(>1 gigabit per second (Gb/s)] recording and retrieval.

Optics in Health Care and the Life Sciences

- The National Institutes of Health (NIH) should establish a study section for RO1 grants devoted to biomedical applications of light and optical technology. An initiative to identify the human optical properties suitable for noninvasive monitoring should also be established.
- The National Science Foundation (NSF) should increase its efforts in biomedical optics and pursue opportunities in this area aggressively. This will require a broader interpretation of the NSF charter regarding health care in order to support promising technologies that bridge the NIH and NSF missions.

Optical Sensing, Lighting, and Energy

- The Department of Energy (DOE), the National Institute of Standards and Technology (NIST), and industry, in cooperation with the technical and professional societies, should pursue a program to enhance the coordination and transfer of optical sensor technology among industry, academia, and government agencies.
- DOE, the Environmental Protection Agency (EPA), the Electric Power Research Institute (EPRI), and the National Electrical Manufacturers Association (NEMA) should coordinate their efforts and create a single program to enhance the efficiency and efficacy of new lighting sources and delivery systems, with the goal of reducing U.S. consumption of electricity for lighting by a factor of two over the next decade, thus saving about $10 billion to $20 billion per year in energy costs.
- World leadership in optical science and engineering is essential for the United States to maintain its dominance in energy-related technologies such as laser-enhanced fusion, laser uranium enrichment, and solar cells. DOE should continue its programs in this area.

Optics in National Defense

- The Department of Defense (DOD) should ensure the existence of domestic manufacturing infrastructures capable of supplying low-cost, high-quality optical components that meet its needs via support for the Defense Advanced Research Programs Agency (DARPA) and the Manufacturing Technology Program.

- A central, coordinated DOD-DOE time-phased plan should be developed and conducted to enable worldwide optical detection and verification of chemical species that threaten civilians and military personnel through hostile attacks.
- A coordinated multiyear DOD plan should be conducted to develop radio-frequency (RF) photonic phased antenna-array technology for radar and communications.
- Key technologies such as high-power laser activities and new optics should continue to be pursued by DOD.

Optics in Industrial Manufacturing

- A multiagency-supported application and test facility should be established in a service center setting using the DARPA-sponsored Precision Laser Machining Consortium as a model for extension of laser materials processing and other optically assisted manufacturing techniques.
- NIST should support development of optical metrology and machine vision systems with improved performance, with the ultimate objective of plug-and-play capability.

Manufacturing Optical Components and Systems

- Government agencies and the optics community should recognize the importance of optics standards, especially their significance in international trade. The U.S. government should participate actively in the setting of such standards. NIST should be given the funding necessary to take the lead in this area.
- Government agencies should continue to support the activities necessary to introduce cost-effective precision aspheric components into both military and commercial products.
- DOD should continue to maintain technology assets and critical skills in optics manufacturing in order to meet future needs.
- The Bureau of the Census should involve representatives of the optics industry in the next revision of the North American Industry Classification System (NAICS) codes.
- Collaborative programs in optics manufacturing should include universities so that students are trained in the latest technical solutions to production problems.
- DOD, NIST, and the DOE national laboratories should establish together a cooperative program that provides incentives and opportunities to develop new ideas into functioning methods for optics fabrication.

Optics Research and Education

- Multiple government agencies should form a working group to collaborate in the support of optics, in a crosscutting initiative similar to the earlier one for high-performance computing and communications systems.
- The Department of Commerce should explicitly recognize optics as an integrated area of knowledge, technology, and industry and should structure its job and patent databases accordingly.
- NSF should recognize the dramatic new opportunities in fundamental research in atomic, molecular, and quantum optics and should encourage support for research in these areas.
- Femtosecond optics and sources offer an opportunity for dramatic impact on science and technology. Agencies should focus attention on this opportunity and encourage innovative work in this cutting-edge field.
- R&D and applications of solid-state lasers are cross-disciplinary and should be supported by a special initiative involving multiple agencies.
- Progress in materials science and engineering is critical to progress in optics. The committee recommends that DARPA coordinate and invest in research on new optical materials and materials processing methods with the goal of achieving breakthrough capability through engineered semiconductor, dielectric, and nonlinear optical materials.
- Multiple agencies with interest in the crosscutting science and technology of extreme ultraviolet (EUV) and soft x-ray optics and techniques should encourage research in this area because of the substantial potential economic payback in the near future.
- NSF should develop an agency-wide, separately funded initiative to support multidisciplinary research and education in optics. Opportunities include fundamental research in atomic, molecular, and quantum optics; femtosecond optics, sources, and applications; solid-state laser sources and applications; and EUV and soft x-ray optics.
- The professional societies should work to strengthen optics as a recognized crosscutting area of science and technology through the recently established Coalition for Photonics and Optics. They should evaluate optics programs and jointly produce an annual guide to educational programs in optics. The professional societies should continue to expand their commitment to professional education in optics.
- Universities should encourage multidisciplinarity in optics education, cutting across departmental boundaries, and should provide research opportunities at all levels, from the bachelor of science to the doctorate and from basic science to applied technology.

Major Recommendations Sorted by Intended Actor

Department of Defense (Including DARPA)

- DARPA should establish a program to seek new paradigms in optical storage that will reach toward the theoretical storage density limit of about 1.0 TB/cm^3, with fast (> 1Gb/s) recording and retrieval.
- DOD should ensure the existence of domestic manufacturing infrastructures capable of supplying low-cost, high-quality optical components that meet its needs via support for DARPA and the Manufacturing Technology Program.
- A central, coordinated DOD-DOE time-phased plan should be developed and conducted to enable worldwide optical detection and verification of chemical species that threaten civilians and military personnel through hostile attacks.
- A coordinated multiyear DOD plan should be conducted to develop RF photonic phased antenna-array technology for radar and communications.
- Key technologies such as high-power laser activities and new optics should continue to be pursued by DOD.
- DOD should continue to maintain technology assets and critical skills in optics manufacturing in order to meet future needs.
- DOD, NIST, and the DOE national laboratories should establish together a cooperative program that provides incentives and opportunities to develop new ideas into functioning methods for optics fabrication.
- Progress in materials science and engineering is critical to progress in optics. The committee recommends that DARPA coordinate and invest in research on new optical materials and materials processing methods with the goal of achieving breakthrough capability through engineered semiconductor, dielectric, and nonlinear optical materials.

National Science Foundation

- NSF should increase its efforts in biomedical optics and pursue opportunities in this area aggressively. This will require a broader interpretation of the NSF charter regarding health care in order to support promising technologies that bridge the NIH and NSF missions.
- NSF should recognize the dramatic new opportunities in fundamental research in atomic, molecular, and quantum optics and should encourage support for research in these areas.

- NSF should develop an agency-wide, separately funded initiative to support multidisciplinary research and education in optics. Opportunities include fundamental research in atomic, molecular, and quantum optics; femtosecond optics, sources, and applications; solid-state laser sources and applications; and EUV and soft x-ray optics.

Department of Commerce (Including NIST)

- DOE, NIST, and industry, in cooperation with the technical and professional societies, should pursue a program to enhance the coordination and transfer of optical sensor technology among industry, academia, and government agencies.
- NIST should support development of optical metrology and machine vision systems with improved performance, with the ultimate objective of plug-and-play capability.
- Government agencies and the optics community should recognize the importance of optics standards, especially their significance in international trade. The U.S. government should participate actively in the setting of such standards. NIST should be given the funding necessary to take the lead in this area.
- The Bureau of the Census should involve representatives of the optics industry in the next revision of the NAICS codes.
- DOD, NIST, and the DOE national laboratories should establish together a cooperative program that provides incentives and opportunities to develop new ideas into functioning methods for optics fabrication.
- The Department of Commerce should explicitly recognize optics as an integrated area of knowledge, technology, and industry and should structure its job and patent databases accordingly.

Department of Energy

- DOE, NIST, and industry, in cooperation with the technical and professional societies, should pursue a program to enhance the coordination and transfer of optical sensor technology among industry, academia, and government agencies.
- DOE, EPA, EPRI, and NEMA should coordinate their efforts and create a single program to enhance the efficiency and efficacy of new lighting sources and delivery systems, with the goal of reducing U.S. consumption of electricity for lighting by a factor of two over the next decade, thus saving about $10 billion to $20 billion per year in energy costs.

- World leadership in optical science and engineering is essential for the United States to maintain its dominance in energy-related technologies such as laser-enhanced fusion, laser uranium enrichment, and solar cells. DOE should continue its programs in this area.
- A central, coordinated DOD-DOE time-phased plan should be developed and conducted to enable worldwide optical detection and verification of chemical species that threaten civilians and military personnel through hostile attacks.
- DOD, NIST, and the DOE national laboratories should establish together a cooperative program that provides incentives and opportunities to develop new ideas into functioning methods for optics fabrication.

National Institutes of Health

- NIH should establish a study section for RO1 grants devoted to biomedical applications of light and optical technology. An initiative to identify the human optical properties suitable for noninvasive monitoring should also be established.

Environmental Protection Agency

- DOE, EPA, EPRI, and NEMA should coordinate their efforts and create a single program to enhance the efficiency and efficacy of new lighting sources and delivery systems, with the goal of reducing U.S. consumption of electricity for lighting by a factor of two over the next decade, thus saving about $10 billion to $20 billion per year in energy costs.

Federal Agencies in General

- Congress should challenge industry and its regulatory agencies to ensure the rapid development and deployment of a cost-effective broadband fiber-to-the-home information infrastructure.
- A multiagency supported application and test facility should be established in a service center setting using the DARPA-sponsored Precision Laser Machining Consortium as a model for extension of laser materials processing and other optically assisted manufacturing techniques.
- Government agencies should continue to support the activities necessary to introduce cost-effective precision aspheric components into both military and commercial products.
- Multiple government agencies should form a working group to collaborate in the support of optics, in a crosscutting initiative similar to the earlier one for high-performance computing and communications systems.

- Femtosecond optics and sources offer an opportunity for dramatic impact on science and technology. Agencies should focus attention on this opportunity and encourage innovative work in this cutting-edge field.
- R&D and applications of solid-state lasers are cross-disciplinary and should be supported by a special initiative involving multiple agencies.
- Multiple agencies with interest in the crosscutting science and technology of EUV and soft x-ray optics and techniques should encourage research in this area because of the substantial potential economic payback in the near future.

U.S. Congress

- Congress should challenge industry and its regulatory agencies to ensure the rapid development and deployment of a cost-effective broadband fiber-to-the-home information infrastructure.

U.S. Optics Industry

- To push CD and DVD technologies to higher effective storage densities and performance levels, U.S. industry should develop multilayer storage media; low-cost optical systems for writing and reading data; and efficient, low-cost techniques for mass replication and assembly of multilayer disks.
- To retain the U.S. technological edge in three-dimensional recording, industry and universities should nurture and accelerate the development of advanced three-dimensional recording media, the design of low-cost optical systems, and the study of systems integration and architectures. It is imperative that these activities be coordinated among university and industrial researchers.
- Congress should challenge industry and its regulatory agencies to ensure the rapid development and deployment of a cost-effective broadband fiber-to-the-home information infrastructure.
- DOE, NIST, and industry, in cooperation with the technical and professional societies, should pursue a program to enhance the coordination and transfer of optical sensor technology among industry, academia, and government agencies.
- DOE, EPA, EPRI, and NEMA should coordinate their efforts and create a single program to enhance the efficiency and efficacy of new lighting sources and delivery systems, with the goal of reducing U.S. consumption of electricity for lighting by a factor of two over the next decade, thus saving about $10 billion to $20 billion per year in energy costs.
- Collaborative programs in optics manufacturing should include universities so that students are trained in the latest technical solutions to production problems.

Universities

- To retain the U.S. technological edge in three-dimensional recording, industry and universities should nurture and accelerate the development of advanced three-dimensional recording media, the design of low-cost optical systems, and the study of systems integration and architectures. It is imperative that these activities be coordinated among university and industrial researchers.
- Collaborative programs in optics manufacturing should include universities so that students are trained in the latest technical solutions to production problems.
- Universities should encourage multidisciplinarity in optics education, cutting across departmental boundaries, and should provide research opportunities at all levels, from the bachelor of science to the doctorate and from basic science to applied technology.

Professional Societies

- DOE, NIST, and industry, in cooperation with the technical and professional societies, should pursue a program to enhance the coordination and transfer of optical sensor technology among industry, academia, and government agencies.
- The professional societies should work to strengthen optics as a recognized crosscutting area of science and technology through the recently established Coalition for Photonics and Optics. They should evaluate educational programs in optics and jointly produce an annual guide. The professional societies should continue to expand their commitment to professional education in optics.

B

Workshop Participants

During the preparation of this report, the committee depended on input from a large number of colleagues. Noted in this appendix are the participants in six workshops that the committee held in 1995 and 1996. Many others helped by commenting on early drafts of the report or providing information in their particular areas of expertise. The field of optics is so broad and diverse that the committee could never have accomplished its task without the assistance of this wider community. The committee thanks all of those who gave so freely of their time and expertise.

Optics in Health Care and the Life Sciences
Portland, Oregon—September 15, 1995

Speakers

W. Grundfest, Cedars Sinai Medical Center
Lasers and Optics in Medicine and Surgery

T. Jovin, Max Planck Institute for Biophysical Chemistry
Lasers and Optics as Developing Tools for Biology

K. Wilson, Applied Biosystems, Inc.
Optics in Biotechnology

W. Webb, Cornell University
Two-Photon and Near-Field Microscopy

J. Pawley, University of Wisconsin at Madison
Confocal Microscopy: From Laboratory to Commercialization

R. Tsien, Howard Hughes Medical Institute, University of California at
    San Diego
    Optical Detection of Biological Signals for Pharmaceutical Screening
E. Gratton, University of Illinois at Urbana-Champaign
    New Developments in Optical Imaging
R. Krueger, Anheuser Busch Eye Institute, St. Louis University School of
    Medicine
    Lasers in Ophthalmology
K. Gregory, Oregon Medical Laser Center
    Lasers in Cardiology
J. Jett, Los Alamos National Laboratory
    Flow Cytometry: From Laboratory to Clinic
R. Rava, Affymetrix
    Optics and High-Density DNA Arrays
R. Johnston, Molecular Dynamics
    High-Throughput Microscopy

Background Paper

M. Elbaum, Electro-Optical Sciences, Inc.

Technical Advisers

T. Baer, Biometric Imaging Systems
A. Yodh, University of Pennsylvania

Optics and Manufacturing
Irvine, California—October 11-12, 1995

Speakers

A. Bergh, Optoelectronics Industry Development Association
    The Manufacturing Infrastructure for Optoelectronics
T. Vorburger, National Institute of Standards and Technology
    Optical Sciences in Metrology
G. Osbourn, Sandia National Laboratories
    Optical Science Issues in Machine Vision
R. Harner, Dow Chemical
    The Chemical Industry
D. Rockwell, Delco Electronics, and D. Roessler, General Motors R&D
    Center
    The Use of Optics-Based Systems in Automotive Manufacturing
R. Withrington, Hughes Aircraft Company
    Aerospace Manufacturing

B. Dorwart, Shannon and Wilson
The Use of Optics in the Construction Industry

M. Fleming, Duplex Products, Inc.
The Future of Optics and Printing Goes Digital

A. Lake, Boeckeler Instruments Company, and R. Breault, Arizona
Optics Industry Association
A Small Company Approach to Future Competitiveness

J. Hanley, Pinnacle Micro, Inc. (invited)
Manufacturing Issues in Laser Disk Optical Storage

M. Morris, Rochester Photonics Corporation
Micro- and Diffractive-Optics Technology: Current Status and Future Impact

R. Haitz, Hewlett-Packard Components Group
Manufacturing Issues in Semiconductor-Based Optoelectronic Components

S. Brueck, Center for High Technology Materials
Opto-Electronic Diagnostics for Semiconductor Manufacturing

P. Reid, Hughes Danbury Optical Systems
Deterministic Manufacturing of Complex Advanced Lithography Optics

S. Depp, IBM T.J. Watson Research Center
Opto-Electronics Issues in Display Manufacturing

D. Kessler, Kodak
Optics in Imaging Systems

D. Belforte, Belforte Associates
Industrial Lasers

Background Paper

T. Montonye, SPIE

Optics in Information Technology Washington, D.C.—November 30 and December 1, 1995

Speakers

T. Li, AT&T Bell Laboratories
Long-Distance Lightwave Telecommunications

P. Shumate, Bellcore
Broad Band Access: Reaching the Customer

M. Phillips, ATx Telecom Systems
Analog Lightwave Transmission

V. Chan, MIT Lincoln Laboratory
Optical Space Communications

M. Kryder, Carnegie Mellon University
Magnetic and Optical Recording: Survey and Future Perspectives

J. Miceli, Eastman Kodak Company
Optical Disk Storage Trends

A. Bell, IBM
Optical Storage Systems and Applications

D. Carlin, Optex Corporation
Advanced Optical Data Storage

C. Brackett, Bellcore
Multiwavelength Networks

D.A.B. Miller, AT&T Bell Laboratories
Photonic Switching

D. Psaltis, California Institute of Technology
Optical Computing

R. Nelson, Motorola
Optical Data Links

W. Worobey, Sandia National Laboratories
Overview of Flat-Panel Display Technology

T. Buzak, Technical Visions, Inc.
The Outlook for Large Flat-Panel Displays

J. Florence, Texas Instruments
Miniature Displays for Projection Imaging Applications: DMDs, LCDs, ELs, and Others

M. Stefanov, Kent Display Systems
Electronic Paper: Survey and Prediction

Background Papers

C. Ryan, GPN AT&T Network Systems
Impacts on Photonic Access and Long-Haul Transmission

P.E. Green, Jr., IBM T.J. Watson Research Center
The Optoelectronic Key to Future Information Infrastructures

D.T. Gall, TWC
Present and Future of Analog Lightwave Transmission Systems in the Cable Television Industry

T.E. Darcie, AT&T Bell Laboratories
Issues and Future Perspective for Analog Lightwave Systems and Technology

J.H. Hong, Rockwell Science Center
Future in Optical Storage: Holographic Mass Memory

B.H. Schechtman, National Storage Industry Consortium
Optical and Magnetic Recording Technologies in the Next Millennium

B. Bell, MOST Inc.
Some Thoughts on Optical Disk Storage

Energy, Space, Environment, and Optical Sensing
Irvine, California—December 14-15, 1995

Speakers

R. Spellicy, Radian Corporation
Optical Environmental Monitoring

R. Menzies, JPL
Active Sensing: LIDAR

T. Vo-Dinh, Oak Ridge National Laboratory
Optical Environmental Biosensors

R. Birk and B. Davis, NASA Stennis Space Center
Space-Based Optical and Laser Sensing of Earth's Resources: Applications in Terrestrial Management

B. Snavely, National Science Foundation
New Telescopes and Astronomical Imaging Techniques

J. Breckinridge, JPL
Optical Engineering for Space Science

R. Withrington, Hughes
Defense Electrooptical Systems: Laser Guides, Sensors

S. Alejandro, Air Force Phillips Laboratory
Future Challenges in Non/Counterproliferation and NBC Defense Remote Sensing

J. Hiller, Oak Ridge National Laboratory
Optical Chemical Sensors: Fiber-Optic Probes, Flow Rates

M.R. Stapelbroek, Rockwell
Detectors and Detector Arrays for Visible and IR Wavelengths

R. Menzel, Texas Tech University
Trends in Examination of Physical Evidence by Optical Techniques

R. Huggins, Boeing
Optical Systems for Aerospace

W. Weber, Ford Motor Company
Optics Research Trends in the Automobile Industry

H. Powell, Lawrence Livermore National Laboratory
Advanced Optical Technology for Inertial Confinement Fusion

P. Iles, Applied Solar Energy
Space Solar Cells

J. Benner, National Renewable Energy Laboratory
Terrestrial Solar Cells

S. Selkowitz, Lawrence Berkeley National Laboratory
Optical Applications for Advanced Lighting Systems

R. Englemann, Oregon Graduate Institute
LED Status and Prospects for High-Efficiency Lighting Applications

Background Papers

J. Bilbro, NASA Marshall Space Flight Center
Center of Excellence for Space Optical Systems

R. Deese, Electro Tech, Inc.
Commercial LED Lighting Systems

N. Pchelkin, U.S. Air Force Phillips Laboratory
Law Enforcement and Security Uses of Lasers and Optics

D. Hartmann, University of Washington
Global Space Based Sensing: Atmospheric/Global Climate Change

F. Faxvog, Honeywell
Ring Laser Gyro Aircraft Sensors

D. Taylor, Walt Disney Imagineering
Entertainment Uses of Lasers/Optics

M. Finger, Lawrence Livermore National Laboratory
Surveillance: Drugs, Explosives, Border Control

T. Amundsen, Metrologic Systems
Barcode Scanner Systems

P. Kelley, Tufts University
Solar Cell and Energy Systems

M. Hutchins, Hutchins International, Ltd.
New Sulfer Dimer Lighting Sources

E. Browell, NASA/Langley
Lidar Atmospheric Remote Sensing

J. Martin, TECO Electric Technology Center
Energy Efficient Lighting

M. Rea, Rensselaer Polytechnic Institute
New Lighting Systems and Light Sources

M. Smith, FLIR Systems, Inc.
New Commercial IR/Thermal Cameras

W. Krupke, Lawrence Livermore National Laboratory
Laser Sensors and New Energy Sources

Optics Research and Education
Irvine, California — January 4-5, 1996
Speakers

H.J. Kimble, California Institute of Technology
Quantum Measurement and Computing

C. Wieman, JILA
Cooling and Trapping

S.J. Smith, Stanford University
Optics in Cellular and Molecular Biology

M. Nuss, AT&T Bell Laboratories
Technology and Applications of Femtosecond Lasers

P. Bucksbaum, University of Michigan
Physics with Femtosecond Sources

R. Hochstrasser, University of Pennsylvania
Lasers in Chemistry

D. Markle, Ultratech Stepper
EUV Optics for Advanced Lithographies

F. Cerrina, University of Wisconsin at Madison
Applications of X-Rays to Materials Science

B. Kincaid, Lawrence Berkeley National Laboratory
Uses of High-Brightness, Partially Coherent Light from Third-Generation Synchrotron Sources

D. Welch, Spectra Diode Labs, Inc.
Semiconductor Diode Lasers

W. Krupke, Lawrence Livermore National Laboratory
Advanced Solid-State Lasers

M. Fejer, Stanford University
Engineering Nonlinear Materials

M. Mansuripur, University of Arizona
Materials for Optical Storage

D. Moore, University of Rochester
Gradient Index Optics

B. Thompson, University of Rochester
Education and Training in Optics: A Philosophical Overview

M.J. Soileau, University of Central Florida
Graduate Education

R.R. Shannon and J.D. Gaskill, University of Arizona
Four-Year Undergraduate Education

R.F. Novak, Monroe Community College
Two-Year Undergraduate Education and Training

D. Hull, Center for Occupational R&D
K-14 Education and Training

Background Papers

H. Winick, LCLS Working Group
A Short-Wavelength Linac Coherent Light Source (LCLS) Using the SLAC Linac

J. Madey, Duke University
Ultraviolet and Extreme Ultraviolet Free Electron Laser Light Sources

G. Kubiak, Sandia National Laboratories
Laser-Produced Plasma EUV Sources

D.L. Matthews, Lawrence Livermore National Laboratory
A Center for the Development and Application of Laser Light Sources

D.G. Hall, University of Rochester
Education in Optics at the B.S., M.S., and Ph.D. Levels at the University of Rochester

C. Joenathan, R.M. Bunch, and B.M. Khorana, Rose-Hulman Institute of Technology
Optics Education at the B.S. and M.S. Levels: Past, Present, and Future

J.O. Dimmock and S. Kowel, University of Alabama, Huntsville
Optical Science and Engineering: A Perspective

Optical Society of America
Continuing Education Report for COSE

SPIE
SPIE Educational Services

Optics and Defense
Washington, D.C.—March 18-20, 1996

Speakers

J. Stichman, Sandia National Laboratories
The Manufacturing Factory of the Future

J. Hurd, Planar Inc.
A U.S. Industry Perspective on Flat-Panel Display Manufacturing

R. Balcerak, Defense Advanced Research Projects Agency (DARPA)
Focal Plane Array Manufacturing

R. Tapp, Optical Imaging Systems
Flat-Panel Display and Sensor Manufacturing for Dual-Use Applications

M. Chang, New Focus Inc.
High-Value, Low-Cost, Low-Volume Manufacturing of Optics

R. Fedchenko, 3D Systems Inc.
Optical Systems for Rapid Prototyping and Manufacturing

H. Pollicove, University of Rochester
   Low-Cost Rapid Fabrication of Optics
L. Marabella, TRW
   Alliances for the Use of Lasers in Manufacturing
K. Thompson, Optical Research Associates
   Optical Design
L. Marquet, Department of Defense
   Weapons-Class Lasers
C. Volk, Litton Industries
   Laser Gyros
L. Buchanan, DARPA
   Defense Needs in Optics
L. Figueroa, Boeing Defense and Space Group
   Optics in Aircraft Manufacturing
M. Ealey, Xinetics
   Advanced Adaptive Optics
I. Ury, Ortel Corp.
   RF Transmission Through Fibers
D. Scifres, Spectra Diode Labs
   Semiconductor Lasers and OEICs: Past, Present, and Future
P. Trotta, Texas Instruments
   Production of Optics for FLIRs and Other Electro-Optic Sensors
G. Riley, Lockheed Martin
   Manufacturing Airborne Focal Plane Array Sensor Systems
J. Bilbro, NASA Marshall
   Optics Metrology
R. Aronno, Tinsley Laboratories
   Precision Asphere Manufacturing: A Small Company Perspective
P. Anthony, Bell Labs
   Packaging of Photonic Components
P. Beauchamp, OCLI
   Optical Coatings
J. Atherton, Lawrence Livermore National Laboratory
   Large Optics and the National Ignition Facility

Society Representatives

J. Pearson, SPIE
D. Hennage, Optical Society of America
J. McMahon, Institute of Electrical and Electronics Engineers/Lasers and
   Electro-Optics Society
P. Baker, Laser Institute of America
B. Hitz, Laser and Electro-Optics Manufacturers Association
D. Wilson, American Precision Optical Manufacturers Association